The Media's Role in Defining the Nation

David Copeland
General Editor

Vol. 5

PETER LANG
New York • Washington, D.C./Baltimore • Bern
Frankfurt am Main • Berlin • Brussels • Vienna • Oxford

David Copeland

The Media's Role in Defining the Nation

The Active Voice

PETER LANG
New York • Washington, D.C./Baltimore • Bern
Frankfurt am Main • Berlin • Brussels • Vienna • Oxford

Library of Congress Cataloging-in-Publication Data
Copeland, David.
The media's role in defining the nation: the active voice /
David Copeland.
p. cm. — (Mediating American history; v. 5)
Includes bibliographical references and index.
1. Press—United States—Influence. 2. Press and politics—
United States—History. 3. Journalism—Political aspects—United States.
4. Mass media—Political aspects—United States. I. Title.
PN4888.I53C58   302.230973—dc22   2009043511
ISBN 978-1-4331-0380-3 (hardcover)
ISBN 978-1-4331-0379-7 (paperback)
ISSN 0085-2473

Bibliographic information published by **Die Deutsche Nationalbibliothek**.
Die Deutsche Nationalbibliothek lists this publication in the "Deutsche
Nationalbibliografie"; detailed bibliographic data is available
on the Internet at http://dnb.d-nb.de/.

Cover concept by Miriam Williamson
Author photo by Brook Corwin.

The paper in this book meets the guidelines for permanence and durability
of the Committee on Production Guidelines for Book Longevity
of the Council of Library Resources.

© 2010 Peter Lang Publishing, Inc., New York
29 Broadway, 18th floor, New York, NY 10006
www.peterlang.com

All rights reserved.
Reprint or reproduction, even partially, in all forms such as microfilm,
xerography, microfiche, microcard, and offset strictly prohibited.

Printed in the United States of America

Table of Contents

Preface .... *vii*

1. *Introduction* *1*

2. *Cooking Up Paragraphs, Articles, Occurrences: Colonial* *11*
   and Revolutionary America

3. *My Pen and Press Are the Only Formidable Weapons I Have* *43*
   Ever Used: The Early Republic

4. *The Great Organ of Social Life, the Prime Element of* *69*
   Civilization: The Antebellum Era and Civil War

5. *We Expect Great Results from This Work: Post-Civil War* *99*
   and Yellow Journalism

6. *There Is Filth on the Floor and It Must Be Scraped Up:* *129*
   The Muckrakers and Press of the Early 20th Century

7. *My Medium Is Everywhere: New Media and the New Century* . *157*

8. *My Notebook Still Carries Bloodstains: The World Wars* *189*
   and the Cold War

9. *The Whole World Is Watching: The Journalism of Change,* *221*
   1950s-1970s

10. *The Scramble to Fill 24 Hours of Air Time: The End of* *257*
    the Twentieth Century

11. *Show-Me Journalism: Media Transformation in the* *277*
    Twenty-First Century

Epilogue *295*

Bibliography *297*

Index *317*

Preface

This is not the first book written on the media and the defining of the nation. It will not be the last because media have played such a crucial role in the development of the United States. The growth and development of the country, in fact, parallels the growth and development of media as the chief sources of information dispersal, entertainment, commerce, and as the center of dialogue among the populace. In 1750, a New York printer remarked that newspapers were indispensable, something that colonials could not do without because they had a taste for news. In other words, they wanted to know what was going on, and the medium of the day for information sharing—newspapers—provided it. By the 1800s, "What's the news" became a standard greeting among Americans.

Maxwell McCombs and Donald Shaw pointed out in their classic 1972 study of agenda setting that Americans closely follow the information media present. And while they don't necessarily believe everything that media tell them, Americans do think about what media say and tend to frame their understanding of information in the way that media present it. After thirty years of study, McCombs, Shaw, and others have concluded that there is more than a push of information to a passive audience. There is push and pull with the audience and the media, and people naturally look for media information that agrees with personal beliefs and ideology. We read the *Washington Times*, watch television commentary programming on FOX, or follow the YouTube and Twitter commentary of Philip DeFranco because what is being said is comfortable to us and reaffirms our beliefs. Those media outlets know that their audience adheres to their points of view, takes that into account, and pushes to its audience the information it wants. This is exactly how America's media have operated from the 1760s into the twentieth century and are operating now. Opposing ideologies turned to media as a means of pushing an agenda. Media became the place to effect change in government, in society, in economics, in practically everything.

Many have, however, promoted objective reporting of news for more than a century rather than a biased presentation, but even objective media information can end up having an agenda and change the course of a city, state, or nation. One might consider Edward R. Murrow objective in his coverage of Senator Joseph McCarthy and that Bob Woodward and Carl Bernstein were just looking for answers to who ordered the break-in at the Democratic National Convention offices in the Watergate building and why. But good watchdog journalism always has an agenda of undoing wrongs and setting them right—effecting change.

This book is not a study of agenda setting nor any of its evolutions. It is, however, a book that looks at the way the nation has turned to, listened to, and debated through media to shape issues and events that, in turn, helped to define the country. Using media has been intentional. Some of the events included here are so significant that all know of them, like the way patriots turned to the press to foster rebellion and independence in the 1760s and 1770s. Others occurred on a much more localized plane, such as Horace Carter's dangerous but successful newspaper crusade to wrest control by the Ku Klux Klan of a rural North Carolina community in the 1950s.

The use of media no matter the societal level, therefore, is important, but we often forget that media have stimulated debate and effected change from the smallest of communities to the entire nation. It's something that no one should forget in the twenty-first century. Technology is providing new means of reaching people, of providing a place to debate, discuss, and shape thought; and anyone who believes that online media reaching out to small groups and individuals did not play a role in the 2008 election should rethink that notion. The same power has always accompanied technological advancements. Look at photography during the Civil War, or at the news surrounding radio and television as they were being developed and then appropriated for use to share information with people.

There is no attempt here to provide a comprehensive history of the United States nor of the history of media. But, this book is both on a selected basis because it deals with the way the nation and media became so entwined that looking at the development of one without the other really does not provide a true picture. This book is an attempt to demonstrate that throughout the course of American history we need to remember how the media have played an active role in nearly all that transpired. The active voice of the people through any medium available has been a driving force, something we have really not been able to do without.

A number of people made this project possible. Mary Savigar and Peter Lang have been supportive of this book as well as others in the Mediating American History series. At Elon University, colleagues Harlen Makemson and Brooke Barnett read parts of the manuscript and offered guidance as to how to approach the broad subject matter. Shannon Martin of Indiana University also provided impetus for this project. Colleague and friend, she always finds the right words to say. As always, Dean Paul Parsons, Associate Dean Connie Book, and Department Chair Don Grady in the School of Communications at Elon University have supported my efforts in any way they could. The University also provided summer research funds that were critical in beginning this project.

As always, it's family that makes the difference in projects such as this. To Holly, Hunter, Victor, and Janet I say thanks for being there. Mostly, I thank my wife Robin for the ability to show to me every day what strength and courage are all about. Her ability to find the best in every situation may be the most active of voices I have ever seen. Finally, a special thank you to Sophie Louise, the newest member of our family. Though she's too young to understand, this book is dedicated to her and the future that she and her generation will shape.

1. Introduction

In 2005, Lu Ann Cahn, a reporter for Philadelphia's NBC affiliate, burst into a Colwyn Borough meeting and took over the microphone set up for public comment. There, with cameras running and officials bewildered, Cahn claimed that officials in Delaware County were using taxpayers' money to fund an illegal bar for borough managers, the police chief, and other elected officials in the fire station of the dry borough. Cahn's guerilla tactics included infiltrating the bar with hidden cameras to support her claim. Armed with that information, she attacked borough officials at the meeting, Cahn's stint at the microphone being the central portion of the story. For her work, Cahn and WCAU-TV won an Emmy for investigative reporting in regional news, and Delaware County's "Dirty Little Secret," as the report was called, was exposed. The bar closed, and county money was no longer diverted into illegal activities.

A little more than a century earlier, John Cockerill, managing editor of New York publisher Joseph Pulitzer's *World*, sent an aspiring reporter named Nellie Bly "undercover" into the city's Women's Asylum on Blackwell's Island to reveal the treatment of those housed there. After Bly spent ten days in the asylum, Pulitzer sent a lawyer to Blackwell's Island with papers to free Bly. She then filled the pages of Pulitzer's paper with stories of the horrors of the asylum. Bly, the reporter, was the central figure in the reporting, but the exposé helped to change the way the city treated patients in mental institutions.

A hundred years before that, according to John Adams, his cousin Samuel and the printers of the *Boston Gazette* sat around "cooking up paragraphs, articles, occurrences, &c., working the political engine,"[1] which resulted in the American Revolution. The "Journal of Occurrences," one of their creations, provided printers throughout America with a detailed account of British activity in occupied Boston. Part fact, principally propaganda, the Journal and other creations by Adams were aimed at changing the direction of the British colonies that would later comprise the United States. Used at first only by the most patriot-minded of printers, by 1774 nearly every American printer eagerly reprinted what Adams and others had to say about the British and began to believe that they were embarking upon "the cause of virtue, of liberty, of God."[2]

In each of these examples—though they may be extreme—media created the news and at the same time reported it. Even though Elizabeth Cochrane, a.k.a. Nellie Bly, accepted Cockerill's assignment in order to land a job at the *World*, there can be little doubt that she believed, just as Lu Ann Cahn and Samuel Adams no doubt believed, that her work would improve conditions or rectify a wrong, thereby making the situation better for the public. The change may have affected only local issues or events, or, as in the case of Samuel Adams, the change had repercussions that altered the course of a nation, perhaps the world.

In 1897, William Randolph Hearst gave this type of reporting a name. For him, it was the journalism of action. "It doesn't wait for things to turn up," Hearst said. "It turns them up."[3] For more than three centuries, the press has been an instrument used to shape the direction of the people it serves. People realized the power that the printed word regularly distributed to the public. It could change the direction of public policy both politically and economically. It could shape the moral compass of a people. It could set the agenda for a nation.

People have called this type of journalism many names. Jock Lauterer refers to it as community journalism, where "citizen-journalists" are "intimately involving themselves in the welfare of the place, in the civic life of their towns, participating as active members of the very community they're covering."[4] Some have begun to call the type of journalism practiced by Cahn and others guerilla journalism or ambush journalism. A century ago, President Theodore Roosevelt classified similar reporting tactics as "muckraking." Whatever the name, journalists taking action have played a pivotal role in the history of the United States. They have told people what to think about as Maxwell McCombs and Donald Shaw pointed out in 1972.[5] But in many cases they have also told people what to think.

Most media practitioners today, however, generally advocate an objective, non-biased form of reporting, which David Mindich has said has been in place since the end of the nineteenth century.[6] At that time, a battle raged in New York City about the way journalism should be practiced. Locked in a contest for circulation, Hearst and his competitor Joseph Pulitzer employed the flamboyant style that would earn the two and their papers the moniker of yellow journalism. Hearst asked, "May a newspaper properly do things, or are its legitimate functions confined to talking about them?"[7] The answer in the last years of the nineteenth century and the beginning of the twentieth for Hearst and

Pulitzer was an unwavering call to action. According to W. Joseph Campbell, that meant that a journalism of action paper employed "activist strategies" and "injected itself as the self-activating participant in solving crime, extending charity, influencing foreign policy, and thwarting what it deemed abuses of municipal government,"[8] which Hearst said represented "the final stage in the evolution of the modern newspaper."[9] Many Americans agreed with Hearst.

Others, however, did not. In 1896, after purchasing the *New York Times*, new publisher Adolph Ochs set in motion the objective, non-biased style that Mindich says became the nation's standard. Deploring the style used by Hearst, the *Times* declared, "We hold that the reader is entitled to have as fully as practicable, and in any case with entire impartiality, the facts as they come out on one side or another of the questions at issue. We especially hold that it is a wrong and an offense to any reader to seek to influence his views either by imperfect or distorted reports or by 'head-lines.'"[10] Ochs added the *Times*' motto, "All the News That's Fit to Print," which meant, according to Campbell, no participatory journalism and an ethical approach to information presentation.[11] Ochs expected his reporters to observe and report what happened. The *Times*' job was to present the facts and the story in an unbiased, fair, and objective manner. Conclusions could be drawn by the reader but not by Ochs' reporters.

While most Americans expect this style of news presentation, it has not been standard press practice for most of the three hundred years the nation has had press outlets. Active involvement by journalists in shaping policy; serving as an agent to change the direction of a town, state, or country; or influencing public opinion was a central function of the press for at least two centuries and has continued. When press outlets grew to gargantuan size in terms of circulation and when new, electronic media added still more voices to the discussion early in the twentieth century, the role of the individual reporter—and perhaps, too, the individual media outlet—often became less of a factor in shaping the way the nation used its news than in the era of the Muckrakers. That did not diminish the power of what was reported, though. Edward R. Murrow, before his ear-defining and shocking radio report from Buchenwald in 1945, announced that he was the least important person there, so he let those in the concentration camp speak. In the post-World War II era, when journalists have increasingly espoused objectivity and fairness following the directives of the Hutchins Commission Report of 1947,[12] much of what has appeared in newspapers and magazines, on radio and television, and now online has the journalist actively involved in creating the story or working to shape ideology. Even objective news presentation can radically shape direction.

Those who craft words into persuasive sentences; sentences into paragraphs; and paragraphs into essays, editorials, news articles, books, radio and television commentary, motion pictures, and even Web logs, intentionally shape their writings to achieve a purpose. It may be simply to inform, but, often, it is to make those who read or view their work think about something and then, perhaps, seek to make changes. When writers assume this purpose, they become actively involved in the issue at hand; their voice assumes an active role in the subject. And, generally, writers express their ideas using active-voice verbs where the subject of the sentence performs the action. Doing so is logical because, as Strunk and White said in their classic work *The Elements of Style*, "The active voice is usually more direct and vigorous than the passive" because the active voice "makes for more forcible writing."[13] It is important that we understand the active way media have been used from the country's colonial foundings through today to solve problems, address issues, and set or change the agenda on any level. The active voice of citizen-journalists and trained journalists has revealed, shaped, and perhaps created the nation's history.

The message is pivotal to change, but so, too, is the medium. As Marshall McLuhan stated in *Understanding Media* in 1964, the medium can become the message. Speaking about television, McLuhan said that the ubiquitous presence of TV sets in homes altered the nature and power of the message, but he also said that any technological innovation that revolutionized information dispersal affected society.[14] The printing press changed the course of Western history because, with it, ideas about religion and speaking one's mind were able to be produced in numbers not imagined or even considered possible by religious or royal power brokers prior to Gutenberg's invention.[15]

For Alexis de Tocqueville, the ability of newspapers to penetrate the nation meant that the medium was able to rally "the interests of the community round certain principles" and that "their influence in the long run becomes irresistible, and public opinion, perpetually assailed from the same side, eventually yields to the attack."[16] Seventy years later, Nathan Stubblefield claimed the same thing for a new medium—radio. The Kentucky inventor boldly proclaimed that radio would soon be everywhere,[17] which is exactly what happened by World War II. Radios, according to the Broadcast Measurement Bureau, could be found in more than 90 percent of all homes.[18] Approximately 80 million Americans remained glued to an estimated 21 million home sets when reporters announced victory first on the European front and then in the Pacific, a remarkable number considering the manufacturing, business, government, and military efforts underway that required people to be working.[19]

Consequently, individual media have had the power to change the course of the nation as those who created the active messages incorporated new and expanding technological advances to share their ideas. Taken together, the message, creator, and the medium have the potential to produce powerful results, and we most notice this power when media are used for ideological purposes.[20]

When scholars offer examples of media's power to effect change, they often point to the press of the Revolutionary era or that of the early republic. Historian David Ramsey said in 1789 that "the pen and the press had merit equal to that of the sword" in causing the Revolution.[21] And, the title of Arthur M. Schlesinger's classic 1957 study sums up the powerful influence of the press during this time period—*Prelude to Independence: The Newspaper War on Britain 1764–1776*. The use of the press during this time period as a tool for ideological change may be the most notable example, but the same conclusions can be reached when looking anywhere along history's timeline in the United States. "What is it that controls the different departments of the government and all the varied industrial and social interests within the limits of the Republic? The answer is, emphatically, public opinion enunciated through the press," the U.S. Bureau of the Census stated in 1866. "The press is the real representative of the people, the great conservative power held by them to guard public and individual liberty."[22]

While this book's purpose is to demonstrate media's pivotal role in defining the nation's direction on multiple levels, not everyone agrees, especially as technological advances have caused changes in media from the last quarter of the twentieth century on. When Tocqueville visited America, citizens depended upon printed media exclusively. At least half of all American homes subscribed to a newspaper where it was read aloud to family members.[23] The average paper had long been shared with other families, ending up at least twenty miles from the place it was purchased,[24] and newspapers and magazines served as the catalyst for public discussion and oration, often following a public reading in a tavern or on the street.[25] The saturation of the content of newspapers through all strata of society, therefore, was extremely high. With the growth of online media and me-type journalism, however, the possibility exists that in an age when everyone has the potential to be a journalist that individual agendas and a multitude of voices might achieve opposite results from the opinion of Tocqueville and the Census Bureau in 1866. Gladys Ganley believes that if we have access to so much information from so many sources at one time that instead of helping in the process of democracy that "the glue of social cohesion" could be lost. "Power to the people could mean that nobody is in control, with unpredictable political consequences."[26] Or, as Neal Postman said, "[O]ur problem is not that we don't have enough information….People don't know what to do with the information. They have no organizing principle."[27]

Even if Ganley, Postman, and others are correct in assuming that an abundance of information from a growing number of media—especially online—

is producing a society with less knowledge about what's going on rather than more (which would seem more logical), this prediction does not diminish efforts by those using the media to voice ideas or to use media outlets as a means to change policy, as Lu Ann Cahn did in Philadelphia with the Delaware County speakeasy and numerous other reporters have done and continue to do with issues of concern to their constituencies.

Sometimes, the journalists become the story. For most people who follow the news, the mention of Matt Drudge immediately calls to mind Bill Clinton and the Lewinsky affair because Drudge's decision to break the story became as revealing about the journalist becoming a part of the story as the potential for online journalism. Edward R. Murrow will forever be linked with the coverage of London during World War II and for undermining Senator Joe McCarthy's "Red Scare." No information about the My Lai Massacre in Vietnam will fail to mention reporter Seymour Hersh, nor will stories about what happened at Abu Ghraib prison in Iraq.

Journalists making change and being a part of the story can be personal, too. This was the case with *ABC World News Tonight* anchor Peter Jennings, who died of lung cancer in August 2005. Jennings announced on-air on April 5 that he was battling the disease and continued an online dialog about the illness through July on the network's *World News Tonight* Web site.[28] The network, three months after Jennings' death, began a month-long series of reports on smoking cessation programs, tips on quitting smoking, public policy issues, and options for the treatment and prevention of lung cancer. Programs also aired on ABC radio and on the network's Web site. An entire Web site, *Quit Smoking ABC*, was also established. It includes dozens of articles and Web logs.[29] The efforts by the network and ABC affiliates continues.

We continue to debate the way media collectively and individually shape information because the voices and faces of media can sometimes become so important that just by reporting information, they can affect the event and change the course of a nation. *U.S. News & World Report* did not call CBS anchor Walter Cronkite "the most trusted man in America" in 1976 capriciously. He had earned the title through years of quality reporting, so when in 1968, Cronkite said that the Vietnam war was unwinnable, media reporting on the war was altered. More importantly, though, so was the nation's view of the war.

This book is structured chronologically and is designed to tell the story of media influence on multiple levels. Within the framework, the technology of creating and sharing media plays a vital role. At certain points, new media became the news. Many elements of media influence are obvious. The press' role in agitating and advocating separation from Britain has already been mentioned, as has Vietnam. The role of the journalists or non-journalists is vital, too, to the story. The idea is not to reveal only the obvious points where media have been the principal tool that affected the nation but to see how media have been employed at every level imaginable to create change. The change can be positive or negative, depending upon perspective. About a third of the people living in America in the 1770s, for example, thought the push to separate from Britain was a bad idea. Bob Woodward and Carl Bernstein were heroes to millions, villains to multiple others in their coverage of the Watergate affair.

What is central to this work is the active nature of the participants and their ability to use media. Samuel Adams wrote using at least twenty-five pseudonyms. He rarely spoke at public events, knowing the printed word in a newspaper had the power to spread his ideas farther than any public oration could.[30] Nellie Bly was always a part of a Nellie Bly story, and Lu Ann Cahn was the principal actor in her Emmy-winning exposé. Even those who are supposed to be the subject of media—elected officials—turn to the media to define and guide the nation. As Scott McClellan said in his 2008 memoir of his time as George W. Bush's press secretary, the administration was in "permanent campaign" mode, using media to hype the rationale for invading Iraq in 2003 and as a means to accomplish anything it wanted by shaping public opinion.[31]

"The mass media serve as a system of communicating messages and symbols to the general populace," according to Noam Chomsky and Edward Herman. Therefore, they work to "inform, and to inculcate individuals with the values, beliefs, and codes of behavior that will integrate them into the institutional structures of the larger society."[32] This is why the study of media's relationship with the nation is so important. President Bush and his advisors knew—just as Samuel Adams, Thomas Jefferson, Horace Greeley, Joseph Pulitzer, Franklin Roosevelt, Martin Luther King, and thousands of others who have used media to spread a message did—that media have the potential to sway the people, to set a nation's agenda, to be a driving force. As Federalist judge Alexander Addison said in 1799, "Give to any set men the command of the press, and you give them the command of the country, for you give them the command of public opinion, which commands everything."[33] Shaping, guiding, and defining, then, can be noble, nefarious, or necessary; it may be a combination of these and other ideals. It is this book's purpose to demonstrate how media have been central to all reasons in defining and shaping the United States.

NOTES

A note on primary sources: In all cases, original spelling, grammar, and presentation have been preserved. As a result, in the sources through the nineteenth century, items may appear in all capitals, italics, bold, and small caps. Words such as today will be spelled to-day. In many cases, all nouns, not just proper nouns, were capitalized.

1. John Adams, *The Works of John Adams, with a Life of the Author, Notes and Illustrations*, ed. C.F. Adams, 10 vols. (Boston, 1850–1856), 2:219, quoted in Philip Davidson, *Propaganda and the American Revolution, 1763–1783* (Chapel Hill: University of North Carolina Press, 1941), 227.
2. *Virginia Gazette* (Williamsburg, Pinkney), December 20, 1775.
3. *New York Journal*, October 13, 1897.
4. Jock Lauterer, *Community Journalism: Relentlessly Local*, 3rd ed. (Chapel Hill: University of North Carolina Press, 2005), preface.
5. Maxwell McCombs and Donald Shaw, "The Agenda-Setting Function of Mass Media," *Public Opinion Quarterly* 36 (1972): 176–187.
6. David T. Z. Mindich, *Just the Facts: How "Objectivity" Came to Define American Journalism* (New York: New York University Press, 1998), 11.
7. *New York Journal*, October 5, 1897.
8. W. Joseph Campbell, "1897: Journalism's Exceptional Year," *Journalism History* 29, 4 (2004): 192.
9. *New York Journal*, October 13, 1897.
10. *New York Times*, October 20, 1897.
11. Campbell, "1897," 194.
12. The Hutchins Commission Report of 1947 is generally called this because of Robert Hutchins, who chaired the Commission on Freedom of the Press. The committee's report was officially titled *A Free and Responsible Press. A General Report on Mass Communication: Newspapers, Radio, Motion Pictures, Magazines, and Books*. The commission proposed five general requirements to maintain a free and responsible press in the United States. The commission said that 1. The media should provide a truthful, comprehensive and intelligent account of the day's events in a context which gives them meaning. 2. The media should serve as a forum for the exchange of comment and criticism. 3. The media should project a representative picture of the constituent groups in the society. 4. The media should present and clarify the goals and values of the society. 5. The media should provide full access to the day's intelligence.
13. William Strunk, Jr. and E. B. White, *The Elements of Style*, revised ed. (New York: Macmillan, 1959), 13.
14. Marshall McLuhan, *Understanding Media* (London: Routledge and Kegan Paul, 1964).
15. David A. Copeland, *The Idea of a Free Press: The Enlightenment and Its Unruly Legacy* (Evanston, Ill.: Northwestern University Press, 2006), 8–9.
16. Alexis de Tocqueville, *Democracy in America*, 2 vols. (New York: Alfred A. Knopf, 1946), 1:187–88.
17. Nathan B. Stubblefield, *St. Louis Post-Dispatch*, January 10, 1902. See also, Trumbull White, ed., *Our Wonderful Progress: The World's Triumphant Knowledge and Works* (1902), 300.
18. *Radio Families–USA–1946* (New York: Broadcast Measurement Bureau, 1946), 1.
19. Kenneth G. Bartlett, "Social Impact of the Radio," *Annals of the American Academy of Political and Social Science* 250 (March 1947): 90. According to the 1940 U.S. Census, approximately 130 million people lived in the forty-eight states.
20. Pam Shoemaker and Stephen D. Reese, *Mediating the Message: Theories of Influence on Mass Media Content* (New York: Longman, 1996).
21. David Ramsey, *The History of the American Revolution*, 2 vols. (1789), 2:319.

22. U.S. Bureau of the Census, *Statistics of the United States* (Washington, D.C.: Government Printing Office, 1866), xiv.
23. William J. Gilmore, *Reading Becomes a Necessity of Life: Material and Cultural Life in Rural New Britain, 1780–1835* (Knoxville, 1989), 193–94.
24. *New York Journal*, February 24, 1795.
25. Richard D. Brown, *Knowledge Is Power: The Diffusion of Information in Early America, 1700–1865* (New York and London: Oxford University Press, 1989), 292. See, also, David Waldstreicher, *In the Midst of Perpetual Fetes: The Making of American Nationalism, 1776–1820* (Chapel Hill: University of North Carolina Press, 1997).
26. Gladys D. Ganley, "Power to the People via Personal Electronic Media," *Washington Quarterly* 10, 3 (1992): 20.
27. In Max Frankel, "Summer Musings," *New York Times Magazine*, June 25, 1995, 24.
28. "Peter Jennings dies of lung cancer," *cnn.com*, August 8, 2005, http://www.cnn.com/2005/SHOWBIZ/TV/08/07/jennings.obit/.
29. *Quit Smoking ABC*, http://www.quit-smoking-abc.com/Have_a_Plan_to_Quit_Smoking_and_Stick_to_It.html.
30. Philip Davidson, *Propaganda and the American Revolution, 1763–1783* (Chapel Hill: University of North Carolina Press, 1941), 5.
31. Scott McClellan, *What Happened: Inside the Bush White House and Washington's Culture of Deception* (New York: PublicAffairs, 2008).
32. Edward S. Herman and Noam Chomsky, *Manufacturing Consent: The Political Economy of the Mass Media* (New York: Pantheon Books, 2002), 1.
33. *Columbian Centinel* (Boston), January 1, 1799.

2. Cooking Up Paragraphs, Articles, Occurrences: Colonial and Revolutionary America

John Adams described their activities as "cooking up paragraphs, articles, occurrences, &c., working the political engine."[1] Toiling late into the night, Samuel Adams, printers Benjamin Edes and John Gill, and others wrote news stories and essays in the office of the *Boston Gazette, and Country Journal* that filled the pages of the most radical newspaper in America and then the columns of newspapers throughout New England and the rest of the thirteen British colonies that would become the United States. Years later, the future second president called that work the "real revolution." If you wanted to understand the American Revolution, he said, you didn't need to study the cannons, muskets, or the battles between colonists and Britons. No, you needed to spend time with the newspapers and pamphlets of the day. There, Adams explained, was where a revolution in the hearts and minds of Americans occurred. The printed word captured the attention of people and swayed public opinion toward separation from Great Britain, which led to revolution and the formation of the United States of America.[2]

THE GOODLIEST TERRITORY

People turned to the press to define America before settlers began arriving from Europe. Seeing the potential for their respective countries, European speculators, eager to populate the New World, subsequently "cooked up" pamphlets, books, and broadsides that described "the mysterious land across the Atlantic" as "alive and beautiful and thrilling." In a sense, the colonization effort of the New World was based in defining it via the printed word. According to Julie Williams, "The printed word was the lure, the promise" that America would provide a land of abundance similar to Europe but much better with opportunities never possible for so many in the Old World. Within the printed word, Europeans—especially the English—discovered maps, directions for successful colonization, and the assurance that the New World was a bountiful land ready for the taking by the industrious. As a result, Williams says, "The first significant role of the press in relation to colonial America was the persistent suggestion that Europeans should settle there. Without the printed word to help lure settlers, Europeans might have had less of an interest in establishing so many settlements so quickly."[3]

Explorer Richard Hakluyt captured the attention of the English monarchy in colonizing the New World when he spoke of lands that "abounde in gold, Rubies, Diamonds" in addition to the most important spices for which Columbus had sought a trade route by sailing east.[4] Ralph Lane wrote that the New World was "the goodliest and most pleasing Territorie of the world: for the continent is of an huge and unknowen greatnesse."[5] Hakluyt, Lane, and others believed that because this land was essentially parallel with the Mediterranean region that it would produce the same agricultural products.[6] In a longer tract, Lane, who joined other Englishmen in their initial explorations along the North Carolina and Virginia coasts and was the first governor of the Roanoke Island settlement, said that the land possessed "the most sweete, and healthfullest climate, and therewithal the most fertile soyle, being manured in the world."[7] His assessment was based as much on the location of the lands explored along the Atlantic as on anything observed.

Explorer John Smith provided England with hundreds of pages of descriptions of the New World. He said that New England possessed "excellent good woods, for building Houses, Boats, Barks or Ships, with an incredible abundance of most sorts of Fish, much Fowle, and sundry sorts of good Fruits for mans vse." Settlers would also find "Quaries of stones," "Free-stone for building, Slate for tyling, smooth stone to make Furnasses and Forges for Glasse and Iron, and Iron Ore sufficient conueniently to melt in them." Ultimately, Smith said, "this is a most excellent place, both for health and fertilitie....I would rather lieu [live] here then any where."[8]

To make emigration appealing to commoners, Smith said that in New England there "are no hard Landlords to racke vs with high rents, or extorting fines, nor tedious pleas in Law to consume vs with their many yeeres disputation for iustice; no multitudes to occasion such impediments to good orders as in popular States: so freely hath God and his Maiestie bestowed those blessings on them will attempt to obtaine them, as here euery man may be master of his owne labour and land, or the greatest part ... and if he haue nothing but his hands, he may set vp his Trade; and by industry quickly grow rich, spending but halfe that time well, which in *England* we abuse in idlenesse, worse, or as ill."[9]

The writings of Hakluyt, Lane, Smith, and others captured the attention of English men and women who wanted an opportunity for a better life. After the first colonization efforts were established, enterprising Englishmen looked for ways to expand settlement. In 1682, Samuel Wilson published *An Account of the Province of Carolina*. Looking to attract colonists for the tract of land owned by the eight Lords Proprietors, Wilson described the colony of Carolina as "a pleasant and fertile Country, abounding in health and pleasure, and with all things necessary for the sustenance of mankind." In order to find out more about emigrating, Wilson invited prospective settlers to visit the Carolina Coffee House in London, which had been established as an information center and travel agency for those interested in leaving England and moving to Carolina.[10] Much later, James Oglethorpe turned to promotional literature to populate his Georgia colony. By the time of the Georgia settlement, however, an extensive number of newspapers existed in Britain, and Oglethorpe and his associates, in order to reach prospective colonists, decided "to cause such Paragraphs to be Published in the said News Papers as may be proper for the promoting of the said Designs."[11]

Promotional literature provided the impetus for emigration to America. It described a land rich in potential. Hardworking people, it said, were ensured a large tract of land and financial success through their endeavors because the land offered everything needed. The literature of colonization worked so well, not just because it described an idyllic place, but because people in England were suffering under oppressive government, disease, and economic depression. Massachusetts Bay founder John Winthrop said that England grew "weary of her Inhabitants."[12] Bad times combined with extensive promotion via the printed word to lay before the people a place that held tremendous promise. The combination sent people to America from England, at first in small groups, but soon by the thousands. An estimated 1,980 colonists lived in what would become the colonies of British colonial America in 1625, but by 1700, that number was more than 250,000. More than 21,000 emigrated during the tumultuous 1630s to New England alone.[13]

SHAPING BELIEF

When colonizing efforts produced stable communities in New England, obtaining a printing press became a priority. The official business of a colony needed to be recorded, and this was one reason for a colony to obtain a press. But there was another. Religion was an integral part of the English immigrants' lives, and colonials saw as part of their purpose in colonization ensuring that Christianity remained central to the life of all in the New World. The biblical message, therefore, became the essential element produced on British colonial America's first printing press, which the Reverend Joseph Glover purchased in 1638 and set up in Cambridge.

Even before colonization efforts began, Europeans believed it was their duty to bring Christianity to New World natives. Englishman John Rastell wrote in 1519 that "a great Meritoryouse dede" could be performed by Europeans if they brought their way of life to the Indians.[14] Or, as John Smith said, "what can hee doe lesse hurtfull to any; or more agreeable to God, then to seeke to convert those poore Salvages to know Christ."[15] That meant bringing the Christian message to them. John Eliot, when he reached New England, assumed responsibility for doing this. He learned the language of the Indians and then carefully translated the Bible into Algonquin, the New Testament first followed by the Old, and had them printed on the Cambridge press.[16] Eliot's efforts, along with those of dozens of other missionaries, converted thousands of Native Americans to Christianity, but the effort was less successful than hoped. If Indians would not become Christians, then they did not fit into the definition of what colonials believed those living in the New World should be. If Indians rejected the message, then it was the duty of colonials under the concept of *vacuum domicilium* to seize the heathens' lands and put it to proper use since the land should be under the control of the righteous.[17] This fact is important in understanding the defining and shaping of America. A belief system based in the religious ideals of the Protestant Reformation became essential. Even though Roman Catholics were tolerated to a certain extent as they were initially in Maryland, settlers needed to fit into a fairly rigid religious belief system. The eradication of Indians could be justified under such a system and so could the expulsion of Europeans who did not subscribe to the understanding of Christianity of specific colonies. The Puritans in New England expelled Anne Hutchinson and Roger Williams because they refused to accept the collective beliefs of the group, not because they rejected the biblical message.

The printed word had the power to define, but it could also corrupt, according to many—especially those who held political, religious, and social power. This fact had been acknowledged in the Western world since Martin Luther's Ninety-Five Theses were published in 1518, and those in power realized that through a mass medium, revolutionary ideas could be disseminated that had the potential to upset the stability of a country or the basic belief system of millions.[18] For that reason, control of the press was vital in the early development of British colonial America. Some, like Virginia's William Berkeley, hoped that an open printing press could be avoided. In his much-quoted statement, the Virginia governor wrote of the way the printed word "brought disobedience, and heresy, and sects into the world."[19] The English government agreed, because in 1686 it passed a law that required each of its American colonies to acquire a press to produce official business, but the law also said that *no* "book pamphlet or other matters whatsoever be printed" without government approval or a license.[20]

The printed word held power and potential that was immediately realized. William Bradford in Philadelphia ran afoul of the law three times for what he published prior to 1690. After the third in 1689, Bradford pleaded with Pennsylvania authorities to release him from arrest. What the Quaker printer had to say was not intended to address the power of the press to define or set an agenda; it was only meant as statement explaining that printing was what Bradford did to make a living. To deny access to his press to anyone meant that he—Bradford—limited his economic security. But what Bradford said pointed out why the government feared the press and why it would become a defining element of America. "If I print one thing to-day, and the contrary party bring me another to-morrow, to contradict it, I cannot say that I shall not print it."[21] By 1689, people had realized the power of the press to effect public opinion. Opposing viewpoints, presented via the printed word, allowed for public debate. Even in a time before a single newspaper existed in British colonial America, printers, public officials, and most citizens understood that open debate could sway public opinion. Capturing public opinion had been central to all that happened in England from the Puritan Revolution of the 1630s forward, and the idea of open debate was based in the writings of John Milton and others. If concepts and beliefs could be presented to the people, Milton posited in *Areopagitica* in 1644, then those that were true would be the ones accepted by the people.[22] One had to hear all the opinions, however, to discern which were based in truth, logic, and religious ideals, according to Milton. Then, ideas could sway public opinion, which is exactly what William Bradford said as he pleaded for Pennsylvania authorities to release him from jail and to return to him the tools of his printing trade, even though he never mentioned anything other than the presentation of opposing ideas. "Printing," Bradford said, "ought rather to be encouraged than suppressed."[23] The reason was obvious to all: With public debate, the direction of a colony or a country could be defined. For some, this was what should occur. For others, an open press was anathema.

INTRODUCTION OF NEWSPAPERS

Printers in America began newspapers for many reasons. Some started papers for purely economic reasons. "I had a prospect of getting a Penny by it … having no other Way to support my Family" was what Boston printer Thomas Fleet said in 1741 when asked about his printing.[24] Benjamin Franklin no doubt was interested in financial gain, but he also felt that a newspaper should have a "Zeal for the Publick Good."[25] To that end, Franklin used his success with the *Pennsylvania Gazette* and *Poor Richard's Almanack* to establish other printers throughout the colonies. Through this network of printers, Franklin was able to gather more news than any other printer, but he was also able to disseminate his ideas more widely because those in his network were obligated to print what Franklin sent them. As a result, he was able to disperse his ideas throughout colonial America for many years. Others were more direct in their use of the press. William Livingstone, William Smith Jr., and John Morin Scott began the *Independent Reflector* in New York in 1752, boldly explaining that their goal was "vindicating the *civil and religious RIGHTS* of my Fellow-Creatures....In a Word, I shall dare to attempt the Reforming the *Abuses of my Country*, and to point out whatever may tend to its Prosperity and Emolument."[26]

The printer of America's first newspaper, Benjamin Harris, listed a number of reasons for starting *Publick Occurrences Both Forreign and Domestick* in Boston in 1690. He proposed "First, *That* Memorable Occurents of Divine Providence *may not be neglected or forgotten.*" "Secondly, *That people every where may better understand the Circumstances of Publique Affairs.*" "Thirdly, *That something may be done toward the* Curing, *or at least the* Charming *of that* Spirit of Lying, *which prevails amongst us.*" The rationale for the paper that was provided by Harris, a printer who had been forced to flee England in 1685 because his publications were considered seditious, appears straightforward and appropriate for the seventeenth century. They were. Religion was pivotal in Massachusetts, which was established by the Puritans. It is logical to assume that all citizens needed to understand what was going on around them in order to be active and knowledgeable in public affairs. Finally, New England was still an unsettled territory. Rumors were easily spread. King Philip's War of 1675 left colonists in constant fear of Indian raids on settlements. Publications like *The Captivity of Mary Rowlandson* (1682) fed these fears, so it was logical for Harris to promise that his paper would lay bare unsubstantiated rumor about Indians or other subjects. In addition, England and France were engaged in war. Known as King William's War, Native American and French attacks in the region had people on edge.

Publick Occurrences also provided its readers with a unique feature. The paper, a single sheet folded into four pages, left the final page blank. Harris may have later done away with it had his newssheet been allowed to survive. But with the first issue, the last page served as a means for readers to add news that was not in the paper and then pass it on. Readers became newswriters—participatory journalists—at the most basic level. The blank sheet allowed readers to comment on the news, too, a seventeenth-century blogging concept that would require three centuries to emerge again.

The situation in Boston was much more complicated than what might appear on the surface or to anyone who might think of *Publick Occurrences* solely in relation to its significance as America's first newspaper. The religious, public affairs, and rumor aspects that Harris outlined in his prospectus also referred to the tumultuous political situation in Massachusetts. The colony was in the midst of a struggle for control that was based in the politics and religion of the period. Harris had strong ties with religious dissenters in England, and in Boston, he quickly aligned himself with Cotton Mather and other Puritans who had been ousted from positions of power in the colony. King Charles II revoked Massachusetts' charter, making it a royal colony in 1684, and in 1686, James II appointed Sir Edmund Andros the colony's royal governor. Andros displaced the Puritans by giving control of the Old South Church to the Anglicans. In the ensuing rebellion, Andros was removed from power, and he and several government officials were jailed. Harris publicly supported these activities as did Mather and other Puritans. In the midst of the upheaval, Harris published *Publick Occurrences*, and members of the governing council reacted. Some members, like Elisha Cooke, strongly opposed the activities of Mather and knew of his close affiliation with Harris.[27] Banning Harris' paper was one way to keep Mather and other Puritans from having a ready mouthpiece. *Publick Occurrences* could also be considered a "disturbance of the peace & subversion of the governmt."[28]

Publick Occurrences' prospectus on its face, then, appears to present a newspaper that claims several noble reasons for its existence. Reading the paper's only issue affirms this, but the news embarrassed the government in its descriptions of the way several Indians were tortured and executed by militiamen. The combination of Harris' activities, which included publishing anti-Andros broadsides, his close association with Cotton Mather and the rebellion against the Andros' government, and the political power struggle taking place surely made it simple for the governing council to declare that Harris published his paper illegally because it was not licensed. But if Harris' past publishing ventures were any indication, *Publick Occurrences* was set to become a news sheet that could help direct the religious and political direction of Massachusetts. Banning it kept one side—that which wanted a return to power—from having a ready voice that could reach the people. The governing council's decree also kept dissenting voices out of the newspaper when postmaster John Campbell turned his handwritten newsletters into America's first continuously published newspaper, the *Boston News-Letter*, in April 1704. Campbell agreed that his paper would be "published by Authority," which means its content was approved by the Massachusetts government before it reached the public. Seventeen years later, however, a group of "concerned citizens" introduced a new newspaper to Boston. Its purpose was to use the printed word to change the course of what was happening in the city and colony.

SAVING LIVES THROUGH PRESS DEBATE

In 1721, the issues differed little from those in 1690; religious and political control of Massachusetts divided colonists. Governor Samuel Shute and the colonial assembly were locked in a power struggle. Anglicans hoped to use the situation to make another change in the colony's power: They wanted to have their religion declared the state church of Massachusetts instead of Puritanism. Headed by John Checkley, a group approached printer James Franklin—also a staunch Anglican—and the *New-England Courant* printed its first issue in August 1721. The writers for the *Courant*, whose names were later revealed by apprentice and younger brother of James Franklin—Benjamin—when he wrote their names on each of the articles of the papers that he saved, attacked Mather and other Puritan leaders. The *Courant*'s writers, they said, "Vilify and abuse the best Men we have, and especially the principal Ministers of Religion in the Country."[29] The *"Hell-Fire Club of Boston"*—as the *Courant*'s writers were called—aggressively attacked Shute and his government, which led on two different occasions to James Franklin's arrest. The Governor's Council, on the first, said that Franklin's paper was a "high affront" to the government and should never be published again unless Franklin agreed to a review of his paper by a government official.[30]

In terms of attempting to shape the direction of a community, however, the *New-England Courant*'s attacks on smallpox inoculation helped create a debate that would last for years in colonial America. The inoculation controversy tied in well with the religious debate, too, because Puritan minister Cotton Mather was the leading proponent of inoculating as a way to lessen the dangers of smallpox. The cadre of *Courant* writers included Dr. William Douglass, the colony's only university-trained physician. Douglass and Checkley led the attack on inoculation, beginning in the first edition of the *Courant*.

Smallpox was perhaps the most deadly of diseases facing people in British colonial America. More than 80 percent of Jamestown's settlers died from the disease, which they brought with them from Europe, during the first two decades of colonization. Smallpox claimed an estimated 90 percent of Native Americans who came in contact with it since the disease was unknown in America before colonization.[31] The 1721 smallpox epidemic marked the seventh time the disease had struck Boston since 1700. About 60 percent of Boston's residents contracted the disease in 1721 with 15 percent of them dying. The medical community, which was almost entirely composed of self-trained healers, had no idea how a person contracted smallpox, and it turned to the typical treatments of the day to treat it. Smallpox began with flu-like symptoms followed by pus-filled blisters. These soon burst and opened the way for other diseases to attack the sick. To treat the disease, the typical doctor of the period might quarantine those who were sick, and he blamed the disease on the "humours" or bodily fluids. The way to make an ill person better was to rid the body of the "bad" fluids such as blood, bile, and urine, which led doctors to bleed patients or to cause them to vomit or have bouts of diarrhea, all of which left the patient weakened and more susceptible to the disease. If smallpox did not kill its victims, it could leave them blind and with brain or kidney damage.[32]

Checkley, Douglass, and others were horrified when Mather enlisted practitioner Zabdiel Boylston to inoculate people during the epidemic. Mather had contracted smallpox during an earlier epidemic. After recovering, he began to search for ways to defeat the disease. From talking with his African slave and reading about how women in the Middle East gave people a weakened form of the disease to build up resistance to it, Mather concluded that inoculation was the only way to prevent a full-blown case of smallpox, and it was the first attempt at immunology in the English-speaking world. This, of course, created a firestorm of reaction with Checkley and Douglass taking their objections to inoculation directly to the people via the *Courant*.[33]

Douglass called inoculation "a Wicked and Criminal Practice" that was "*imprudent*," and "*notoriously false.*" He referred to Mather as "*the Town Lyar*" and Boylston as a "quack."[34] All three of Boston's papers were used to debate the merits or folly of inoculation. Douglass pointed out how no trained physicians in Europe practiced inoculation. Where it had been tried, those inoculating ceased the practice because of the dangers involved. In addition, inoculation could not be trusted since one did not receive the disease in "the ordinary way," which meant one did not contract the disease naturally.[35] Another *Courant* financier, Doctor George Steward, pointed out that inoculation violated all sensible medicine since it injected a person in perfect health with a deadly disease—one the person might never contract.[36] In the *Boston Gazette*, the pro-inoculation forces outlined precisely the successes of inoculation in Boston during the current epidemic. They listed the number of people inoculated and the number of them that died—one. They explained how people reacted to inoculation. Finally, they concluded by calling inoculation "a most merciful and wonderful Work of GOD."[37]

Winter helped bring an end to the smallpox epidemic, but the newspaper debate provided colonials with concrete proof of the power of inoculation. In 1725, the *American Weekly Mercury* announced that Louisa, daughter of George II, had been inoculated for smallpox.[38] In 1730 during another outbreak of smallpox in Boston, the printers of the *New-England Weekly Journal* published each week the number of deaths that occurred from the disease. They broke the tally down by those who succumbed following inoculation and those who died from the disease contracted in the "natural" way. One in four who contracted the disease in the natural way died, the paper said, while "those that have taken it by *Inoculation*, there are more than *one hundred*, who are through the Mercy of GOD recovered, and *four* have died."[39] Twenty-five percent casualties versus less than 4 percent seemed to the printers of the *Journal* to make inoculation the obvious choice for people, but they published the list of those inoculated, too, so that people could look through and discern which important citizens or friends might have chosen to be injected with a weakened dose of smallpox.[40] By 1730, inoculation was generally seen as a viable method of fighting smallpox. According to the *Pennsylvania Gazette*'s reprint of an article from *Chambers's Dictionary*, "The Practice seems to be useful....Advantages impossible to be had when the Distemper is caught in the natural Way. It has also been constantly observed ... the Danger next to none, the Recovery easy, and that the Patient is equally secured from the Distemper for the future, as he would be by having gone thro' it in the natural Way."[41] Though inoculation was limited by legal action into the 1760s, the practice was accepted in America, and the debate in the press made inoculation the logical choice for people.

PUBLIC ISSUES, PUBLIC DEBATE, PUBLIC SPHERE

As the colonies thrived, people became more concerned with the issues of everyday life in their communities. While all the British settlements along the Atlantic seaboard were not free from the danger of attacks by Native Americans or enemies of the British, most of the longest established towns and cities were because they were insulated—at least from the west—by expansion into the backcountry. Danger still persisted at sea, and wars with France and Spain created concern and hardships throughout the colonial period. Other issues, though, affected everyday life, and the press became the principal vehicle to address them. In a period when the number of printers and print shops was limited, it is amazing how much people turned to the press to sway opinion. Printer William Goddard said that newspapers were "necessary & important Alarms,"[42] not in reference to everything that appeared in newspapers but to those items that affected the welfare of people closely. People used the printed word to discuss any thing they believed affected them in some way. Usually, the topic that created the most debate revolved around political issues, but legislation could involve many subjects that related more to everyday life than to governing, including religious and economic issues.

In 1750, about 1.7 million people lived in British colonial America.[43] A dozen newspapers served the colonies and were published in the major population centers, though none of the major cities' populations exceeded fifteen thousand. The total number of papers published by individual printers varied, but the typical printer produced about 600 papers each week. Despite what seems to be a small number of newspapers, their ability to disseminate information into the general population was much greater than their circulation figures. Newspapers were easily shared, not only within the town they were published but with other areas of the colonies. Printers scheduled publication of their papers to coincide with the schedules of ships and post riders so that the latest issues of their papers could be dispersed as quickly as possible. Printers sought the position of postmaster in order to send their publications throughout the colonies postage free. Printers also struck deals with postmasters in order to obtain this franking privilege. In 1758, Benjamin Franklin, working with Williamsburg printer William Hunter as co-postmasters, devised a plan to make sure papers moved freely or for very little money.[44]

As a result, newspapers from across the colonies were available in all population centers, taverns, and lodging houses where they were read aloud and their content discussed and debated. Most newspaper subscriptions, though, were sold to households. There, they would have been read to the household and then shared with other families, perhaps as many as four or five, who would have done the same.[45] As a New Jersey citizen explained, "one subscriber would read it on the evening of its arrival and pass it over to his neighbor the next morning."[46] Papers were considered "the gen'ral source throughout the nation, of ev'ry modern conversation."[47] By the middle of the 1750s, people of all economic and social levels were using them to take part in discussion in the public debate. "How common is it to see a Shoemaker, Taylor, or Barber, haranguing with a great deal of Warmth on the publick Affairs?" a New Yorker observed in 1756. "He will condemn a General, Governor, or Province with as much Assurance as if he were of the Privy Council, and knew exactly wherein they had been faulty:—He gets his Knowledge from the News-Papers, and looks upon it undoubtedly true because it is printed."[48] Even though literacy rates in America were relatively high, the fact that papers were read aloud and their content debated and discussed in public meant that those unable to read could still be a part of the conversation or at least know the issues at hand.

Printing and newspapers, therefore created what German sociologist Jürgen Habermas called the public sphere.[49] For a public sphere to exist, printers had to be relatively free to print whatever they received on any subject. For the most part, such a situation existed in colonial America by the middle of the eighteenth century. In 1735, John Peter Zenger, who had been held in jail for six months on charges of seditious libel brought by Governor William Cosby, was found not guilty by a New York jury even though libel law of the time demanded that the printer of the *New-York Weekly Journal* be found guilty. In the rare cases when printers were ordered to stop printing material colonial governments found contentious, they tended to ignore the constraints. In Boston in 1748, a young firebrand named Samuel Adams turned to the press to attack Governor William Shirley, whom he and others felt had usurped the authority of the colony's legislature. Though Adams and the *Independent Advertiser* leveled caustic attacks on the governor, they faced no repercussions. In fact, following Zenger's trial, no American printer of the colonial period was ever convicted of libel, even though some were charged with the crime.[50]

With the colonial press relatively unencumbered by any restraints on what was published, the public sphere afforded people the opportunity to debate the issues that most affected them. Or as John Adams said in 1765, it became "easy and cheap and safe for any person to communicate his thought to the public."[51] Rarely did all people agree on any issue, so they argued their points in the papers in hopes of creating a consensus that could then turn the item of concern into a political issue that would then be voted on by the colony's legislature. A number of issues became lightning rods of contention. Most were specific to an issue affecting a single colony. Others were issues that required action by individual colonies but were general issues of concern up and down the Atlantic seaboard. The debate over the establishment of a public school in New York with religious oversight provides an example of an issue that was initially local in nature. Later, however, it would become an issue throughout America as people struggled with basic rights, many demanding that their religious freedoms be protected and leading to the religion clauses of the First Amendment.

In 1746, New Yorkers initiated discussions about beginning a college in the colony. Four colleges already existed in America: Harvard; William and Mary; Yale; and the Log College in Princeton, New Jersey. Rapidly growing New York believed it should have an institution, too. As the colony moved toward making the school a reality, two factors affected discussion: the general purpose of a college and the fear that the college might lead to a "state church" in New York. Most Americans wanted their clergy to be well educated, and training young men for the ministry was one of the main purposes of a college. Even though the existing colleges accepted students of various religious affiliation, they trained ministers for specific denominations—for example, Yale for Congregationalists and the Log College for Presbyterians. Issues surrounding religion had been debated in the newspapers since the Great Awakening of the late 1730s and 1740s. At that time, a number of religious groups in America split, and issues of religious freedom were constantly discussed in newspapers as will be seen shortly. People became concerned about the possible restriction of their rights to worship as they pleased. A college could greatly affect religious freedom, some said, especially in a colony like New York.[52] The Anglican Trinity Church in New York City existed because of a royal charter and received financial support from taxes through a 1693 law.[53] If the Anglican Church ran the college, therefore, many believed that it would only be a matter of time before New York would establish Anglicanism as the colony's official religion. All one had to do, they said, was look to the east and south. In Connecticut, Congregationalism was supported by the government, and New Jersey was already leaning toward Presbyterianism since the establishment of the Log College. With the issues before the people, debate ensued.

The *New-York Mercury*, printed by Hugh Gaine, served as the press outlet for those who supported the Anglican-controlled college. The *Independent Reflector* provided the bulk of the initial opposition to a church-run, government-supported school. Not by chance, Gaine was Anglican, and the men behind the *Reflector*—Livingstone, Smith, and Scott—were Presbyterian, though they favored no religious affiliation for the school.[54] With the sides firmly set, the debate surrounding the college began. One Anglican supporter said that the whole purpose of an education was "to teach and engage the Children to *know God in Jesus Christ*, and to love and serve him."[55] Another said "that *national Establishments* can alone diffuse, thro' a Country, the full social Advantages arising from Religion....But if the Wisdom of a country has established one religion, and that the best they could frame or devise, we are sure to have this one Religion flow down thro' the Ages."[56] Yet another writer pointed out that the Church of England was already established in New York, and "what is already establish'd in the Province" should be the religious affiliation of the college.[57]

William Livingstone led the attack on the plans for the college. He did not oppose the colony establishing a school; he simply objected to state support of the institution combined with oversight by a religious denomination. Livingstone very much anticipated the premise of the separation of church and state that was understood in the wording of the Constitution and in the First Amendment[58] and whose disconnect President Thomas Jefferson deemed divided by a "wall of separation" in 1802.[59] On August 9, 1753, Livingston published an essay titled "*The Absurdity of the civil Magistrate's interfering in Matters of Religion*" in the *Independent Reflector*. There, he said, "Government was instituted for the Establishment and Preservation of our civil Interests, it is an Abuse to suppose the Magistrate can have any Right to interfere in Matters of religious Belief." He believed that the establishment of the college under Anglican direction and funded by the state was "*the height of Tyranny and Oppression*" and that such an entwining was an "Absurdity to suppose" because to do so would mean that government would have the power to "*enslave* the *Consciences of Men!*" For six weeks, Livingstone laid out his reasoning as to why the new college should not be directed by a religious body if it was state supported. He concluded by saying, "I exhort, I beseech, I obtest, I implore you, to expostulate the Case with your Representatives, and testify your Abhorrence of so perilous, so detestable a Plot."[60] Ultimately, King College opened in 1754 under Anglican direction but without direct government support.

RELIGIOUS REVIVAL = BIG NEWS

In the 1730s and 1740s, religious revival swept through the American colonies. The Great Awakening, as the period would later be called, affected all parts of British colonial America, and few if any of the more than 900,000 people living in the thirteen colonies at the time escaped the effects the series of revival meetings had on American religion. Just as dissenters in England had done a century before during the Puritan Revolution, the revivalists in America—and those who opposed their ideas on religion—turned the printing press into one of their greatest tools for the spreading of their message. At the heart of the controversy and the use of the press was an itinerant preacher from England named George Whitefield. According to Harry Stout, Whitefield pioneered the use of the press among all strata of American society. Whitefield developed a new form of communication, Stout says, "in which people were encouraged—even commanded—to speak out concerning the great work of grace in their souls."[61] Whitefield became, according to America's first media historian Isaiah Thomas, "the common topic of conversation from Georgia to New Hampshire. All the newspapers were filled with paragraphs of information respecting him, or with pieces of animated disputation pro or con."[62]

Whitefield arrived in America in November 1739, but newspaper stories about his activities in Britain began appearing in American newspapers as early as 1737, when John Peter Zenger copied an article from the *Gentleman's Magazine* that talked about Whitefield's plans to begin an orphanage in Georgia.[63] In April 1739, stories about Whitefield's preaching in England started running continuously in papers. News items about ministers were not uncommon, but the stories about Whitefield claimed that he "preach'd to 4 and 5000 People" regularly and to an unbelievable "10,000 People, at Kennington-Common" on May 4.[64] On May 8, Whitefield preached again at Kennington Common, and those in attendance numbered an astounding twenty thousand, according to the news account. Whitefield, following his sermon, was warned never to preach there again "for his own security" because of the divisions his preaching were causing among people.[65] Whitefield was an ordained Anglican minister, but he tended to ignore the tenets of the Church of England, promoting, instead, the ideas of "the New Birth," which adhered to the concept of individuals as being reborn solely through the grace of the Messiah.[66] As a result, Whitefield "like a comet drew the attention of all classes of people."[67] His powerful voice and his controversial message were the perfect combination for the first newspaper story to capture the attention of nearly all Americans regardless of where they lived.

Controversy, then, fueled the interest in Whitefield along with his strong religious message. No one tended to remain ambivalent where the man called the "Grand Itinerant" was concerned, and Whitefield's actions only added to his story. Often, he attacked local ministers who refused to allow him access to their pulpits, which led to disputes in newspapers. In the South, he chastised slave owners. Everywhere, he challenged religious orthodoxy, which led to numerous splits in America's Protestant denominations. Whitefield, however, fueled the news about himself and even stirred up controversy because traveling with him was William Seward, who probably served as the first public relations specialist in American history.[68] Seward, just as he had done in Britain, prepared notices about where Whitefield would be preaching as he toured the colonies. Seward sent those announcements ahead of Whitefield to newspapers of the respective colonies in order to put Whitefield on the radar of locals. After revival services, Seward prepared news stories about them. He always included in his write-ups the number of people who were in attendance. And just like the stories from England that initiated interest in Whitefield, those in attendance always numbered in the thousands. Fifteen thousand attended Whitefield's final sermon in New York City, Seward's press release stated.[69]

Because Whitefield was a lightning rod of a figure, people responded in the press to all that Seward wrote and to what Whitefield preached and did. One angry writer claimed that the number of people reported as attending a Whitefield sermon were "always exaggerated, being often doubled and sometimes trebled."[70] Others were less polite, calling Whitefield and Seward liars. When Seward wrote a lengthy story about how Whitefield was able to close down two "dens of iniquity" in Philadelphia—the dancing school and the concert hall—one writer pointed out that both men had "little Regard to Truth" and were simply saying whatever was needed to spread Whitefield's fame and ideology. The story forced Benjamin Franklin to admit that he ran the article "at the Request of Mr. Seward" and that "the Article allow'd to be literally true, yet by the Manner of Expression 'tis thought to insinuate something that is not."[71]

By the time Whitefield left America in 1741, religious views of many had been forever changed. A century earlier, members of what were considered radical religious sects had been hanged in Massachusetts, but after all of the debate and discussion surrounding the Great Awakening and Whitefield, Quakers—among those executed—were granted permission to use Boston's Faneuil Hall, and in the next decade Jews were granted religious liberty in Pennsylvania. Even though many colonies still had government-supported churches, the Great Awakening and the preaching of Whitefield changed the standing order of religion in America to a great extent because the press allowed unrestricted debate on religious issues.[72] The renewed interest in religion led to hundreds of publications. In 1738, the year before Whitefield arrived in America, printers produced a total of 133 pamphlets, and about 42 percent of them were on religion. By 1741 and Whitefield's departure, printers produced 146 pamphlets on religion; they made up more than 60 percent of all tracts published and represented more than a 142 percent increase in religious publications.[73] The Great Awakening and Whitefield's preaching, which forced people to ask themselves "What must I do to be saved," laid the groundwork for evangelical Christianity in America. It made the printed word and the newspaper especially central to debate and discussion because Seward's press releases always found their way into print. Or as Benjamin Franklin said after the dancing hall fiasco in Philadelphia, "any one is entitled to the Use of it [the press], who thinks it necessary to offer his Sentiments on disputable Points to the Publick."[74] People in New England knew of Whitefield's activities in South Carolina and Georgia. People in those colonies stayed apprised of Whitefield's actions in New England, the Middle colonies, and everywhere else. Colonials realized that events that occurred in America hundreds of miles from them could have serious implications—just how much those events could affect them was about to be discovered, and the press would again be the link for Americans.

WAR, REVOLT, TAXES, AND MORE WAR

From 1750 to 1760, British colonial America's population grew by 36 percent. The number of newspapers that served that population grew twice as fast.[75] Americans did have a taste for news, which printer James Parker noted in 1750. He said that a newspaper was "an Amusement we can't be without."[76] The increase in the number of newspapers in America in the 1750s was a reaction to events that threatened the colonies' existence, and it was simply the beginning of a series of crises—expounded upon and fueled by the press, the number of papers growing by more than 263 percent from 1754 to 1775—that would formulate an American identity and lead to revolution. In the mid-1750s, though, America faced what historian Fred Anderson called, "the most important event to occur in eighteenth-century America," and historian Frank Luther Mott deemed "the great running story" of the colonial era.[77] They described the French and Indian War, a part of a global confrontation between Britain and France called the Seven Years' War.

The French and Indian War forced America's newspaper press, which was barely fifty years old, to mature. The French and Indian War was the fourth declared war between Britain and France that involved fighting in North America and in the waters of the Atlantic and Caribbean, but it was the first that threatened colonies in all regions of British colonial America at roughly the same time. This meant that newspapers had to react more quickly than before because an attack on Virginia could have repercussions for New England. As we have seen, a system of free exchanges of newspapers was set up in the 1750s using established postal routes. The post moved regularly and quite rapidly among the major cities of Philadelphia, New York, and Boston, which greatly enhanced the dispersal of information. Timeliness was measured in days and weeks in the 1750s, but news during the war moved about as fast as physically possible. As an example, the *New-Hampshire Gazette* printed a story on August 29, 1760, about fighting that occurred in the South Carolina backcountry less than three weeks earlier. Because the implications of a French takeover of the American colonies meant a change in religion, government, and probably a way of life, printers found ways to pass on as much information as possible to people. Between their regular publication dates, they produced supplements or "extraordinaries" whenever information that would be of value arrived at their offices. Some printers announced that they were reducing the advertisements in the paper in order to provide more war news.

During the colonial period, newspapers would print stories that might not be true since the verification of information was, at best, difficult because of the speed it moved and the manner in which it was obtained. During the French and Indian War, printers tried to authenticate reports they received, especially when the information had far-reaching implications. Printers started to trust what had been printed in other newspapers more than stories that circulated on the street, in pubs, or in hastily written correspondence from soldiers. For example, the *Boston Evening-Post*'s printers heard rumors in September 1760 that the French had surrendered Canada. The capitulation of Canada meant for all practical purposes that the fighting in America would end. To report such news of great import erroneously would be irresponsible, they felt. They set the initial story in type on the third page of their paper and waited. Soon, more reports arrived from a different source, and the printers added another paragraph that began by saying, "The above account is confirmed" and the verification via the other stories added following the printers' notice.[78] To ensure accuracy, some papers hired "News-writers" to send information back to the paper. Reporters would not become a regular feature of American journalism until the 1840s, but stories in New England papers during the invasion of French Canada in 1759 and 1760 contained references to "the weekly news writer" and the "Writer of News." In one story in the *Boston Gazette*, the news writer prefaced his report with this comment: "We have spared no pains in making the above accounts as brief, and at the same time as accurate as possible."[79]

The war also caused the colonies to begin thinking more as one entity, though the transformation to a collective from individual units would take many more years to become a common way of thinking for the majority of Americans. Here, newspapers played an essential role by providing an outlet for numerous voices. One of them, known as the *Virginia Centinel*, produced a series of newspaper essays during the war to remind colonials of the dangers they faced. "FRIENDS, COUNTRYMEN!" the *Centinel* cried. "AWAKE! ARISE! When our Country, and all that is included in that important Word, is in most threatening Danger....If Liberty, Property, Religion, Life, are Terms of any Significancy ... certainly, you must be alarmed, when YOUR COUNTRY IS IN DANGER."[80]

The move toward uniting the colonies was a byproduct of another effort: keeping the Iroquois Confederation known as the Six Nations as allies and trade partners with the British colonists. Most Native American groups west of the Appalachians and in Canada had long been closely associated with the French. As one of the two most powerful groups of Indians east of the mountains (the Cherokee in the South being the other), London wanted to ensure that the confederation maintained its ties to the colonists. Governor James De Lancey of New York was ordered to arrange a meeting with the Six Nations at Albany and representatives of all colonies from Virginia northward. For Benjamin Franklin and several others, this became the opportune time to push for a union of American colonies. Franklin had long felt that such a union was essential to deal with trade, defense, and affairs with Native Americans. He wrote to New York printer Hugh Gaine in 1751, "Were there a general Council form'd by all the Colonies, and a general Governor appointed by the Crown to preside ... every Thing relating to Indian Affairs and the Defence of the Colonies might be properly put under their Management."[81] The planned Albany Congress provided Franklin with the perfect catalyst to promote his plan of American union, and on May 9, 1754, about a month before delegates left for Albany, the *Pennsylvania Gazette* ran what would be America's first political cartoon.[82] "JOIN, or DIE" featured a snake severed into parts, each representing the colonies, and surrounded with heavy black border, the symbol of something tragic or of the death of someone important. The message was self-evident, and newspapers throughout America either reprinted Franklin's story, mentioned his editorial cartoon, or reproduced their own version of the disjointed rattlesnake.[83] With the woodcut, Franklin observed that "the present disunited State of the British Colonies, and the extreme Difficulty of bringing so many different Governments and Assemblies to agree in any speedy and effectual Measures for our common Defence and Security" almost surely ensured "the Destruction of the British Interest, Trade and Plantations in America." The Albany Congress failed to produce a strong Native American alliance, but "The Commissioners from the several Governments were unanimously of Opinion, That an Union of the Colonies was absolutely necessary in order to defeat the Schemes of the French," the *Boston Gazette* reported.[84]

In the months after the congress, newspaper printers, along with colonial governors and some ministers, became the strongest advocates for a union of colonies. They pushed for it, declaring in the pages of their papers that it was the only way to ensure American survival. Boston Congregational minister Jonathan Mayhew, who would preach numerous sermons on American unity, declared "What UNION can do, we need only look towards those Provinces, which are distinguished by the Name of THE UNITED, to know....Ye cannot be saved from the Storm you are now threatened with, yea, which is already begun, except ye are at UNION AMONGST YOURSELVES; and exert your Strength together, for your common Interest. Upon this Condition, you are safe, even without a Miracle; otherwise, nothing short of one can save you...."[85] Printer Hugh Gaine appended his thoughts to an anonymous essay on "the PRESENT STATE of this Continent," saying: "*I hope, and pray the Almighty, That the British Colonies on this continent, may cease impolitically and ungenerously to consider themselves as distinct States, with narrow, separate and independent Views, pursue temporary and ineffectual Expedients, and sink their public Wealth into private Emoluments; that they will unite like Brother Protestants, and Brother Subjects ... and secure to themselves and their Posterity, to the Ends of Time, the inestimable Blessings of Civil and Religious Liberty, and the Possession and Settlement of a great Country, rich in all the Fountains of human Liberty.*"[86]

Though printers advocated and royal governors prodded, the elected legislatures of the various colonies refused to give up what they considered their individual rights. While not voting union, legislatures enacted legislation that provided the defense and security that the Albany Plan of Union advocated. Mayhew seemed to know that representatives were not ready to create a union of colonies and said, "But, in the mean time, each Government ... will undoubtedly concur in such Measures as are necessary and practicable for the Common Safety."[87] And, they did. Colonial assemblies provided money and troops to fight the French and Indians without hesitation. Even though New York, for example, had to constantly worry about French and Indian aggression from Canada, the colony readily sent aid to Virginia to help that colony fight the French and Indians in the Ohio Valley.[88] Colonial governments may have rejected the Albany Plan of Union, but newspaper coverage of calls for union demonstrates that a collective consciousness was developing in British Colonial America, led to a great extent by those who produced the weekly newssheets. News about the French and Indian War was truly "the Main Subject of the Public Attention,"[89] and the extensive amount of space provided for war news proved that the press could keep the public informed. Printers enhanced news presentation, news sharing, and service to communities. They proved the press was ready for what lay ahead, though no one in 1760 had any idea of what would happen next or the role the press would play in defining America. But John Holt, who began working for newspapers in 1754 summed it up well: "It was by means of News papers that we receiv'd & spread the Notice of the tyrannical designs formed against America and kindled a spirit that has been sufficient to repel them."[90]

The Stamp Act of 1765, William Smith Jr. observed, "lost Great Britain the affection of all her Colonies."[91] Historian Arthur Schlesinger said the period after the French and Indian War represented "the newspaper war" on Great Britain where Americans "exhibited extraordinary skill in manipulating public opinion" as people were excited "in one community by newspaper accounts of activities in other places."[92] Many people, however, were already thinking this way. The press of the 1750s had created this mindset through coverage of the French and Indian War, so people's ideas about colonial relations were already changing. "All the colonies," as one writer said in 1757, "have a natural Connection and Interest in one another, and in the same Places."[93] In the 1760s and 1770s, this use of the press to disseminate information widely and to use it as a propaganda tool became essential. Considering America as a separate, independent entity from Britain was becoming more and more accepted, and thinking of those who lived in the British colonies as Americans, not Britons, seemed more and more natural.

The events and people of the 1760s and 1770s have become the fabric of what many consider America to be. How the events played out in the press, and how the key players used the public prints and were discussed in them have defined what it means to be an American. The Stamp Act crisis of 1765 became the first in a long list of events. David Ramsey, a historian contemporary with the era, said, that printers "provided their united zealous opposition" to the tax. [94] The Stamp Act was a direct affront on their livelihood because a penny tax was placed on any whole sheet of paper, and fines of up to ten pounds would be levied to any printer who might violate the law. It was only natural, therefore, that printers opposed the Stamp Act, but other opponents quickly pointed out that the tax was not just a tax upon those who used paper as a part of their occupation but an attack upon all Americans. Society—any society—existed for the public good and economic well-being of individuals in order to enhance the community as a whole, both now and in the future, they believed. Within this concept, which developed from Enlightenment thought including the idea of John Locke that government existed principally to promote the welfare of the governed, the polemicists of rebellion framed their rhetoric.[95] "AWAKE! Awake, my Countrymen," the anonymous "B.W."—who was probably Samuel Adams—wrote in the *Boston Gazette*, "defeat the designs of those who enslave us and our Posterity.…This is your Duty, your burden, your indispensable Duty. Ages remote, Mortals yet unborn, will bless your generous Efforts, and revere the Memory of the Saviours of their Country."[96]

America's reaction to the Stamp Act included riots, tarring and feathering stamp agents, and the closing of a number of papers in protest of the tax. No paper ever printed on a sheet of taxed paper, and in March 1766, Parliament repealed the Stamp Act, proving that a newspaper-led protest could change—at least temporarily—London policy toward the colonies. But Britain felt that Americans should pay their just part of the Seven Years' War regardless of opposition, and taxes were the way to do it. In 1767, London levied a duty on imported tea, paint, lead, glass, and paper through the Townshend Acts. It was becoming obvious that colonials were dividing along irreconcilable lines, one saying taxes were inevitable since the colonies were a part of Britain and the other side saying Britain had no right to tax since Americans were not represented in Parliament. Eventually, this side would push for separation and independence, dropping the idea of no taxation without representation and replacing it with Franklin's "JOIN, or DIE" moniker and promoting the concept that British rule meant tyranny, oppression, and slavery for all Americans.[97] If Britain intended to tax items imported into America, then Americans could do without them or make them themselves. One writer suggested, "let us unanimously lay aside foreign superfluities, and encourage our own manufacture. SAVE YOUR MONEY AND YOU WILL SAVE YOUR COUNTRY!"[98]

Americans continued their newspaper war on Britain and joined it with this economic one. In reaction to the Stamp Act, *Boston Gazette* printer Benjamin Edes, and eight of the city's merchants formed the Loyal Nine. Soon, they opened their group to like-minded individuals, and the Sons of Liberty were born. Quickly, Sons of Liberty groups formed throughout America to boycott British goods and foment resistance to British authority. Acts of opposition by the various groups throughout the colonies were important, and as opposition to British authority spread, news of the activities of the Sons of Liberty became essential elements of news coverage of events. But the press as a means of resistance had worked so well with the Stamp Act that it became the critical tool for those who believed it was time to sever ties with Britain. In Boston, reaction to the Townshend Acts was so volatile that British troops were sent to occupy the city in September 1768. That's when Edes, Samuel Adams, and others began "cooking up paragraphs, articles, occurrences, &c., working the political engine," as John Adams described it.[99] They created the "Journal of Occurrences" to describe what was happening in Boston. The first Journal carried a dateline of September 28, the same day the occupation began and appeared in the Patriot *New-York Journal*. It said that Boston was surrounded by fourteen ships of war "at a Time of profound Peace." It described how British troops "landed under Cover of the Cannon of the Ships of War, and marched into the Common, with Muskets charged, Bayonets fixed … with the Train of Artillery upwards of 700 Men." After more descriptions of the events in Boston, the first Journal closed with these words: "*The above Journal you are desired to publish for the general Satisfaction, it being strictly Fact.*"[100] The "Journal of Occurrences" continued its diatribe against the British through August 1, 1769, and it appeared throughout the colonies under that title and was also called "Journal of the Times" and "Journal of Transactions."

While Sam Adams and company were at work in Boston, Philadelphia lawyer John Dickinson provided more newspaper ammunition against the British. Using the earthy pseudonym "Pennsylvania Farmer," Dickinson attacked the Townshend Acts. In a series of twelve letters that appeared in almost every paper in America, Dickinson made the case for opposition to British duties on commodities. In his last letter, he said, "Let these *truths* be indelibly impressed on our minds—*that* we *cannot be* HAPPY, *without being* FREE—that we cannot be FREE *without being* SECURE IN OUR PROPERTY, that, *we cannot be* SECURE IN OUR PROPERTY, *if, without our consent, others may as by right, take it away*," adding "'SLAVERY IS EVER PRECEDED BY SLEEP.' *Individuals* may be *dependent* on ministers, if they please. STATES SHOULD SCORN IT."[101] Numerous others joined the newspaper debate on both sides of the issue, and nonimportation of British goods became a rallying issue in newspapers. Women banded together to spin thread in order to decrease dependence on British cloth, papers reported, while others promised not to drink English-imported tea.[102] What made these articles so powerful, besides their content, was the way in which printers throughout the colonies reprinted them. Colonials realized, just as they had during the French and Indian War, that what was happening in Boston would affect them eventually, no matter where they lived in the thirteen colonies. And, as Arthur Schlesinger said, "it aided materially in providing the basis for common attitudes and common actions."[103]

The press only intensified its coverage of the events of the 1770s. The Boston Massacre of March 5, 1770, provided the opportunity to demonize British soldiers in the "inhuman tragedy" that took the life of five Bostonians. Visuals were essential to press coverage. The *Boston Gazette* provided a woodcut of coffins with the initials of those killed, and Paul Revere produced "The Bloody Massacre perpetrated in King Street," a colorized image that went on sale three weeks after the shooting. The massacre's anniversary became a means "that kept it burning with an incessant flame" among Americans, according to David Ramsey.[104] None of those anniversary speeches was more incendiary than the one made by Joseph Warren just six weeks before shots were fired at Lexington and Concord. Calling it "the bloody Tragedy," Warren asked who spilled American blood: "haughty France or cruel Spain," "the grim Savage." "No," he answered. "The arms of George our rightful King have been employed to shed that blood which freely would have flown at his command.…Having redeemed your country, and secured the blessing to future generations, *who*, fired by your example, shall emulate your virtues, and learn from *you* the heavenly art of making millions happy."[105]

As the riff between the colonies and Britain grew, the papers provided more and more information that pointed toward schism. When Patriots dressed as Indians dumped British tea into Boston harbor in protest of the continued tax on the drink, many viewed the action as one writer in Boston did. "The News of the destruction of the Tea, as it was the ONLY way left prevent the chains prepared for us," he said, "gave satisfaction to all the Friends of American Liberty, who heard of it."[106] As the situation worsened, efforts were made to control the press, and those who advocated neutrality or who believed that an impartial press—or one that allowed Loyalist views to be espoused—were threatened into silence. James Rivington said, "THE TRUE SONS OF LIBERTY … proclaim the deeds" of all even "TRAITORS to the CAUSE OF FREEDOM."[107] By 1774, however, the same newspapers, printers, and everyday citizens who had constantly advocated that "The freedom of the press is the great bulwark of Liberty" when people had tried to silence them on controversial subjects, now thought like printer John Holt, who believed that the cause of freedom was too important to allow the press "to give a fair hearing to such performance.… I disdain such publications."[108] Others provided a sterner warning, saying "if you print, or suffer to be printed in your press anything against the rights and liberties of America, or in favor of our inveterate foes, the King, Ministry, and Parliament of Great Britain, death and destruction, ruin and perdition, shall be your portion."[109] Printers like Rivington were hanged in effigy, arrested, had their shops ransacked and presses destroyed, and/or forced to flee the country. This left one predominant voice in the newspapers—that of those who favored independence.

Not all Americans, however, believed that independence was the right course of action for the colonies even though open confrontation had existed since April 1775. Those people kept speaking out, but because of the mindset of printers like Holt, their outlets for discussion grew increasingly limited. Then, in January 1776, an anonymous pamphlet titled *Common Sense* appeared. "*Common Sense* was the work of an 'original genius,'" one New Yorker claimed.[110] The anonymous writer turned out to be a newly immigrated Briton named Thomas Paine who had grown sympathetic to the American plight. After a year in America working for the *Pennsylvania Magazine* and writing anonymous letters that claimed independence was God's will for America, Paine penned his pamphlet. There, he advocated that it was only "common sense" for America to initiate complete separation from England via a declaration of independence. His words struck a chord with Americans, and after one month *Common Sense* had sold a half-million copies and was in its third printing.[111] "I find Common Sense is working a powerful change in the minds of men," George Washington said.[112] The commander in chief was correct. Paine structured the idea of declaring independence into a logical argument that made it superfluous to remain colonies of Great Britain and subservient to a tyrannical monarch. Silas Deane, who helped negotiate America's alliance with France but was later accused of being a traitor for calling the Revolution hopeless, explained in a letter in September 1776 how this press-driven message affected the people. Even "the very poorest," he said, "are not an ignorant unprincipled rabble, heated and led on to the present measures by the artful & ambitious few." Instead, Deane explained, they form their opinions "with Gazettes & political publications, which they read, observe upon and debate in a Circle of their Neighbors."[113]

Paine, like thousands of other Americans, put his words into action, enlisting in the colonial army. The colonies had declared their independence in July, but the war effort was going poorly. So, Paine began writing again. "These are the times that try men's souls. The summer soldier and the sunshine patriot will, in this crisis, shrink from the service of their country; but he that stands it *now*, deserves the love and thanks of man and woman. Tyranny, like hell, is not easily conquered; yet we have this consolation with us, that the harder the conflict, the more glorious the triumph."[114] For the next six years, a principal goal of newspapers was to bolster the morale of the military and civilian population, of which "The Crisis" series served as a template. When the British surrendered at Yorktown in 1781, newspapers were viewed as co-equals with the military in winning the war. As David Ramsey said, "the pen and the press had merit equal to that of the sword."[115]

***

The press played a key role in defining America even before settlers began arriving from Europe. Colonization literature painted a portrait of America as a place that offered emigrants all they needed to survive and to become social and economic successes. Even though the press developed slowly in America because the colonies lacked large population centers that could support public prints, by the middle of the eighteenth century, multiple newspapers served the major urban centers of Philadelphia, New York, and Boston. The combined circulation of Boston papers in 1754, for example, created a population-to-newspaper-issue ratio of seven to one, identical to that of London.[116] Because of the oral tradition that existed, newspapers were read aloud in homes, taverns, coffeehouses, and other public places, which opened their content to a broad spectrum of people who then discussed and debated issues. Newspapers, sold exclusively by subscription, traveled throughout British colonial America via the mails. They were shared, so their reach greatly exceeded their subscription numbers.

Printers actively pursued contributions to their newspapers. Benjamin Franklin made the call for correspondents to send him news of "*every remarkable Accident, Occurrence, &c. fit for public Notice.*"[117] As a result, citizens actively turned to newspapers to air their opinions, grievances, and compliments. Public prints became, as many printers claimed, "The gen'ral source throughout the nation, of ev'ry modern conversation."[118] Medicine, education, religion—all issues of importance to a community—were discussed in papers, broadsides, and pamphlets. As time passed, the public sphere that printed information created helped people throughout the colonies realize that there were some issues common to all. The religious matters that the Great Awakening addressed and created and the issues of survival during the French and Indian War produced two distinct results: They forced the press into a heightened sense of service to the public as the principal source of information throughout colonial America, and they helped colonials realize that issues that affected one part of British colonial America could have repercussions in others. Jack Greene has concluded, "a powerful process of social and cultural convergence" occurred that "rendered the differences among them [colonials] less and less pronounced."[119] This was the work of the press as printers picked up news from one newspaper and shared it with their readers.

In the 1760s and 1770s, especially, Americans voiced actively their opinions on taxes, representation, and their relationship with Britain. The debate was intense. Actions by groups like the Sons of Liberty combined with newspaper essays to produce a debate that ended in a declaration of independence and war. In this process, polemists cooked up paragraphs to inflame via emotion, and they employed reason to argue the validity of their points of view. The press gradually assumed a partisan bent that would continue for a century, and some would argue has never disappeared. The press of early America revealed the power of the written word. The public prints were a new and evolving medium. As such, they were powered by printers who accepted submissions of essays and information from people who wished to influence public opinion. By the middle of the 1750s, these essay/editorials were quite powerful. They were able, according to Andrew Eliot, the pastor of Boston's New North Church, "to awaken and arouse" the people to action.[120] This use of the press to set the agenda and shape the direction of colonial America was critical in the process of creating a new nation. As Americans worked to figure out the next steps of the United States of America following the Revolution, the press would become even more critical as the young nation looked toward the formulation of a new government and how to hold together a fledgling country.

NOTES

1. John Adams, *The Works of John Adams, with a Life of the Author, Notes and Illustrations*, C. F. Adams, ed., 10 vols. (Boston, Mass., 1850–1856), 2:219; quoted in Philip Davidson, *Propaganda and the American Revolution, 1763–1783* (Chapel Hill: University of North Carolina Press, 1941), 227.

2. John Adams to Thomas Jefferson, August 24,1815, *The Works of John Adams, with a Life of the Author, Notes and Illustrations*, C. F. Adams, ed., 10 vols. (Boston, Mass., 1850–1856), 10:172.
3. Julie Hedgepeth Williams, *The Significance of the Printed Word in Early America: Colonists' Thoughts on the Role of the Press* (Westport, Conn.: Greenwood, 1999), 30.
4. Richard Hakluyt, *Divers Voyages Touching the Discovery of America and the Islands Adjacent* (1598; reprint, London: The Hakluyt Society, 1850), 33.
5. Richard Lane, "Letter to M. Richard Hakluyt Esquire, and another Gentleman of the Middle Temple, from Virginia," in Richard Hakluyt, *The Principal Navigations Voyages Traffiques & Discoveries of the English Nation*, 12 vols. (London, 1598; reprint, Glasgow, 1903–1905), 8:299.
6. Earle, "Pioneers of Providence," 486.
7. Ralph Lane, *Discourse on the First Colony*, in John Smith, *Writings with Other Narratives of Roanoke, Jamestown, and the First English Settlement of America* (New York: Library of America, 2007), 847.
8. John Smith, "The Description of New England," in *The General Historie of Virginia, New England, and the Summer Isles, 1624*" (London, 1624), 208–09, in *Travels and Works of Captain John Smith*, ed. Edward Arber (Edinburgh: John Grant, 1910), 707–08.
9. Smith, "The Description of New England," 210, in *Travels and Works of Captain John Smith*, 709.
10. H. Roy Merrens, "The Physical Environment of Early America: Images and Image Makers in Colonial South Carolina," *Geographical Review* 59, 4 (October, 1969): 532.
11. Allen D. Candler et al., comps., *The Colonial Records of the State of Georgia* (Atlanta, 1904), 2:3, quoted in Rodney M. Baine, "James Oglethorpe and the Early Promotional Literature for Georgia," *William and Mary Quarterly*, 3rd series 45,1 (January 1988): 103.
12. John Winthrop, "General Observations for the Plantation of New England," May 1629, *Winthrop Papers*, 2:14, quoted in David Hackett Fischer, *Albion's Seed: Four British Folkways in America* (New York and Oxford: Oxford University Press, 1989), 16.
13. The 1625 figure is based on Williams, *The Significance of the Printed Word in Early America*, 46. The 1700 figure comes from *Historical Statistics of the Unites States: Colonial Times to 1970* (Washington, D.C.: U.S. Department of Commerce, 1975), 1168. The 1630 immigration figures are attributed to Edward Johnson, *Johnson's Wonder–working Providence 1628–1651*, ed. J. F. Jameson (New York, 1910), 58, quoted in Fischer, *Albion's Seed*, 17.
14. John Rastell, *Interlude of the Four Elements* (London, 1519), quoted in David A. Copeland, *Colonial American Newspapers: Character and Content* (Newark: University of Delaware Press, 1997), 44.
15. John Smith, *A Description of New England* (1616), in Smith, *Writings*, 152.
16. Williams, *The Significance of the Printed Word*, 59.
17. James Axtell, *Beyond 1492: Encounters in Colonial America* (New York and Oxford: Oxford University Press, 1992), 29; Alfred A. Cave, "Canaanites in a Promised Land: The American Indian and the Providential Theory of Empire," *American Indian Quarterly* 112 (Fall 1988): 279.

18. A. G. Dickens, *Reformation and Society in Sixteenth–Century Europe* (London: Thames and Hudson, 1966), 51.
19. Quoted in William Waller Hening, *The Statutes at Large Being a Collection of All the Laws of Virginia (1619–1792)*, 13 vols. (Richmond, Va.: Samuel Pleasants, 1809–1823), 2:517.
20. Quoted in A. C. Goodell, *Proceedings of the Massachusetts Historical Society* (June 1893), 173.
21. Quoted in Anna Janney DeArmond, *Andrew Bradford: Colonial Journalist* (Newark: University of Delawar Press, 1949), 5.
22. John Milton, *Areopagitica: A Speech of Mr. John Milton for the Liberty of Unlicens'd Printing, to the Parliament of England* (London, 1644), *The Works of John Milton*, 20 vols. (New York: Columbia University Press, 1931), 4:346–47.
23. Quoted in DeArmond, *Andrew Bradford*, 5.
24. *Boston Evening-Post*, March 30, 1741, 1.
25. The Busy Body (Benjamin Franklin), *American Weekly Mercury* (Philadelphia), February 4, 1728/29, 1.
26. Z, "*The* INTRODUCTION, *or Design of this Paper*," *Independent Reflector* (New York), November 30, 1752, 2.
27. Wm. David Sloan and Julie Hedgepeth Williams, *The Early American Press, 1690–1783* (Westport, Conn.: Greenwood, 1994), 2–6.
28. *The Andros Tracts: Being a Collection of Pamphlets and Official Papers* (Boston: Prince Society, 1868–1874), xxxv, 77–78, quoted in Sloan and Williams, *The Early American Press*, 6.
29. *Boston Gazette*, January 15, 1721/22, 2.
30. Isaiah Thomas, *The History of Printing in America* (1810; reprint edition, New York: Weathervane Books, 1970), 237–39.
31. James H. Cassedy, *Medicine in America: A Short History* (Baltimore, Md.: Johns Hopkins University Press, 1991), 4–5.
32. David A. Copeland, *Debating the Issues in Colonial Newspapers* (Westport, Conn.: Greenwood, 2000), 13.
33. Sloan and Williams, *The Early American Press*, 25.
34. *Boston News-Letter*, July 24, 1721, 1; *New-England Courant* (Boston), August 21, 1721, 1.
35. *New-England Courant* (Boston), August 21, 1721, 2.
36. *New-England Courant* (Boston), December 18, 1721, 1.
37. *Boston Gazette*, October 20, 1721, 3.
38. *American Weekly Mercury* (Philadelphia), July 22, 1725, 2.
39. *New-England Weekly Journal* (Boston), April 6, 1730, 2.
40. *New-England Weekly Journal* (Boston), April 20, 1730, 1.
41. *Pennsylvania Gazette* (Philadelphia), May 28, 1730, 1.
42. Quoted in Ward L. Miner, *William Goddard, Newspaperman* (Durham, N.C.: Duke University Press, 1962), 126.
43. *Historical Statistics of the Unites States*, 1168.
44. Richard B. Kielbowicz, *News in the Mail: The Press, Post Office, and Public Information, 1700–1860s* (Westport, Conn.: Greenwood Press, 1989), 15–17.
45. Charles E. Clark, *The Public Prints: The Newspaper in Anglo-American Culture, 1665–1740* (New York and Oxford: Oxford University Press, 1994), 259.
46. Quoted in Kielbowicz, *News in the Mail*, 26.

47. This line of verse appeared in numerous American papers. They include *Virginia Gazette* (Williamburg, Purdie and Dixon), 22 January 22, 1770; *New-York Gazette: or the Weekly Post-Boy*, April 16, 1770; *New-York Journal; or the General Advertiser*, April 19,1770; *New-London Gazette*, May 25, 1770; *Providence Gazette; and Country Journal*, July 7, 1770.
48. *New-York Weekly Post-Boy*, November 8, 1756, reprinted in *Boston Evening-Post*, November 22, 1756, 1.
49. Jürgen Habermas, *The Structural Transformation of the Public Sphere: An Inquiry into a Category of Bourgeois Society*, trans. Thomas Burger (Cambridge, Mass: Harvard University Press, 1989).
50. David A. Copeland, *The Idea of a Free Press: The Enlightenment and Its Unruly Legacy* (Evanston, Ill.: Northwestern University Press, 2006), 180.
51. Quoted in Michael Warner, *The Letters of the Republic: Publication and the Public Sphere in Eighteenth-Century America* (Cambridge, Mass.: Harvard University Press, 1990), 71.
52. Copeland, *Debating the Issues*, 154–55.
53. Milton M. Klein, ed., *The Independent Reflector* (Cambridge, Mass.: Harvard University Press, 1963), 26.
54. Alfred Lawrence Lorenz, *Hugh Gaine: A Colonial Printer-Editor's Odyssey to Loyalism* (Carbondale: Southern Illinois University Press, 1972), 7–8.
55. *New-York Mercury*, June 3, 1754, 1.
56. X.Z.&., *New-York Mercury*, July 23, 1754, 1.
57. *New-York Mercury*, July 30,1753, 2.
58. See Frank Lambert, *The Founding Fathers and the Place of Religion in America* (Princeton, N.J. and Oxford: Princeton University Press, 2003).
59. Thomas Jefferson, "To Messrs. Nehemiah Dodge and Others, a Committee of the Danbury Baptist Association, in the State of Connecticut," January 1, 1802, in *The Writings of Thomas Jefferson*, ed. Andrew A. Lipscomb and Albert Ellery Bergh, 20 vols. (Washington, D.C.: Thomas Jefferson Memorial Association, 1905), 16:281–82.
60. *Independent Reflector* (New York), April 26, 1753, 90.
61. Harry S. Stout, *The New England Soul: Preaching and Religious Culture in Colonial New England* (New York: Oxford University Press, 1986), 193.
62. Thomas, *The History of Printing in America*, 568.
63. *New-York Weekly Journal*, June 13, 1737, 2.
64. *Boston Weekly News-Letter*, April 19, 1739, 4; *Pennsylvania Gazette* (Philadelphia), July 19, 1739, 1; *New-York Weekly Journal*, July 23, 1739, 3.
65. *New-England Weekly Journal* (Boston), July 24, 1739, 1.
66. *New-England Weekly Journal* (Boston), July 17, 1739, 1.
67. Thomas, *The History of Printing in America*, 568.
68. See Frank Lambert, "'Pedlar of Divinity': George Whitefield and the Great Awakening," *Journal of American History* 77 (1990): 812–37; Copeland, *Colonial American Newspapers*, 215–23.
69. *Pennsylvania Gazette* (Philadelphia), October 23, 1740, 2.
70. *Pennsylvania Gazette* (Philadelphia), May 1, 1740, 1.
71. *Pennsylvania Gazette* (Philadelphia), May 8, 1740, 2.
72. Copeland, *The Idea of a Free Press*, 176–77.
73. Figures taken from the bibliographic listings in Charles Evans, *American Bibliography*, 14 vols. (Chicago, 1904), 2:109–326; and Roger P. Bristol, *Supplement to Charles*

Evans' American Bibliography (Charolottesville: University Press of Virginia, 1970), 58–78.
74. *Pennsylvania Gazette* (Philadelphia), July 24, 1740, 1.
75. David Copeland, "'Join, or Die': America's Press during the French and Indian War," *Journalism History* 24:3 (1998): 118.
76. *New-York Gazette Revived in the Weekly Post-Boy*, January 22, 1750, 1.
77. Fred Anderson, *Crucible of War: The Seven Years' War and the Fate of Empire in British North America, 1754–1766* (New York: Alfred A. Knopf, 2000), xv. Frank Luther Mott, *American Journalism, A History: 1690–1960*, 3rd ed. (New York: MacMillan, 1962), 52.
78. *Boston Evening-Post*, September 22, 1760, 3.
79. *Boston Gazette, and Country Journal*, August 6, 1759, 1. For another reference to news writers, see *New-Hampshire Gazette* (Portsmouth), March 23, 1759, 1.
80. "The Virginia Centinel, no. 1." The "Centinel" wrote a series of essays on the dangers of the invading French and Indians to the American colonies. The essay was first printed in the *Virginia Gazette* (Williamsburg), April 30, 1756, which is no longer extant. Other papers throughout the colonies ran the warnings, as well. This version comes from the *Maryland Gazette* (Annapolis), August 12, 1756, 1 (emphasis included). The same essay also appeared in the *New-York Gazette*, June 14, 1756; *Boston-News Letter*, June 24, 1756, and July 1, 1756; *Connecticut Gazette* (New Haven), June 26, 1756; and *Boston Evening-Post*, June 28, 1756.
81. "Benjamin Franklin to James Parker," March 20, 1750/1, in Benjamin Franklin, *Writings* (New York: Library of America, 1987), 444.
82. Sinclair Hamilton, "The Earliest Device of the Colonies and Some Other Early Devices," *Princeton University Library Chronicle* 10 (1948–49): 118, notes that the "JOIN, or DIE" woodcut was "the first device to appear in this country symbolizing or suggesting the union of the colonies."
83. See, for example, *New-York Mercury*, May 13, 1754, 2; *Boston Gazette, or Weekly Advertiser*, May 21, 1754, 3; *South-Carolina Gazette* (Charleston), August 22, 1754, 2.
84. *Boston Gazette, or Weekly Advertiser*, July 23, 1754, 2. For the text of the Albany Plan of Union, see Henry Steele Commager, ed., *Documents of American History*, 8th ed. (New York: Appleton-Century-Crofts, 1968), 1:43–45.
85. Jonathan Mayhew, *Pennsylvania Gazette* (Philadelphia), August 29, 1754, 1.
86. *New-York Mercury*, September 23, 1754, 2 (emphasis included).
87. *Pennsylvania Gazette* (Philadelphia), August 29, 1754, 1.
88. *Pennsylvania Gazette* (Philadelphia), July 11, 1754, 2 and July 23, 1754, 2.
89. *Boston Evening-Post*, March 15, 1757, 2.
90. John Holt to Samuel Adams, January 29, 1776, quoted in Arthur M. Schlesinger, *Prelude to Independence: The Newspaper War on Britain 1764–1776* (New York: Random House, 1958), 284.
91. Quoted in Gordon S. Wood, *The American Revolution: A History* (New York: Modern Library, 2002), 24.
92. Schlesinger, *Prelude to Independence*, 20.
93. *Pennsylvania Journal and Weekly Advertiser* (Philadelphia), October 13, 1757, 1.
94. David Ramsey, *The History of the American Revolution*, 2 vols. (Philadelphia, 1789), 1: 61–62.

95. Peter C. Messer, *Stories of Independence: Identity, Ideology, and History in Eighteenth-Century America* (Dekalb: Northern Illinois University Press, 2005), 8.
96. *Boston Gazette, and Country Journal*, October 7, 1765, 1.
97. Schlesinger, *Prelude to Independence*, 34.
98. T. H. Breen, *The Marketplace of Revolution: How Consumer Politics Shaped American Independence* (Oxford: Oxford University Press, 2004), 197–98. *Boston Evening-Post*, November 9, 1767, quoted in Breen, 208.
99. John Adams, *The Works of John Adams, with a Life of the Author, Notes and Illustrations*, C. F. Adams, ed., 10 vols. (Boston, Mass., 1850–1856), 2:219; quoted in Philip Davidson, *Propaganda and the American Revolution, 1763–1783* (Chapel Hill: University of North Carolina Press, 1941), 227.
100. *New-York Journal; or, the General Advertiser*, October 13, 1768, 2.
101. A Farmer, "Letter XII," *Pennsylvania Chronicle, and Universal Advertiser* (Philadelphia), February 15, 1768, 20.
102. *Boston Evening-Post*, May 29, 1769, 1; February 12, 1770, 4.
103. Schlesinger, *Prelude to Independence*, 245.
104. Ramsey, *The History of the American Revolution*, 1:91.
105. Joseph Warren, "An ORATION; delivered March 6th, 1775, at the Request of the Inhabitants of the Town of BOSTON; to commemorate the bloody Tragedy of the fifth of March, 1770," *Massachusetts Spy Or, Thomas's Boston Journal*, March 17, 1775, 3.
106. *Massachusetts Spy Or, Thomas's Boston Journal*, December 30, 1773, 3.
107. *Rivington's New-York Gazetteer*, July 14, 1774, 3.
108. *New-York Journal; or, the General Advertiser*, January 5, 1775, 3.
109. Quoted in Leonard W. Levy, *Emergence of a Free Press* (New York and Oxford: Oxford University Press, 1985), 175.
110. *New-York Journal; or, the General Advertiser*, February 22, 1775, 1.
111. Mary Margaret Roberts, "Introduction: Paine's *Common Sense*," in *Pamphlets and the American Revolution: Rhetoric, Politics, Literature, and the Popular Press*, ed. G. Jack Gravlee and James R. Irvine (Delmar, N.Y.: Scholars' Facsimilies & Reprints, 1976), i.
112. Quoted in Sloan and Williams, *The Early American Press*, 172.
113. Silas Deane, "Memoire," September 24, 1776, quoted in Robert M. Weir, "The Role of the Newspaper Press in the Southern Colonies on the Eve of the Revolution: An Interpretation," in *The Press and the American Revolution*, eds. Bernard Bailyn and John B. Hench (Boston: Northeastern University Press, 1981), 100.
114. "The American Crisis, No. 1," *Dunlap's Pennsylvania Packet or the General Advertiser* (Philadelphia), December 27, 1776, 1.
115. Ramsey, *The History of the American Revolution*, 2:319.
116. Figures based on Schlesinger, *Prelude to Independence*, 303; and Sidney Kobre, *Development of American Journalism* (Dubuque, Iowa: Wm. C. Brown, 1969), 28.
117. *Pennsylvania Gazette* (Philadelphia), October 16, 1729, 4.
118. *Virginia Gazette* (Williamsburg, Purdie and Dixon), January 22, 1770, 2; *New-York Gazette: or the Weekly Post-Boy*, April 16, 1770, 4; *New-York Journal; or the General Advertiser*, April 19, 1770, 3; *New-London (Conn.) Gazette*, May 25, 1770, 4; and *Providence (R.I.) Gazette; and Country Journal*, July 7, 1770, 4.

119. Jack Greene, *Pursuits of Happiness: The Social Development of Early Modern British Colonies and the Formation of American Culture* (Chapel Hill: University of North Carolina Press, 1988), 171.
120. Quoted in Davidson, *Propaganda and the American Revolution*, 225.

3. *My Pen and Press Are the Only Formidable Weapons I Have Ever Used: The Early Republic*

In 1808, the printer of Philadelphia's most partisan of newspapers—the *Aurora*—wrote a letter to Stephen R. Bradley, president *pro tem* of the Senate and Republican representative from Vermont. In that letter, William Duane told Bradley, "My pen and my press are the only formidable weapons I have ever used."[1] For a decade, Duane had been at "war," and his newspaper had been the battlefield. Joining the man who began the *Aurora*—Benjamin Franklin Bache—Duane quickly became one of the most despised men in America by the Federalist party. He, as much as Thomas Jefferson or James Madison, according to historian Jeffrey Pasley, deserves credit for the founding of the Democratic-Republican party in the United States and for its eventual wresting of power away from the Federalists, the party of George Washington.[2]

When Duane described his writings and his printing press as weapons, he was correct. The time period following the American Revolution was a volatile one. Americans rejoiced when the war ended and Britain conceded to its former colonies. The "GLORIOUS NEWS" that Cornwallis had surrendered to Washington at Yorktown "was ushered in by ringing of bells, discharging of cannon, displaying of colours, attended with the shouts of a grateful populace." It was, printer Isaiah Thomas declared, "an event that must affect every patriotick American with joy and pleasing sensibility."[3] But, the war of independence soon yielded to another kind of war, one over the direction of the new nation. At the heart of this confrontation, just as Duane said, was the press, though this was nothing new. The press served as agitator leading up to the Revolution. In the 1770s and during the war, the press provided information on how the fight for independence progressed, and it served to rally the young nation in support. In the 1780s, the press became the repository for information about the direction of the new nation. In the 1790s and into the first decade of the nineteenth century, the press truly became a weapon as the battle for the new nation's direction became a principal issue everywhere as newly formed political parties debated their agendas in public prints.[4] Samuel Miller, a Presbyterian minister who later became a professor at Princeton Theological Seminary, wrote that "never was there given to man a political engine of greater power" than newspapers at the end of the eighteenth century, and "never were they actually perused by so large a majority of all classes since the art of printing was discovered."[5]

Thomas, who compiled the first history of the press in 1810, simply quoted from Miller's *Brief Retrospect of the Eighteenth Century* to explain the value of the press in the years after the war in *The History of Printing in America*. Newspapers "have become the vehicles of discussion, in which the principles of government, the interest of nations, the spirit and tendency of public measures ... are all arraigned, tried, and decided... they have become the welfare of the state, and deeply involving both its peace and prosperity.... By means of this powerful instrument, impressions on the public mind may be made with a celerity, and to an extent, of which cannot but give rise to the most important consequences in society."[6]

From the end of the Revolution until 1820, newspapers grew at an astonishing rate, which helps, in part, to explain their position within society. Noah Webster, author of the first work to attempt a standardization of American English—*Grammatical Institute of the English Language*—and editor of the *American Minerva*, grasped the significance of newspapers to Americans and in his paper's inaugural issue stated, "Of all these means of knowledge, Newspapers are the most eagerly sought after, and the most generally diffused. In no other country on earth, not even in Great-Britain, are newspapers so generally circulated among the body of people, as in America."[7] The number of newspapers in America grew so quickly in the last fifteen years of the eighteenth century and the first ten of the nineteenth because of the importance of what was happening in the young republic and its effect on the people. When the American Revolution ended, thirty-five newspapers were published in America. That number grew to ninety-two in 1790, an increase of almost 162 percent in seven years. During the volatile 1790s, the number of papers swelled to 234, an increase of 154 percent for the decade. By the time Thomas Jefferson left office in 1809, America had 329 newspapers, more than 40 percent more than had been published nine years earlier and a whopping 840 percent more than existed in 1783. In 1783, papers were printed once a week, but Americans needed information more often than that. Consequently, newspapers in larger cities began publishing more frequently, twice and three times a week with twenty-four dailies in operation by 1800.[8] By 1820, 512 newspapers served the nation,[9] an increase of more than 55 percent in eleven years, more than 1,362 percent more papers than served the nation when the Treaty of Paris was signed. Just as newspapers in the eighteenth century were shared and read aloud in public places and in homes, newspapers of the early republic were likewise an everyday part of society, able to penetrate the country through the rapidly increasing construction of post roads for mail delivery. By the end of the War of 1812, nearly 44,000 miles of roads crossed the country. Because of the large number of newspapers, the postal service enlisted stagecoaches just for mail service.[10] Obtaining a newspaper almost anywhere in America was not too difficult so long as a post road ran nearby. Regions in the South and in the West had fewer roads, making the availability of newspapers more difficult than from Baltimore northward. Still, nearly 90 percent of what traveled via the mail in the United States consisted of newspapers.[11] Throughout the nation, then, newspapers provided regular reading matter and subjects for public debate, leading one Boston paper to say, "Almost the total reading of at least half the people of this country, and a great part of the reading of the other half, is from newspapers." The paper then explained that Americans' desire to know what was happening "has given rise to a general form of salutation on the meeting of friends and strangers: *What's the news?*"[12]

Newspapers of the early republic were about as far from objective in the political arena as could be imagined. Objectivity and fairness were not considered necessary at all. In fact, editors scoffed at the pair. Isaac Hill, owner of the *New-Hampshire Patriot*, said, "we view the idea of an impartial paper as preposterous; for there is no man in his senses who does not view one of the great parties ... as being essentially right, and the other wrong." He then added that newspapers "had the greatest effect in producing the change and opposition to our general government."[13] Hill's comments address what made the issues of the times so critical to nearly every American and why the press became the "weapon," to use William Duane's analogy, to define and shape the young nation. *Boston Courier* printer Benjamin True summed up the power of the press in 1805 when he said, "The American press is now generally acknowledged to be an engine of great influence."[14]

DISASSEMBLING AND CREATING GOVERNMENT

With the ratification of the Articles of Confederation in 1781, a writer in the *Providence Gazette* declared, "A Union, begun by necessity, cemented by oppression and common danger, and now finally consolidated into a perpetual confederacy of these new and rising States; And thus the United States of America, having, amidst the calamities of a destructive war, established a solid foundation of greatness."[15] Many people, however, including most newspaper printers, quickly discovered the weaknesses of the confederation. During the next six years, dissatisfaction with the Articles of Confederation increased, and a concerted effort to rid the new nation of confederation status grew. Sentiment that America should create a different form of government swelled from the pages of newspapers. In October 1786, the *New-Haven Gazette* introduced an epic poem that described America in crisis. Titled "American Antiquities" but called *The Anarchiad* since in the first installment of "The Antiquities" that was the name of the alleged ancient poem that had been recently discovered and would be presented. What *The Anarchiad* had to say, its introduction explained, ranked "amongst the productions of human genius."[16] In the tenth installment of a series that ran from October 26, 1786, through September 13, 1787, the poem harkened back to the 1754 warning by Benjamin Franklin that unity was the only way to ensure the security of a country. *The Anarchiad* reminded Americans that under their present governmental system—the Articles of Confederation—the new nation was in danger because its people were not unified and the national government had no power to ensure the country's future. The poem warned:

> "Ere death invades and night's deep curtain falls,
> Thro' ruin'd realms the voice of Union calls,
> On you she calls! attend the warning cry,
> 'YE LIVE UNITED, OR DIVIDED DIE.'"[17]

Clearly, the weaknesses of the Articles of Confederation, which limited the power of the federal government, especially in the areas of trade and taxes, jeopardized the nation. Writers used the press to emphasize this. "Thus we see that in the course of national affairs, the liberty of the people is in a more hazardous situation when the sinews of government are unstrung, than when they are drawn too tight," a Massachusetts citizen explained. "I have been led to these reflections, by considering the present state of our federal government, which presents a melancholy picture of a headless body, where the tremulous motion of the severed nerves, is the only sign of remaining life." The writer concluded by pointing out that "the confederation, that appeared so perfect in its original state, is become a loose, incompleat agreement."[18]

As a result, influential individuals—led by James Madison—pushed for a meeting to review the Articles. Review turned into revision, and in May 1787 delegates from all states except Rhode Island gathered in Philadelphia. Just as they had when the Albany Congress of 1754 proposed its Plan of Union, most printers supported the idea of creating a stronger government. They believed, however, that the work that was being done was so critical that it needed to be done without daily reports of the delegates' deliberations and agreed not to publish any information that might be leaked from the sessions. They did, however, discuss the delegates and their credentials, the hopes for the new government, and ideas about what form a new government might take. They made sure that "the eyes of the whole continent" focused on Philadelphia.[19] Once the Constitution was finished, the *Pennsylvania Packet* published it in its entirety on September 19, two days after the delegates ratified it. Every American newspaper followed and printed the new Constitution.

Once the Constitution was completed, however, the work of ratification by the states began, and here, the press became the essential tool for approval by the states. According to David Redick, a member of Pennsylvania's Supreme Executive Council, "The new Constitution for the United States seems now to engross the attention of all ranks."[20] The press truly served as the place for public debate because the battle for ratification would not be easy, nor would it occur swiftly. Three-fourths of the states needed to give the new governmental document an affirmative vote in order for it to take effect. Nearly a year after the Constitution was approved in convention, New Hampshire, the ninth state, ratified it. But four states, including New York and Virginia—two of the most populous and powerful—had yet to okay it, though they would soon. According to Saul Cornell, in the battle over ratification, "Planter aristocrats, middling politicians, and backcountry farmers were bound together by the tenuous connection provided by the world of print."[21] Once something was published in one paper, it was often picked up and used by others. Cornell says that once a writing entered the public sphere it often assumed a life of its own, that "authors no longer controlled how they [their essays] were read."[22]

People responded to these writings in different ways, depending on where they lived and their personal situations. "Let us be of one heart, and of one mind. Let us seize the golden opportunity to secure a stable government, and to become a respectable nation," the *Independent Chronicle* said in support of the Constitution. "Let us remember our emblem, the twisted serpent, and its emphatical motto, *unite or die*! This was once written in blood; but it is as emphatical now as then. A house divided against itself, cannot stand. Our national existence depends as much as ever upon our union; and its consolidation most assuredly involves our posterity, felicity and safety."[23] This became a principal rationale of the Federalists, those who favored the Constitution. A strong union was essential for the survival of the United States, they believed. They also were convinced that it was essential for the survival of the nation to ensure that the proper people ran the country. For that reason, those supporting ratification employed a triumvirate of writers who laid out a logical case for supporting the Constitution. Alexander Hamilton, James Madison, and John Jay—three well-known and influential leaders who were critical to the activities at the Constitutional Convention—created a set of papers called *The Federalist*.

Writing independently, but publishing all they wrote under the pseudonym "Publius," Madison, Hamilton, and Jay developed a rational, Enlightenment-inspired series of eighty-five essays, the first presented in the New York *Independent Journal* in October 1787 and the others rotating among three other New York papers until May 1788, to lay out why the nation should adopt the Constitution as presented. The essays were also reprinted in pamphlet form. The selection of New York as the focus of Publius' essays was not coincidental, either. Hamilton, especially, knew that his home state would be a difficult one to persuade toward ratification. Using the principal papers of New York City to introduce *The Federalist* helped the press to focus attention on the battle for ratification there. Though three-fourths of the states had approved the Constitution before New York did the same at the end of June 1788, without New York, and Virginia, which ratified a week earlier, the country would have likely not survived.

In "The Federalist, No. 1," Hamilton—the chief writer of *The Federalist* essays—explained that the papers were intended to provide an analysis of the Constitution and answer the complaints that were being raised about the new document. Publius said he was prepared to discuss, "*The utility of the UNION to your political prosperity—The insufficiency of the present Confederation to preserve that Union—The Necessity of a government at least equally energetic with the one proposed to the attainment of this object—The conformity of the proposed constitution to the true principles of republican government—Its analogy to your own state constitution*—and lastly, *The additional security, which its adoption will afford to the preservation of that species of government, to liberty and to property.*"[24]

Opposition, however, to the new government was considerable, and the press, naturally, provided a way for those who opposed it to unite and present a common set of criticisms.[25] The Anti-Federalists pointed out what the Constitution lacked. Elbridge Gerry of Massachusetts and a delegate to the Constitutional Convention succinctly summed up Anti-Federalist objections to the Constitution in its present form: "My principal objections to the plan, are, that there is no adequate provision for the representation of the people—that they have no security for the right of election—that some of the powers of the Legislature are ambiguous, and others indefinite and dangerous—that the Executive is blended with and will have an undue influence over the Legislature—that the judicial department will be oppressive—that treaties of the highest importance may be formed by the President with the advice of two thirds of a *quorum* of the Senate—and that the system is without the security of a bill of rights. These are objections which are not local, but apply equally to all the States."[26]

Gerry's message obviously struck a sympathetic note and took on a life of its own. His address appeared in forty-six different Anti-Federalist newspapers and was turned into a pamphlet. Some Federalist papers ran Gerry's objections, too.[27] The issue that seemed to permeate objections of people everywhere was Gerry's last—a bill of rights. "[T]he omission of a bill of rights [is] a defect in the proposed constitution," the *Pennsylvania Herald* declared.[28] "The jury to question is, whether it is a plan of government formed upon *revolution* principles, and the liberty and happiness of the people fully and effectually secured?" a writer called A Confederationalist said two weeks later. "I say the question ought to be whether the plan, if adopted, will secure to the people the blessings of a free and equal government? I say that a declaration of those inherent and political rights ought to be made in a BILL OF RIGHTS, that the people may never lose their liberties by construction."[29]

The idea of a free press became a central argument for a bill of rights in some parts of the United States, though it was of less concern in areas where state constitutions already granted such a protection. The argument reinforced the centrality of the press to free speech for people and a public sphere for debate, discussion, and the ability to effect public opinion and thereby the direction of the nation. Without this right—of speaking freely via the printed word—Anti-Federalists were saying, "what controul can proceed from the fœdral government to shackle or destroy that sacred palladium of national freedom?"[30] Without it, one of the people's most important tools would be lost, and many who opposed ratification of the Constitution felt that without a bill of rights, Americans would give up one of their inherent rights. "If the liberty of the press be an inherent political right, let it be so declared, that no despot however great shall *dare to gain say it*. If it is not so declared it may be denied. Declare it to be an inherent and political right, and that it ought to be held sacred, and we then shall by certain upon what ground we stand," A Confederationalist demanded.[31] Federalist arguments ensured the adoption of the Constitution, but the print war by Anti-federalists ensured that a means of ratification existed and that what people considered essential rights—freedom of religion, speech, press, assembly, and the ability to petition the government concerning grievances—would forever be protected.

CHANNELING IDEOLOGY

Hamilton and Madison worked together to create *The Federalist Papers* and believed the Constitution was essential to the survival of the United States. But, as Publius, they were, according to Stanley Elkins and Eric McKitrick, a "split personality" on exactly how the elements of the new government might be applied in relation to the individual states' rights. Hamilton believed that the powers of the federal government were "unconfined" and extensive. Madison understood federal powers to be "few and defined."[32] As a result, how the new government would operate immediately became a central issue, and the press, even more than it had been a tool in the battle over adoption, served as the means to set the political ideology of the new nation. According to Donald Stewart, the press of this period both molded public opinion as it mirrored it better than perhaps it ever has in American history.[33] Because printers of the period realized that they had been central to the debate over the Constitution, they elevated the printed word's role in defining the new nation. Most had agreed that the Constitution deserved ratification, but they were not in agreement in its interpretation. As a result, partisanship that went well beyond that of the revolutionary 1760s and 1770s developed. The ideology of the country's developing political parties—the Federalists and the Democratic-Republicans—required a channel that reached directly to the people. During the 1790s and into the nineteenth century, newspapers, especially where editors and printers adopted the philosophies of one of the parties, became a tool of influence beyond what anyone might have imagined. It became a time when, Jeffrey Pasley says, the influence of editors developed into "the tyranny of printers."[34]

In 1789, the new government led by George Washington and the Federalist party employed selected newspapers to publish laws, orders, and resolutions of the national government. In return, the printers of those papers were paid for their work. Disseminating this information to the public was essential for the new government, but government patronage soon became political patronage because government officials realized that they could reward printers who supported a particular political agenda with lucrative government contracts. As a result, Culver Smith says, "printing, politics, and patronage became intermingled." Doing this seemed logical, considering the nature of the press and public attitudes toward newspapers at this time. Most Americans thought of newspapers as part of the political system.[35] Hamilton, who had turned to the press with *The Federalists Papers* in 1787, quickly grasped the possibilities for a new paper in New York City, which served as the nation's temporary capital. The *Gazette of the United States* was begun by John Fenno, who considered himself an editor of the paper, not its printer. His goal was to create a national newspaper that could support the policies of the federal government, and his political stance and goals as an editor fit perfectly with those of Hamilton. Hamilton served as the nation's first secretary of the treasury, and in that role he maintained direction of the Federalists and soon gave Fenno printing contracts for the Treasury Department and eventually provided Fenno with a "loan" of two thousand dollars.[36] By the time that the capital moved from New York to Philadelphia in 1790, the *Gazette of the United States* was a critical component of the Federalist political machine, sending copies to thousands of people throughout the United States weekly.

The use of Fenno's paper to promote the Federalist party stimulated a response by the Democratic-Republicans. Madison and fellow Virginian Thomas Jefferson assumed the philosophical reins of the party. Jefferson was Washington's secretary of state, so he possessed the same power as Hamilton in terms of offering patronage to a printer. The Virginians began looking, and they found Philip Freneau, a newspaperman without a paper who was often called the "Poet of the Revolution" for his verse "The Rising Glory of America." Freneau and Madison had also been roommates at Princeton, making his selection even easier.[37] Freneau printed his first *National Gazette* on October 31, 1791. Jefferson began providing Freneau with State Department publishing contracts and assorted information for the paper. Jefferson claimed that he never wrote for a newspaper, but many Federalists doubted it, one in Boston saying that the *National Gazette* was "published under the eye of that established patriot and republican Thomas Jefferson."[38] Before the end of the year, Freneau wrote, "PUBLIC opinion sets the bounds to every government, and is the real sovereign in every free one."[39]

During the first half of the 1790s, America's two political parties were developing with editors gradually aligning themselves based on patronage and ideology. While printer-editors may have believed strongly in the respective viewpoints of the Federalists and Republicans, they were, first and foremost, hardworking people trying to make a living with a printing press. The most common name for an American paper in the 1790s included the word "advertiser." Printers had to sell enough advertising to support their livelihoods since Americans were notoriously lackadaisical in their efforts to pay for newspaper subscriptions. So, for some printers, ideology was enough, but practically all printers sought out political patronage. Patronage ensured that a printer could produce a paper and not worry about having enough money to stay in operation. It also ensured that editors would support the political agenda of those who paid them and provide an avenue to address the public. The average printer-editor during the last decade of the 1790s and first decade of the 1800s earned, at best, about $10 a week, but lucrative government printing contracts produced enormous salaries for some printers, especially if the printer were aligned with the party in power nationally or locally. Freneau claimed that his rival Fenno was paid "*two thousand* or twenty five hundred dollars a year." The Federalists "cannot otherwise than have some sort of influence on the Editor of the Gazette of the United States," Freneau said.[40] Early in the nineteenth century, Samuel Harrison Smith and the *National Gazette* held multiple federal printing contracts, earning Smith $2,000 per year.[41] By 1809, more than 83 percent of all papers acknowledged some sort of party affiliation.[42]

Sometimes patronage was not necessary for a printer to fight without reservation for a political party. Benjamin Russell of Boston's *Columbian Centinel* was totally dedicated to the Federalist cause. Though the Boston editor did receive some government contracts in his long tenure with the paper, they were of secondary concern for Russell. He believed fully in the direction the Federalist party was guiding the young nation. Russell's antithesis was Philadelphia's Benjamin Franklin Bache, grandson of early America's preeminent printer, scientist, and politician. Despite advice to the contrary, Bache decided to follow in his grandfather's profession, having received the best of training as a printer while living with his namesake when Franklin served as America's emissary to France. In 1790, Bache began a typical American paper, the *General Advertiser*, but being enamored with France, he increasingly embraced the Republican political position and started attacks on George Washington's administration, saying to the disbelief of many that "If ever a nation was debauched by a man, the American nation has been debauched by WASHINGTON"[43] after his administration signed Jay's Treaty, which limited American trade to Britain only, while Britain made a few concessions in the Northwest Territory. Bache was soon joined by other printers like Vermont's Matthew Lyon and New York's John Daly Burk in a never-ending series of attacks on the Federalists.

Jay's Treaty with Great Britain averted another war with Britain, but this, in turn, angered France, which had supported the United States since the Revolution and had operated under a treaty with America since 1778. As a result, France began raids on American shipping. In 1797, President John Adams, in an effort to stop those raids, tried to negotiate in secret with the French. Their agents, however, demanded extortion money and hinted that America might be ripe for a bloody revolution similar to the one in France. Known as the XYZ affair, Republican papers publicized the clandestine activities while angry Federalist sheets responded by associating the Republicans with the violence of the French Revolution, calling them traitors to America.

In an effort to silence opposition they feared was out to destroy the Constitution, Federalists in Congress passed the Alien and Sedition Acts. Perpetuation of the union was the rationale Federalists gave for the acts, but the acts were designed explicitly to silence criticism of the government, the kind that Republican papers were increasingly voicing. "Every independent Government has a right to preserve and defend itself against injuries and outrages which endanger its existence," Harrison Gray Otis informed Congress.[44] The Alien and Sedition Act said that anyone who wrote, uttered, printed, or published any false, scandalous, or malicious comments against the government was to be punished by fine of up to $2,000 and a possible two years' imprisonment. The act mentioned specifically the president and all members of Congress. Interestingly, it made no mention of the vice president who happened to be the Republican Jefferson. About twenty-five people were arrested, with fourteen indictments handed down during the acts' duration.[45] The acts were repealed in 1802, and for the most part, attempts at federal control of the press ended, although some states continued to charge editors with libel.

Federalist editors praised the Alien and Sedition Acts. "Whatever American *opposes* the Administration is an anarchist, a Jacobin, and a Trator [*sic*]," the *Massachusetts Mercury* of Boston declared. "It is *Patriotism* to write in favor of our Government—it is Sedition to write against it."[46] Bache was, naturally, a principal target of the acts and was arrested for seditious libel (criticism of the government) in June. His trial was set for October, and in the interim, Bache continued his attacks on the Federalists. He never stood trial, however. In September, Bache contracted yellow fever and died. Burk, who was an Irish immigrant, faced jail because of the Sedition Act and deportation because of the Alien Act. He, too, continued his attacks on the Federalists even though he had been indicted for seditious libel. Burk found it necessary to go into hiding to avoid jail time and deportation. Lyon, who had used his newspaper as a means to fight the Federalists and to get himself elected to Congress, was the first person arrested under the Sedition Act. He spent four months in jail and was fined $1,000. His incarceration did not keep Lyon from being re-elected to the House of Representatives. In total, about twenty-five people were arrested, with fourteen indictments handed down during the duration of the Alien and Sedition Acts.[47]

The work of editors like Bache, Lyon, and Burk was critical to the Republicans because Jefferson resigned his post in the Washington administration in 1793. The *National Gazette* could not survive without his political and financial support, and the Federalists still maintained control of the national government and the majority of newspapers. The 1796 election, however, proved that the Federalists' hold on the national government was weakening. Most Americans rightly assumed that after George Washington announced he would not seek another term as president that either Adams or Jefferson would become the nation's chief executive. "THOMAS JEFFERSON & JOHN ADAMS will be the men," an anonymous Philadelphian predicted and added, "whether we shall have at the head of our executive a steadfast friend to the Rights of the People, or an advocate for hereditary power and distinctions, the people of the United States are soon to decide."[48] While Republican supporters saw Jefferson as the candidate for "the Rights of the People," Federalists believed the election of Jefferson would quite likely end the United States. "If Mr. Jefferson is elected President of the United States," John Fenno predicted, "In 15 months we will be involved in a ruinous war, which will terminate in the fall of the present fabric of government, and a disunion of the states."[49] Adams won the election by three electoral votes, making Jefferson the vice president. Jefferson was now back in a position that afforded him the opportunity to provide patronage to Republican-leaning editors, and he, Madison, and the growing number of Republicans learned from the election that the power to wrest the government from Federalist control lay with the presses of a growing group of sympathetic printers.

THE ENGINE IS THE PRESS

All forms of printed material—newspapers, pamphlets, and broadsides— were important in the dissemination of information as the 1800 election approached, but much of what appeared in pamphlets, for example, ended up shared throughout America via newspapers. Broadsides were a quick way to get the printed work out, too, but they were limited in how much information could be printed on a single sheet. When local politicians in Albany, New York, wanted to promote their candidates in 1799, they produced a broadside listing the candidates they supported. The broadside, however, referred its readers back to the *Albany Gazette* for more in-depth information on the candidates and their selection.[50] Jefferson saw the press as the means to win the election of 1800, despite the fact that the Alien and Sedition Act had become a tool of the Federalists to silence the most outspoken Republican editors. In a letter to Madison, Jefferson said, "The engine is the press. Every man must lay his purse and his pen under contribution."[51] Later, Jefferson would say that the Republican press "has unquestionably rendered incalculable services to republicanism through all its struggles with the federalists, and has been the rallying point for the orthodoxy of the whole Union."[52]

Federalists agreed with Jefferson's assessment of the importance of the press to the 1800 election. "Give to any set of men the command of the press," Federalist judge Alexander Addison warned in his charge to a jury in a Pennsylvania sedition trial, "and you give them the command of the country; for you give them the command of public opinion, which commands everything."[53] William Cobbett, one of the most vicious of Federalist editors, promoted his paper and the role of the editor in this ideological battle, saying in his *Porcupine's Gazette and United States Daily Advertiser*, "Professions of *impartiality* I shall make none." He added that any editor that did not actively involve his newspaper in the political battle of the day was "a poor passive fool, and not an editor."[54] New York editor Charles Pierce agreed: "Every native *American* has but *two choices* to make. He must *rank* with the *friends* or the *foes* of *America. Neutrality* is *criminal.*"[55]

The Republicans used Jefferson's election and the Alien and Sedition Acts to their advantage. Even though the acts were initially an attack upon Republican editors and Republicans allowed the acts to run their course without repealing them, the Republicans did not have to remind many members of the press which party had put legislation into place to limit the voice of newspapers. Jefferson, and the Republicans elected after him, enlarged the political patronage system that was already in place in America, and it became invaluable to the promotion of Republican ideals within the public sphere. Republicans voted in 1804 to add newspapers in the territories to publish governmental proceedings, and Madison—when he was elected in 1808—rewarded newspapers that had worked to elect both himself and Jefferson with government printing.[56] Despite these facts, partisanship in newspapers continued. The rise of the Republican party in the American political system following the election of 1800 and the reciprocal decline of the Federalist party in terms of representation on the national level, ironically, was not reflected in the same proportions in the press. Newspapers that backed the Federalist party remained active throughout the nation. Printers and editors founded more than one hundred papers supporting Federalist policies after the 1800 election and before the end of the War of 1812.[57] But, the use of the press to argue for whether the nation should declare war on Great Britain in 1812 demonstrated that the Republicans possessed the most voices in the nation, both in politics and in the press. Federalist editors, however, were anything but docile or quiet.

INTERNATIONAL PROBLEMS AND ANOTHER WAR

The United States faced numerous crises in the decade following Jefferson's election.[58] The British had stopped the U.S.S. *Chesapeake* in 1807 and demanded to board it. When its captain refused, they fired on the ship for fifteen minutes. The British then boarded the ship and carried off four sailors accused of being British deserters. Four other sailors died, and another eighteen were wounded. Americans were outraged, many demanding the nation declare war. "Whenever matters of dispute between nations is settled by force, then it is *war*," Joseph Gales, editor of the *Raleigh Register*, declared.[59] "There has rarely occurred an outrage, equally aggravated," a New York writer explained.[60] "The honour and independence of our nation insulted beyond the possibility of further forbearance" is how one Virginian described what had happened.[61] Another editor, warned "that America can be to Britain one of her most formidable foes."[62] "Indignant Republican editor William Alexander Rind declared, "All parties, ranks, and professions were unanimous in their detestation of the dastardly deed, and all cried aloud for vengeance."[63] "Have we had kicking enough?" another newspaper editor asked, and added, "we must protect ourselves."[64] But, President Jefferson was not ready for war, and he ignored the demands from multiple voices around the country being made in Republican papers.

Other recurring events, however, would lead the nation to war. Britain and France had established a series of blockades along the European coast, seizing the cargo of any ship trading with the other. America, a neutral in this European conflict, saw its ships' cargoes confiscated. Britain and France continued to escalate their interference with American shipping. By 1812, nearly nine hundred American ships had been taken by the two European nations. Britain was by far the worse offender since its blockades involved many more ships than that of the French navy, which was tiny in comparison to that of Britain. In addition, sailors on American ships were often impressed—declared to be citizens of the nation boarding the vessel and then forced into that country's naval service. From 1803 to 1812, Britain removed about six thousand Americans from assorted merchant ships.[65] Add to this the fact that the

British in Canada were believed to be keeping Native Americans in the Northwest Territory supplied with weapons and pushing them toward open hostilities with Americans, and the British became the greater of the evils in almost all parts of the United States except New England. The *National Intelligencer*, the mouthpiece of the Republicans in Washington, later explained about the Indians "that these barbarians are in British pay" and encouraged to attack Americans in order to cause a war.[66]

Increasingly, a group of American political leaders who became known as War Hawks pushed for the nation to declare war on Britain. Most of the approximately twenty War Hawks were from the Western and Southern states, and all were Republicans. Using the bully pulpit of Congress and the growing number of Republican newspapers, the War Hawks laid out their agenda. The War Hawks placed the blame squarely on Great Britain for the Indian war on the frontier, the seizure of American merchant ships, confiscation of American goods on those ships, and impressment of American sailors. They also blamed Britain for the economic depression created by all of these and the enactment of the assorted embargoes the nation felt it had to use to respond to Britain's belligerent maritime policies. Many Americans were convinced that the nation's only option was war. "We think ourselves bound to state as a *fact*, that we have not seen, we have not *heard* of a single Republican," Richmond printer Thomas Ritchie proclaimed, "who is not opposed to submission to G. Britain—who does not approve the Report on Foreign Relations—and who is not for a war with G. B. unless she recedes from her aggressions."[67]

From the time of the *Chesapeake* affair through the beginning of the War of 1812, editors on both sides of the political spectrum worked to solidify a network of information dispersal for their respective parties. Calls for war by the Republicans gave Federalist editors fodder for their papers. As the inevitability of war grew, Federalist papers openly opposed it. In New England, where the Federalists remained the strongest, antiwar advocates began to suggest secession from the Union was the only answer. The *Columbian Centinel* noted about six months into the war that unless ways around war could be found that "the chain will break" that holds the nation together.[68] Later, the paper's editor declared that it was time for New England to break "the *cobwebs* of a compact which has long since ceased to exist."[69] Republicans, however, possessed the most vocal politicians and editors. Congress declared war, and on June 18, 1812, President James Madison signed the declaration. Its passage in Congress, however, represented the political divisiveness that existed in the United States. Only slightly more than 60 percent of the representatives supported the war. All thirty-nine Federalists voted against it, and forty-nine of the 128 Republicans, about 38 percent, did as well.[70]

The War of 1812 would become the final struggle nationally in the political battle between Federalists and Republicans. Though the war did not go smoothly for the United States, the British, who were embroiled in war in Europe with France, did not turn their full attention toward America until Napoleon was defeated at Waterloo. By then, the British were too tired to give similar attention to the United States. On Christmas Eve, 1814, the warring parties reached a settlement in Ghent, Belgium. The Treaty of Ghent basically returned relations between the two countries to pre-war status. Impressment did stop however, but America gained nothing politically. It did, however, gain respect on the world stage, and Republican papers proclaimed, "This second war of independence has been illustrated by more splendid achievements than the war of the revolution."[71] Republican editors and politicians hammered this concept home, and at least on a national level, the two-party system disappeared. The Federalists could only muster electoral votes from New England in the 1816 presidential election, being labeled "the arch-enemy of mankind" in Republican newspapers.[72] In 1820, the Federalists did not even run a candidate. Some of the staunchest of Federalist editors simply accepted the demise of their party. Benjamin Russell of Boston's *Columbian Centinel* dubbed the new national unity as the "Era of Good Feelings."[73] The increasing strength of Republican papers, fostered by the patronage system that Madison and later James Monroe expanded, gave the party "command of public opinion" on the national stage, which Alexander Addison had warned in 1798 to be the true power of the nation's press.[74]

CAUSING RIOTS, SUPPRESSING FREE SPEECH

Americans, especially Republicans when the Alien and Sedition Acts went into effect in 1798, railed about their speech rights. The Virginia General Assembly passed the Virginia Resolutions in December, saying that in the commonwealth "the Liberty of Conscience and of the Press cannot be cancelled, abridged, restrained, or modified by any authority of the United States." The Alien and Sedition Acts, therefore, were unconstitutional.[75] One Virginia representative, Richard Hay, went even further. "A man may say every thing which his passions suggest; he may employ all his time, and all his talents, if he is wicked enough to do so, in *speaking* against the governments matters that are false, scandalous, and malicious."[76] Even Federalist Alexander Hamilton argued that the rights of the press "consists in the right to publish, with impunity, truth, with good motives, for justifiable ends, though reflecting on government, magistracy, or individuals."[77] Americans, in hundreds of essays and letters in newspapers from when the Federalists proposed the Alien and Sedition Acts until their expiration in 1802, complained about infringement upon what they considered an essential right and "one of the choicest blessings a free people can enjoy." Silencing newspapers was compared to "*star chamber tyranny*,"[78] in reference to the court begun by Henry VIII to quiet and punish all dissenting voices.

Despite the consistent and continuous calls that elevated freedom of the press to an unalienable right, Americans were also quick to advocate the silencing of newspapers when their message contradicted that of a political group. Patriots hanged printer James Rivington in effigy before the Revolution and then destroyed his press and had him arrested when he continued to print items they felt did not gel with patriot rhetoric. Silencing opposition voices in the press became a way to define the nation's direction as much as the continual repetition of specific ideology in public prints. Days after the United States declared war on Britain, the *Federal Republican* in Baltimore attacked the move, calling it "unnecessary, inexpedient, and entered into from a partial, personal and as we believe, motives bearing upon the front marks of undisguised foreign influence."[79] Baltimore, the nation's third largest city with a population in excess of 41,000, was the home of thousands of Irish and Scottish immigrant shippers and sailors who despised Britain and advocated the Republican agenda. In many places, angry Republicans threatened Federalist editors who openly opposed the war. Few Federalist papers or their editors faced what Alexander Hanson and Jacob Wagner did in Baltimore, however. War supporters were so outraged at what the pair said in their paper that an angry pro-war mob destroyed Hanson's and Wagner's print shop and its contents.

The sudden demise of the *Federal Republican* meant that no strong Federalist voice existed in Baltimore. Following the lead of printers who moved out of Boston when it was occupied by the British during the Revolution in order to continue printing their paper, Hanson and Wagner did the same. The pair of Federalist editors traveled to Georgetown and found someone sympathetic to their cause who would print their paper. Five weeks after the destruction of their Maryland office, on July 27, Hanson returned to Baltimore with a new issue of the *Federal Republican*. Using Wagner's home on Charles Street as the paper's temporary offices, supporters of the *Federal Republican* came to the house with Hanson, including General Henry Lee, Revolutionary War hero. That night, a crowd gathered around the house and began throwing rocks at the window. During the ensuing hours, angry members of the mob rushed the house and broke down the door in an attempt to remove Hanson and his supporters. Those in the house fired on the intruders, killing several. By morning, thousands of people calling for the blood of Hanson and his cronies filled Charles Street.

Finally, those in the house agreed to leave under protective custody and were placed in the jail for safety. On the night of the 28th, with no guards around, the crowd entered the jail, took those in it out, and proceeded to beat them mercilessly. The mob piled those they attacked in the street outside the jail, assuming they were dead. Only because Republican doctors pleaded with the crowd to turn the bodies over to science were all not murdered. As it was, one man was killed. Others received life-changing injuries, including Lee, who suffered a fractured skull, and Hanson, who had several broken bones and a spinal injury.

The scene in "Mobtown," as Baltimore was soon derogatorily nicknamed by Federalists, was praised by some Republican editors. In nearby Wilmington, Delaware, one editor declared that the people of Baltimore had the right to "break to atoms the Trumpet of Sedition"—that is, the *Federal Republican*.[80] But, rather than silence the *Federal Republican*, the riots, ironically, saved it. Federalists throughout America subscribed to the paper, and it continued as a thorn in the side of the Madison administration as a national voice for the Federalists rather than a local one. Hanson was even elected to Senate in 1816.

SHAPING THE LOCAL AGENDA: THE FEMALE VOTE

Even though the debate within the public sphere had been won by the Republicans, Americans continued to use newspapers to shape public opinion, and political parties continued to operate actively on local and state levels. The editor of the West Chester, Pennsylvania, *Village Record* explained that "parties exist, and as much zeal is exhibited [in Pennsylvania], as in any state where the old party names still prevail."[81] Political in-fighting, therefore, continued. In New Jersey, the right of women to vote became the focus of the state's Federalists and Republicans. In 1776, the New Jersey constitution declared that "all inhabitants" of the state could vote if they met property and residence requirements. Whether the framers of the state constitution intentionally worded the voting clause to give women the right to vote is arguable. More than likely, it was a statement of rebellion, just as the Declaration of Independence announced that "all men are created equal." Regardless, the wording opened the door for females to vote.[82] As a result, according to William Whitehead, women voted throughout New Jersey in the pivotal elections at the end of the eighteenth century and beginning of the nineteenth.[83] The women's vote was targeted by the Federalists first, who emphasized "the suitability and desirability of female political participation" in an effort that aided their candidates to capture a number of state elections.[84] Papers noted that "the Females asserted the privilege granted them by the laws of this state, and gave in their votes for members to represent them."[85]

Women eventually became a political tool in New Jersey for both parties on the state level. "Were not the republicans in Essex County drove to the necessity at the last Congress election, of bringing out the women, &c. to vote, on account of their opponents (the federalists) having practiced it for many elections previous," the editor of the Centinel of Freedom explained, the Federalists would have carried a local election.[86] Similarly, a writer to the paper claimed that despite the strong female vote for Federalists that the Republicans still won at the polls. The writer, who called himself "A Young Republican of Elizabeth Town," created poetry that both belittled the female vote and its potential: "Although reinforc'd by the petticoat band, True Republican valor they could not withstand; And of their disasters in triumph we'll sing, For a petticoat faction's a dangerous thing."[87] According to Judith Apter Klinghoffer and Lois Elkis, as New Jersey moved into the nineteenth century, the political parties began to re-evaluate female voting. The Federalists concluded that state Republicans now stood to gain more from the female vote than they did because of Republican success on the national level in 1800. They were wrong because in 1802, Federalists still faired well and resulted in an equal number of delegates elected from both parties. Republicans blamed the outcome on bogus female votes.[88] Until that point, Republicans had pushed strongly to garner female support for candidates. "Our daughters are the same relations to United States as our sons," the Republican Genius of Liberty decreed. "The contrary idea originated in the same abuse of power, as monarchy and slavery, and owes its little remaining support to stale sophistry."[89] The ability of women to vote—as well as that of free blacks—became a liability. Women, who had a right in New Jersey that existed nowhere else in the United States, lost it. Late in 1807, a new election law passed the assembly. It provided "that no person shall vote at elections, unless he be a free white male citizen of this state and of the United States, of 21 years of age, worth fifty pounds ... and resided in the county one year."[90]

DOGS AND HOGS: CLEANING UP THE CITY

Though political issues and a second war with Britain occupied much of the news space in the press during the pivotal decades following the ratification of the Constitution, citizens and editors also used the press to deal with problems confined to specific places. In New York City, one paper—the *New-York Evening Post*—and its editor—William Coleman—began a crusade to improve the city's sanitation. According to Patricia Dooley, New York in the 1810s was plagued with wild dogs that attacked children and with pigs that ran loose, eating the piles of garbage that could be found on streets throughout the city.[91]

Rabies among the animals on the streets was a prime concern, and Coleman asked in one of his editorials, "Are the lives of our citizens to be perpetually endangered? ... Every dog now found in the street should be instantly killed."[92] Swine, Coleman said, created a similar health problem, noting "hogs [*have*] the *freedom of the city*" and might spread rabies just as easily as dogs.[93] Coleman wrote that the people of New York constantly complained about dogs and pigs roaming the streets and that city officials simply ignored the problem. "*The Swinish Multitude* are still permitted to infest our streets, notwithstanding all that has been written and printed against so great a unisance [*nuisance*]. Why is not the voice of our citizens regarded? All cry out against the disgusting evil, and yet it is suffered to exist."[94]

On July 27, 1818, Coleman provided information for New Yorkers on how to treat "*the bite of a mad dog*" and a week later reported that "a mad dog made his appearance in the lower end of town, and in his route destroyed nine chickens, one cat, and one pig, besides biting six dogs!" All dogs that were bitten and the rabid one were destroyed along with any canine that was not wearing a collar.[95] The rabies epidemic continued, and Coleman told how reports confirmed that one dog had bitten several people with many others probably not reporting attacks. "There is no telling how many others he may have bitten," Coleman said. "Why is this? Of what value are all the dogs in the state, when compared to that of the meanest individual?"[96] The combination of rabid dogs and free-roaming hogs led Coleman to ask, "how long will it be before our streets will be full of mad dogs and hogs?" In one advertisement in another city paper, Coleman reported that a butcher advertised forty "*agreeable street companions*" for sale. "If every petty grocer keeps upon the town an equal number, (and they have a right so to do) the *swinish multitude* of NEW-YORK will exceed all calculation!"[97]

Readers joined in. One reader asked that the city clear the streets "of the throngs 'of mongrels, curs, and dogs of low degree,' that infest the streets." He then implored Coleman not "to leave any thing which regards the well-being of the city" untouched when it comes to the welfare of its citizens.[98] When members of the meat-packing industry in the city began to question the *Evening-Post*'s motives for its attacks on animals in the streets, Coleman said he only did so as the voice for the "universal indignation [*that*] prevailed the whole city," which called out in the unanimous "cry of Shame! Shame!" It was his duty, Coleman explained, to make the city a better place for its citizens. "It is true that we have expressed ourselves with freedom, and considerable warmth, upon the subject of the armies of dogs which *to this* day run at large in our city, to the imminent hazard of the lives of our citizens; and also in relation to the droves of hogs which are suffered to infest our streets, maugre all decency, and to the great annoyance of all cleanly and civil people."[99]

Coleman's assault on dogs and hogs did not stop. He noted in 1819 that Philadelphia had passed an ordinance that prohibited dogs from running free in the city's streets, saying he hoped authorities in New York might follow the example.[100] Other New York City papers joined in, and in July 1819, Coleman reported the "The trustees of the village of Brooklyn have passed an ordinance, prohibiting dogs from running at large."[101] He also reported every time another American city passed an ordinance against animals running loose in the streets. In May 1821, Coleman reported that at long last, "*The Common Council* have lately passed an ordinance relating to dogs running at large in the city," which provided for fines against owners. Coleman noted, though, "we have not seen it promulgated except in a single" instance.[102] Even though Coleman's efforts did not immediately change New York's street conditions, they did widen the debate. By 1820, New York physician David Hosack proposed that the city clean up its streets, dispose of sewage and wastes through a system, and improve the city's water supply system in an effort to improve sanitation in the city. Doing so, he said, would make for healthier living.[103] The city's government did have health administrators, but they were generally corrupt, something Coleman often acknowledged indirectly. By the 1830s, the city was actively working to remove sewage in the streets and the epidemic diseases it caused. Coleman's battle to clean up New York's streets raised awareness, widened the debate over sanitation, and brought new voices into the discussion, but it took decades for the city to establish its Department of Public Health, which happened in 1867.[104]

***

Fifty years after the signing of the Declaration of Independence, the United States had witnessed mammoth changes, all chronicled in the press and most debated in support or opposition by media that very much took sides on all issues. The early republic was, according to Thomas Leonard, the first society in history that provided a source that contained news for all. It worked because Americans had a higher literacy rate than any country in the world.[105] The wide dissemination of information also was successful because of the long English tradition that was transferred to America of the right to express oneself publicly and the fact that the issue of free speech and a free press became essential issues in the debate over the ratification of the Constitution. Declaring free speech as a right inherently granted had long been the way Americans started their discussion of any controversial issue, especially one that might rankle those in political, social, economic, or religious power.[106] It became a central issue in the debate over the ratification of the Constitution, making a bill of rights essential to many Americans before they would consider operating as a nation under the new form of government.

    Another factor played a vital role in the press' growth and its importance to and influence on the nation, too, and that was the rapid growth of advertising via newspapers. "The *advertisements*, moreover, which they daily contain, respecting new books, projects, inventions, discoveries and improvements, are well calculated to enlarge and enlighten the public mind, and are worthy of being enumerated among the many methods of awakening and maintaining the popular attention, with which more modern times, beyond all preceding example, abound," Samuel Miller said.[107] "Advertiser" became a standard word in nearly all newspaper nameplates during the period of the early republic. Advertising represented a symbiotic relation between the press and society. Advertisements were obviously important to printers and editors because they paid the bills and were the only way to make money unless political patronage could be obtained since Americans were notorious for not paying for their newspaper subscriptions. A society that was growing, expanding, and finding that it had revenue to acquire "things" depended upon the press to let them know what was available. Within those advertisements, which looked very much like contemporary classified ads and often included a generic woodcut of the item for sale, there really was something for everyone. The press was essential in shaping the capitalistic nature of society, and this trend would only expand throughout the nineteenth century.

For Americans in the last two decades of the eighteenth century and the first two decades of the nineteenth century, the press helped foster a sense of community because so many issues on the national level directly affected life on the local level. As these issues were being debated in public with the press as the catalyst, Americans began to develop an idea of what the United States would be. Newspapers, one writer said, "have become the vehicles of discussion in which the principles of government, the interests of nations, the spirit and tendency of public measures, and the public and private characters of individuals are all arraigned, tried and decided."[108] The press of the period also offered a glimpse into the future of print media in the United States and into the direction society would take. Religious issues and religious publications began to appear and would reach a peak in dissemination by the middle of the nineteenth century. Questions about slavery started to surface with the introduction of the first abolitionist publications after the War of 1812. Publications targeted solely at women were started.

The diversity of information that appeared in newspapers and would help define the nation expanded proportionally with the rapid growth of the press. In the next forty years, however, one topic would increasingly infiltrate all elements of public discussion and debate. It would influence the stances taken by editors and politicians and then by most citizens. Slavery would become the defining element of the nation in its antebellum era, and it would increasingly become the element that drove the press and, therefore, the conversation of the nation.

NOTES

1. William Duane to Stephen R. Bradley, November 10, 1808, quoted in Richard N. Rosenfeld, *American Aurora: A Democratic-Republican Returns* (New York: St.

Martin's, 1997), 3. Stephen R. Bradley information is found in Jacob G. Ullery, ed., *History of Vermonters and Sons of Vermont* (Holyoke, Mass.: Transcript Publishing, 1894), 104–106.
2. Jeffrey L. Pasley, *"The Tyranny of Printers": Newspaper Politics in the Early American Republic* (Charlottesville: University Press of Virginia, 2001), 195.
3. *Massachusetts Spy: Or, American Oracle of Liberty* (Worcester, Mass.), November 8, 1781, 3. Numerous papers published the events at Yorktown as "glorious news." See, for example, *New-Hampshire Gazette; or State Journal, and General Advertiser* (Portsmouth), October 27, 1781; *Boston Gazette, and Country Journal*, November 5, 1781; *Pennsylvania Packet or the General Advertiser* (Philadelphia), November 3, 1781; *The Freeman's Journal: or, The North-American Intelligencer* (Philadelphia), November 14, 1781.
4. See Carol Sue Humphrey, *The Revolutionary Era* (Westport, Conn.: Greenwood Press, 2003), x–xi.
5. Samuel Miller, *A Brief Retrospect of the Eighteenth Century*, 2 vols. (New York: T. and J. Swords, 1803), 2:253. Quoted in Isaiah Thomas, *The History of Printing in America* (1810; reprint, New York: Weathervane, 1970), 19.
6. Thomas, *The History of Printing in America*, 18–19. Quoted directly from Miller, *A Brief Retrospect of the Eighteenth Century*, 2:251–55.
7. *American Minerva* (New York), December 9, 1793, 1.
8. David A. Copeland and Carol Sue Humphrey, *The War of 1812* in *Greenwood Library of American War Reporting*, ed. David A. Copeland, 8 vols. (Westport, Conn.: Greenwood Press, 2005), 2:11.
9. Alfred McClung Lee, *The Daily Newspaper in America* (New York: Macmillan, 1937), 711.
10. Richard B. Kielbowicz, *News in the Mail: The Press, Post Office, and Public Information, 1700–1860s* (Westport, Conn.: Greenwood Press, 1989), 46–47.
11. Frank Luther Mott, *American Journalism, A History: 1690–1960*, 3rd ed. (New York: Macmillan, 1962), 194.
12. *Daily Advertiser* (Boston), April 7, 1814, 2.
13. "Impartial Papers," *New-Hampshire Patriot* (Concord), April 10, 1810, 3.
14. "Prospectus," *Boston Courier*, June 13, 1805.
15. *Providence Gazette and Country Journal*, March 24, 1781, 3.
16. "American Antiquities," *New-Haven Gazette, and the Connecticut Magazine*, October 26, 1786, 287.
17. *New-Haven Gazette, and the Connecticut Magazine*, May 24, 1787, 106.
18. A Bostonian, "A View of the Federal Government of America. Its Defects, and a proposed Remedy," *Independent Chronicle and the Universal Advertiser* (Boston), August 3, 1786, 2. The writer continued his critique in the August 10 issue. The essay appeared in other papers, including, for example, *New-Hampshire Mercury and the General Advertiser* (Portsmouth), August 16, 1786, 1; *Newport Mercury*, August 28, 1786, 1; *Pennsylvania Packet, and Daily Advertiser* (Philadelphia), November 11, 1786, 2.
19. *Independent Gazetteer, or, the Chronicle of Freedom* (Philadelphia), August 7, 1787, 3; *New-York Packet*, August 14, 1787, 2; *New-Jersey Journal* (Elizabethtown), August 15, 1787, 3; *Massachusetts Gazette* (Boston), August 17, 1787, 3; *Connecticut Courant* (Hartford), August 20, 1787, 2; *Maryland Chronicle or the Universal Advertiser* (Fredericktown), August 22, 1787, 2; *Massachusetts Centinel* (Boston)

August 22, 1787, 178; *Cumberland Gazette* (Portland, Maine), August 23, 1787, 3; *New-Hampshire Gazette, and General Advertiser* (Portsmouth), August 25, 1787, 2; *Newport Herald*, August 30, 1787, 1; *Columbian Herald or the Patriotic Courier of North-America* (Charleston), August 30, 1787, 2.
20. David Redick to William Irvine, September 24, 1787, quoted in Saul Cornell, *The Other Founders: Anti-Federalists & the Dissenting Tradition in America, 1788–1828* (Chapel Hill: University of North Carolina Press, 1999), 20.
21. Cornell, *The Other Founders*, 10.
22. Cornell, *The Other Founders*, 11.
23. "On the National Constitution," *Independent Chronicle and the Universal Advertiser* (Boston), June 5, 1788, 2. The article appeared in other papers. See, for example, *Newport* Herald June 19, 1788, 2.
24. Publius, "The FŒDERALIST. No. I," *Independent Journal: or, the General Advertiser* (New York), October 27, 1787, 3.
25. Cornell, *The Other Founders*, 28.
26. E. Gerry, *Massachusetts Centinel* (Boston), November 3, 1787, 54–55.
27. Cornell, *The Other Founders*, 28.
28. *Pennsylvania Herald, and General Advertiser* (Philadelphia), October 10, 1787, 2.
29. "A CONFEDERATIONALIST," *Pennsylvania Herald, and General Advertiser* (Philadelphia), October 27, 1787, 2.
30. *Pennsylvania Herald, and General Advertiser* (Philadelphia), October 10, 1787, 2.
31. "A CONFEDERATIONALIST,"*Pennsylvania Herald, and General Advertiser* (Philadelphia), October 27, 1787, 2.
32. Stanley Elkins and Eric McKitrick, *The Age of Federalism* (New York and Oxford: Oxford University Press, 1993), 103.
33. Donald H. Stewart, *The Opposition Press of the Federalist Period* (Albany: State University of New York Press, 1969), 4–5.
34. Pasley, *"The Tyranny of Printers,"* 285.
35. Culver H. Smith, *The Press, Politics, and Patronage: The American Government's Use of Newspaeprs 1789–1875* (Athens: University of Georgia Press, 1977), xi, 3.
36. Smith, *The Press, Politics, and Patronage*, 14.
37. Pasley, *"The Tyranny of Printers,"* 63.
38. *Independent Chronicle and the Universal Advertiser* (Boston), September 6, 1792, 3.
39. "Public Opinion," *National Gazette* (Philadelphia), December 19, 1791, 59.
40. *National Gazette* (Philadelphia), August 15, 1792, 330.
41. Smith, *The Press, Politics, and Patronage*, 17, 28.
42. Copeland and Humphrey, *The War of 1812*, 13.
43. *Aurora General Advertiser* (Philadelphia), December 23, 1796. John Fenno also ran Bache's comments the same day. See "THE DEMONIAC–No. III," *Gazette of the United States* (Philadelphia), December 23, 1797, 3. In 1794, Bache added *Aurora* to the name of his paper. People called it the *Aurora*, dropping *General Advertiser*. Bache kept it in the nameplate, however.
44. *Annals of Congress*, U. S. House of Representatives, 5th Congress, 2nd Session (Washington, D.C., 1834), 8:2164.
45. James Morton Smith, *Freedom's Fetters: The Alien and Sedition Laws and American Civil Liberties* (Ithaca, N.Y., 1956), 185–87.

46. *Massachusetts Mercury* (Boston), October 5, 1798, 2. The story also appeared in the *New-Hampshire Gazette* (Portsmouth), October 9, 1798, 2 and the *Gazette of the United States* (Philadelphia), October 26, 1798, 1. Publication Date: October 9, 1798
47. Smith, *Freedom's Fetters*, 185–87.
48. *Aurora General Advertiser* (Philadelphia), September 13, 1796, quoted in Carol Sue Humphrey, *The Revolutionary Era* (Westport, Conn.: Greenwood Press, 2003), 218.
49. *Gazette of the United States* (Philadelphia), October 26, 1796, quoted in Humphrey, *The Revolutionary Era*, 299.
50. Michael Schudson, *The Good Citizen: A History of American Civic Life* (New York and London: Oxford University Press, 1998), 78.
51. Jefferson to Madison, 5 February 1799, in Paul L. Ford, *The Works of Thomas Jefferson* (New York, N.Y., 1892–1899), 7:344.
52. Quoted in Stewart, *The Opposition Press*, 634.
53. Alexander Addison, "Liberty of Speech, and of The Press," *Columbian Centinel* (Boston), December 29, 1798, supplement, 2.
54. "To the Public," *Porcupine's Gazette and United States Daily Advertiser* (Philadelphia), March 4, 1797, 1. Cobbett ran this prospectus for several issues following his initial offering on March 4.
55. Charles Pierce, "PROPOSALS FOR EDITING, PRINTING AND PUBLISHING The Oracle of the Day," August 4, 1798, quoted in Robert W. T. Martin, *The Free and Open Press: The Founding of American Democratic Press Liberty, 1640–1800* (New York: New York University Press, 2001), 135.
56. Smith, *The Press, Politics, and Patronage*, 42–43.
57. Pasley, *The Tyranny of Printers*, 351.
58. This section on the period leading up to the War of 1812 is taken from Copeland and Humphrey, *The War of 1812*, 16–19.
59. "British Outrages," *Raleigh Register, and North-Carolina Weekly Advertiser*, July 2, 1807, 2.
60. *New York Spectator*, quoted in *Connecticut Courant* (Hartford), July 8, 1807.
61. *Norfolk Gazette and Publick Ledger*, June 24, 1807.
62. *National Intelligencer and Washington (D.C.) Advertiser*, July 13, 1807, 1.
63. *Washington (D.C.) Federalist*, July 3, 1807.
64. "More yet," *The Balance and Columbian Repository* (Hudson, New York), July 14, 1807, 219.
65. Donald R. Hickey, *The War of 1812: The Forgotten War* (Urbana: University of Illinois Press, 1989), 19, 11.
66. *National Intelligencer* (Washington, D.C.), September 26, 1812, 2.
67. "*TO WHOM IT MAY* CONCERN," *Enquirer* (Richmond), 14 December 14, 1811, 3.
68. *Columbian Centinel* (Boston), January 13, 1813, 2.
69. *Columbian Centinel* (Boston), December 21,1814, 1.
70. U. S. Congress, *Annals of Congress: Debates and Proceedings in the Congress of the United States, 1789–1824*, 42 vols. (Washington, D. C., 1834–1856), 12th Congress, vol.1, 297.
71. "Peace," *National Advocate* (New York), February 20, 1815, 2.
72. *American Watchman and Delaware Republican* (Wilmington), October 7, 1815, 3.
73. "Era of Good Feelings," *Columbian Centinel* (Boston), July 12, 1817, 2.

74. Alexander Addison, "Liberty of Speech, and of The Press," *Columbian Centinel* (Boston), December 29, 1798, supplement, 2.
75. Virginia Resolution, December 24, 1798, *The Founders' Constitution*, Vol. 5, Amendment I (Speech and Press), Document 19, http://press-pubs.uchicago.edu/founders/documents/amendI_speechs19.html.
76. Richard Hay, *An Essay on the Liberty of the Press* (1799), quoted in Leonard W. Levy, *The Emergence of a Free Press* (New York and Oxford: Oxford University Press, 1985), 313.
77. *People v. Croswell*, 3 Johnson's (New York) Cases, 336, 337–339, at 358.
78. *Time Piece* (New York), July, 30, 1798, 2.
79. *Federal Republican* (Baltimore), June 20,1812. The offending information was reprinted in the July 27, 1812 edition.
80. *American Watchman and Delaware Republican* (Wilmington), June 26, 1812, 2.
81. *Village Record* (West Chester, Pa.), November10, 1819.
82. Irwin N. Gertzog, "Female Suffrage in New Jersey, 1790–1807," in Naomi B. Lynn, ed., *Women, Politics, and the Constitution* (New York: Haworth Press, 1990), 48–49.
83. William Whitehead, A Brief Statement of the Facts connected with the Origin, Practice and Prohibition of Female Suffrage in New Jersey, January 21, 1858, New Jersey Historical Society, Proceedings, viii, 101, in *The Historical Magazine* VII (January 1863): 167.
84. Judith Apter Klinghoffer and Lois Elkis, "'The Petticoat Electors': Women's Suffrage in New Jersey, 1776–1807," *Journal of the Early Republic*, 12, 2 (Summer, 1992): 172.
85. "Rights of Women," *New-Jersey Journal* (Elizabethtown), October 18, 1797, 3.
86. *Centinel of Freedom* (Newark), September 15, 1801, 3.
87. A Young Republican of *Elizabeth Town*, "REPUBLICANISM TRIUMPHANT: or, The 'Well-Born' Defeated," *Centinel of Freedom* (Newark), October 23, 1798, 3.
88. Klinghoffer and Elkis, "'The Petticoat Electors,'" 177, 183.
89. *Genius of Liberty* (Morristown), August 7, 1800, quoted in Patricia L. Dooley, *The Early Republic* (Westport, Conn.: Greenwood Press, 2004), 195.
90. "New Election Law," *Trenton Federalist*, November 16, 1807, 3.
91. Dooley, *The Early Republic*, 305.
92. *New-York Evening Post*, August 25, 1818, 2.
93. "Hogs," *New-York Evening Post*, February 16, 1818, 2.
94. *New-York Evening Post*, June 5,1818, 2.
95. "*Mad Dogs*," *New-York Evening Post*, August 5, 1818, 2.
96. "*Dogs*," *New-York Evening Post*, August 21, 1818, 2.
97. "H*ogs*," *New-York Evening Post*, August 25, 1818, 2.
98. F., "MAD DOGS," *New-York Evening Post*, August 27, 1818, 2.
99. *New-York Evening Post*, September 2, 1818, 2.
100. *New-York Evening Post*, June 12, 1819, 2.
101. *New-York Evening Post*, July 3, 1819, 2.
102. *New-York Evening Post*, May 7, 1821, 2.
103. George W. Smillie, *Public Health, Its Promise for the Future* (New York, Macmillan Co., 1955), 168.

104. Israel Weinstein, "Eighty Years of Public Health in New York City," *Bulletin Of The New York Academy Of Medicine*, 23, No. 2 (1947): 221–37; reprint, *Journal Of Urban Health: Bulletin of the New York Academy of Medicine*, 77, No 1 (March 2000): 138.
105. Thomas C. Leonard, *News for All: America's Coming-of-Age with the Press* (New York and Oxford: Oxford University Press, 1995), 29.
106. See David A. Copeland, *The Idea of a Free Press: The Enlightenment and Its Unruly Legacy* (Evanston, Ill.: Northwestern University Press, 2006).
107. Miller, *A Brief Retrospect of the Eighteenth Century*, quoted in Thomas, *The History of Printing in America*, 18–19.
108. *American Watchman and Delaware Republican* (Wilmington), February 22, 1814, quoted in Humphrey, *The Press and the Young Republic*, 140.

4. The Great Organ of Social Life, the Prime Element of Civilization: The Antebellum Era and Civil War

In the initial issue of the New York *Herald*, James Gordon Bennett proclaimed that the press was "the great organ of social life, the prime element of civilization."[1] In antebellum America, he was correct. The press drove national discussion in a continuation and accelerated capacity from the first decades of the nineteenth century because of the rapid increase in the number of newspapers and in the expansion of the press to meet the needs of various groups. Abolitionists, African Americans, religious denominations, women, and others created media outlets to address issues. Technology played a role, too, with the invention of the telegraph, new and faster printing presses, and visual images. The issues that the nation faced, however, often transcended the multitude of outlets that developed even though they had additional items on their agendas. Sectional divisions grew increasingly, and slavery was the issue that lay at the root of these divisions even though many claimed that states' rights or popular sovereignty were the reasons that the country was slowly pulling apart and moving toward disunion.

In 1820, 512 newspapers were published regularly in America with a circulation of slightly less than 300,000. By 1860, about three thousand newspapers were regularly published, with more than 350 daily papers with their circulation reaching nearly 1.5 million. Magazines grew at an even more phenomenal rate. A dozen magazines were published in 1800. By 1860 that number grew to one thousand.[2] Visitors to the United States observed the power of the press. Alexis de Tocqueville explained its power when he wrote following his early 1831–1832 tour of the nation that the press "rallies the interests of the community round certain principles and draws up the creed of every party."[3] And what were the interests of the nation? Slavery, moral and social reform, women's rights, burgeoning immigration, religion, economic depression, urbanization, public education, westward expansion, the desire to hold onto a more agrarian lifestyle, and what might happen to a republic that was becoming increasingly polarized because of these issues that dominated the lines of newspapers and the conversations of many.

Newspaper growth had other effects on antebellum America. Prior to the Revolution, voting rates were low, with only 10 to 15 percent of eligible white males doing so in 1775.[4] But, newspapers continued to grow, prosper, and discuss the issues that affected the direction of the nation. By the 1820s, people in various levels of society turned to papers to voice opinions. By the time of Andrew Jackson's presidency in 1829, more than 50 percent of American households subscribed to a newspaper; and approximately 44 percent of eligible voters participated in the 1832 election.[5] Involvement in public debate by Americans through an expanding press and the rise in people voting cannot be coincidence. Jackson acknowledged this when elected president.

In 1824, Jackson won the majority of popular votes in the United States but failed to capture enough votes in the Electoral College to claim the White House. Deals between candidates gave John Quincy Adams the election in the House of Representatives. Many editors, especially those outside New England, felt political corruption had taken the election of president away from the people. One even said that "public opinion will eventually be respected by the election of the General."[6] As a response to what they viewed as a Shanghaied presidency, editors mounted a campaign for change. Jackson noted the backlash to the election in the papers and followed the lead of America's press. He allowed editors to set the agenda. "The recent demonstration of public sentiment inscribes on the list of Executive duties, in characters too legible to be overlooked, the task of reform," Jackson said in his 1829 inaugural address.[7] Jackson's press secretary Martin Van Buren, who succeeded "Old Hickory" as president in 1837, added, "Without a paper, we may hang our harps on the willows."[8] Van Buren meant that unless those in political power or who desired it had a venue to address the people, all their efforts would be futile. The newspaper editor in the nineteenth century, therefore, increasingly became the person who directed political activities on all levels—national, state, and local.[9]

The rapid growth of newspapers, of course, did not occur in a vacuum. It coincided with the mercuric rise in America's population. The antebellum period began with twenty-two states. By 1860 there were thirty-three. Even more dramatic was the growth of the nation's population, which increased by more than 210 percent. In 1820, America's population was slightly more than 10 million. By 1860, the population surpassed 31 million. Despite efforts to limit slavery through the Missouri Compromise, the Compromise of 1850, the Kansas-Nebraska Act of 1854, and other legislation, the number of slaves grew, too. In 1820, census figures placed the slave population at more than 1.5 million. By 1860 it had swelled to nearly 4 million, an increase of more than 158 percent. Free African Americans by 1860, however, numbered fewer than 500,000.[10] Immigration in part fueled America's population explosion after 1840. More than 1.7 million people entered America in the 1840s, and in the 1850s, that number jumped to more than 2.5 million.[11]

An estimated three thousand newspapers and one thousand magazines for a population of 31 million people do not seem adequate, nor does a 1.5 million daily circulation, but these numbers are misleading. Daily circulation figures naturally omit weekly papers from the tally and their reach into society. The population contained a huge component of young people, too. According to census figures, 50 percent of all white Pennsylvanians in 1860, for example, were 19 or younger, a ratio reflected in most states.[12] Newspaper—and now magazine—sharing continued among people. If half of all American families subscribed to a newspaper by around 1830 as some scholars suggest and those papers were shared with just one family, then the potential existed for 100 percent saturation of newspapers. When you consider that William Gilmore says that newspapers regularly ended up twenty-five miles from where they were purchased, more than one family no doubt saw each issue.[13] Americans continued to read aloud newspapers in homes, and they did the same in taverns and places of lodging. Group reading naturally facilitated discussion and debate. One newspaper's contents, in this setting, could be heard by dozens of people. Looking at newspaper circulation through the lens of shared copies and the reading of them in public in relation to America's population helps explain why, in 1825, a Virginian who proposed a new paper noted that "A thirst for newspaper reading prevails among all ranks of society throughout our country."[14] Finally, when one takes into account that an estimated 888 million copies of all newspapers circulated throughout the United States by the Civil War period,[15] it sheds light on why Tocqueville could say with confidence that newspapers in the United States "drop the same thought into a thousand minds at the same moment."[16]

Understanding newspaper penetration in society only explains part of the reason why the press played such a crucial role in defining the nation and setting the agenda for debate during the antebellum era. The press of the first two decades of the nineteenth century was forced to deal with the critical issues of a new country working hard to establish an innovative form of government. The core values of how that government was to operate under the Constitution were understood by opposing political ideologies very differently, and the result was a very partisan press. Even though the Federalists eventually died out, the evolving and growing nation would not maintain a single national voice for long. But, with the United States fifty years past the Declaration of Independence and thirty-plus years past the ratification of the Constitution, other issues would become vital to public debate. Politics would continue to be a critical issue for all Americans, but a great number of moral, social, and ideological issues arose. These concerns combined with the rapid growth in the numbers of newspapers and their circulations to produce the ideal avenue for public debate often within the public sphere. Because these issues often had great repercussions for people, they turned to media as the means to insert themselves into public discussions. They discovered that the press was the way to do this because it was the only venue that ensured discussions could take place on every level of society.[17]

A DIVIDING NATION

The pivotal issues that the United States faced from the 1820s until the Civil War played out on a national stage, but they had dramatic results locally, too. Efforts to balance regional concerns with national stability meant that nearly everything that became legislation in Washington had repercussions within the states. The press naturally served as the sounding board for discussion and as a place to shape the nation's direction. It did this, generally, via the partisan arrangement between editors and political parties that developed in the 1790s and that would remain entrenched in America until after the Civil War. Newspapermen became politicians. Politicians, through patronage, became editors. Editors became congressional lobbyists.

Both editors and politicians approached issues from an increasingly sectional perspective. Generally, Americans in the South and west of the Appalachians distrusted the national bank, for example, while states above the Mason-Dixon line and east of the mountains supported it. When it was time to renew the bank's charter, President Andrew Jackson, the nation's first president directly tied to the lands beyond the Appalachians, helped kill a new charter, but not before the debate played out on the national, regional, and local media stage. Even in New England, unanimity concerning the bank did not exist. "A bank we must have," Massachusetts senator Isaac Bates stated in Worcester's *Massachusetts Spy*. "The People look to it for redemption from the thraldem in which their business is involved. They wait for it as for the dawn of a bright and glorious morning after a disturbed, dark night of distress and suffering."[18] In nearby Hartford, Connecticut, however, a meeting of local citizens painted a different picture: "The powers of the Bank over the circulating medium are dangerous....The Bank receives the deposits of the nation, without making any adequate compensation for the many millions constantly standing to the credit of the Government." The citizens' resolution concluded by stating, "there exists no necessity for a bank with such powers at the present."[19]

Sectionalism fueled by slavery ultimately became the driving force behind almost every political issue on which the press focused. Increasingly, it became impossible to read newspapers as they discussed issues Americans faced on any level where slavery did not underscore debate. Slavery was a moral issue, and as such an advocacy press directed by abolitionists railed on the evils of it. On a more pervasive level, however, slavery turned into a pivotal issue of national expansion, and the westward movement of Americans in the 1800s seemed always to return to the role that slavery would play in the nation's territorial lands. The press, courtesy of the *United States Magazine and Democratic Review*, gave Americans a name and a divine right for their desire to inhabit all the land between the Atlantic and Pacific oceans. The *Review* in 1845 said, "Our manifest destiny is to overspread the continent allotted by Providence."[20] The move by Americans into lands acquired by the Jefferson administration through the 1803 Louisiana Purchase began long before the coining of manifest destiny, though, and when portions of that territory requested admission into the union as a state, the issue of slavery played the crucial role.

In 1819 and 1820, the first major legislation designed to appease Americans on the issue of slavery and new territory erupted in the pages of the nation's newspapers. Now slavery, which had been an issue of concern as the Founding Fathers grappled with the Constitution and produced the Three-Fifths Compromise as a way to deal with state population and slaves and who put in the document a date when the importation of slaves into the country would cease, became the controlling factor of state expansion. Since the ratification of the Constitution, eight states had joined the Union, four allowing slavery and four deemed free states. Geographically, each fit naturally with the understood division between slave and free states, Mason and Dixon's line between Maryland and Pennsylvania and the Ohio River. Missouri, which was requesting admission as a slave state, was situated above where the Ohio flowed into the Mississippi River. Missouri's admission would also create an imbalance between free and slave states, which stood at eleven of each. This element of the conundrum disappeared, though, when Maine requested admission to the Union in 1820. A part of Massachusetts, Maine now sought sovereignty. If Maine entered the Union as a free state, then Missouri could join as a slave state, thus preserving regional balance in the Senate.

Many Americans, however, saw the issue of slavery in the Louisiana Territory as an issue that transcended regional legislative balance, and the issue became a constitutional one. Rufus King of New York said that, legally, Congress could keep slavery out of American territories, declaring that "it has been established in the States north-west of the river Ohio."[21] But, the Louisiana Purchase itself stated that when its territory was incorporated into the United States that all its citizens were granted "the enjoyment of *all the rights*, advantages and immunities" of every U.S. citizen. That meant, one Virginian pointed out in the Richmond *Enquirer*, any citizen in the territory had the right to own slaves because the Constitution granted the right to own slaves.[22] Missourans believed the question went beyond slavery. Admission to the United States under such terms meant the state would not only have to disallow slavery, but it meant the United States was asking its citizens to "abandon your rights, and suffer an earthly power to dictate the terms of your constitution," Thomas Hart Benton's St. Louis paper explained.[23]

The debate in the papers continued for months. Those who opposed Missouri entering as a slave state built upon the basic argument that slavery was "an inveterate disease, adverse to the very life of liberty, and a contradiction of the sacred declaration, upon which was established resistance to tyranny, emancipation from subjection; and the assertion of the equal rights of all mankind."[24] When Senator Jesse Thomas of Illinois offered a compromise to allow Maine and Missouri enter as free and slave respectively, angry Americans on both sides of the slavery issue denounced the compromise. "A constitution warped from its legitimate bearings, and an immense region of the territory closed for ever against the Southern and Western people—such is a 'sorry sight' which rises to *our* view," Richmond editor Thomas Ritchie noted.[25] Northerners called their elected representatives who supported the Missouri Compromise "political hypocrites," declaring, "*The freedom of our country has been sacrificed in the house of its friends.*"[26]

Most Americans disliked the Missouri Compromise. Thomas Jefferson called it "a fire bell in the night,"[27] but it helped the country avoid schism. The *National Intelligencer* said that with the compromise "we look now only for harmony and conciliation on all sides."[28] Joseph Gales Jr. and William Seaton were wrong. Issues surrounding slavery, territory, and state expansion were only beginning.

WAR, A PROVISO, AND MORE COMPROMISE

Every time the nation faced a national crisis in the years following the Missouri Compromise, it seemed as if at the root one found the issue of slavery. When Texas, a part of the Mexican state of Coahuila, sought independence from Coahuila and was refused by the Mexican government, revolt ensued, which ultimately led to an independent Texas. Approximately 90 percent of the inhabitants of Texas emigrated from the Southern states of the United States, and petitioned the U.S. government to become a state in the Union.[29] Even though Mexico had made slavery illegal, Americans in free states believed that Southerners, despite the fact that they had been living in a territory that banned slavery, would soon introduce the practice if Texas became a state. After all, it was well below the 36-degree, 30-minute line of demarcation between free and slave territory that the Missouri Compromise established for territory west of the Mississippi River. Texas was a massive land mass, and many Northerners had long believed Texas would be subdivided into multiple slave states. "If Texas should be added to the United Sates, it is not an extravagant supposition that there may in process of time be twelve or fifteen additional Slave States incorporated into the Union," editor and abolitionist Timothy Dwight explained. Dwight went on to say that Southern politicians always garnered more power than the size of their constituencies warranted. With a numerical advantage, too, he believed the free states would have "to surrender themselves" to the will of the slave states.[30] As a result, Texas was denied admission into the United States, but doing so was only temporary. Manifest destiny was a buzzword for the nation. New York *Tribune* editor Horace Greeley, borrowing from an Indiana editor in 1837, repeatedly advocated that Americans "go west," and when Southerner James K. Polk—running on an expansion platform—was elected as the eleventh president in 1844, it was only a matter of time, many Northerners lamented, before Texas would join the Union and tilt the balance of power to the South. They were correct, Polk was inaugurated and Congress annexed Texas within the same week.

The American public argued the legitimacy of war with Mexico in newspapers. James Buchanan, the secretary of state, painted the war as a humanitarian effort that increased U.S. stature worldwide: "Heaven has smiled upon the just cause; and the character of our country has been illustrated by a rapid succession of brilliant and astonishing victories. The exploits of our army have elevated our National character, and shed a luster upon our name throughout the civilized world."[31] Others, though, argued that the war was being waged in order to add to the nation's territory. Weston Gales, editor of the *Weekly Raleigh Register, and North Carolina Gazette*, claimed the Mexican War was one that the president initiated solely for personal political purposes. It was a war of conquest to gain land and that there was no legitimate reason for it.[32] Still others said the land grab had one purpose: to broaden the territory open to slavery in the United States, and that thought led one junior American congressman to propose legislation that focused discussions in the United States for years.

Fighting a war requires money, and President Polk needed $2 million about three months into the war with Mexico. Texas was already admitted as a slave state, and Polk, it was generally acknowledged, wanted to add the Mexican territories of New Mexico and California to the United States. Newspapers were arguing the validity of the war and about taking other Mexican territories as spoils of war. Penny editor Arunah S. Abel of Baltimore would even suggest that all Mexicans should become U.S. citizens as part of the paper's "All Mexico" movement.[33] During all the wrangling in the press and in Congress, a first-year representative from Pennsylvania, David Wilmot, made a jarring proposal: "That as an express and fundamental condition to the acquisition of any territory from the Republic of Mexico by the United States by virtue of any treaty ... neither slavery nor involuntary servitude shall ever exist in any part of said territory except from crime, whereof the party shall first be duly convicted."[34] Wilmot's words were amended to the House proposal for the funds that Polk requested. The House of Representatives passed the amendment on the day it was introduced. The Senate, however, put off the bill until the last day of the legislative session. The president wanted Wilmot's amendment stricken from the bill, but Senator John Davis of Massachusetts staged a filibuster to prevent any action by the Senate. By the time he finished, the House had adjourned, and the Senate did the same—without discussing the amendment or passing the war funding bill.[35]

The Wilmot Proviso, as the amendment was called, immediately drew the national discussion away from the war and back to the issue of slavery, or at least to how the war would affect the nation in terms of slavery in its territories in addition to renewed debate on the evils of slavery. The debate in the press generally formed on sectional lines, and the Wilmot Proviso became one of the most debated pieces of legislation in the press never to be enacted, coming to a vote dozens of times before 1850. Nearly fifteen years before shots at Fort Sumter would begin the Civil War, the Wilmot Proviso caused the nation to face the prospect that the United States might not continue so long as slavery created a divide among its people. With amazing premonition, one Boston writer in 1846 observed, "As if by magic, it [the Wilmot Proviso] brought to a head the great question which is about to divide the American people."[36] Though the Proviso never became law, the questions it raised created a stalemate following the Mexican War and affected political action through the 1850s. It even threatened to end the nation until the Compromise of 1850 provided a temporary fix. Henry Clay proposed that California be admitted as a free state and the other territories gained from Mexico free to choose whether or not to allow slavery. More was needed, though, and Illinois senator Stephen A. Douglas finally worked out a deal that killed the Wilmot Proviso but added the Fugitive Slave Act to the compromise. The law mandated that runaway slaves be returned to the South, and failure to do so could mean jail time and a stiff fine to anyone harboring runaways.

Newspapers raged in the North about the immorality of the act. Newspapers, North and South, talked openly that the Fugitive Slave Act would lead to the end of the Union. "The fugitive slave law, as it now stands, can no longer be enforced without jeopardizing the public tranquility to an alarming extent," Horace Greeley explained in the New York *Tribune*. Greeley, one of the most powerful press voices of the age added that "unless the country is to be precipitated upon insurrection and perchance civil war," Congress had to revoke the Fugitive Slave Act.[37] Gamiliel Bailey, the editor of the abolitionist-leaning *National Era*, said, "Our business is, to redeem the Federal Government from vassalage to the Slave Power, and all responsibility for

Slavery, and, through its constitutional exercise, foster a sentiment of Liberty in the slave states."³⁸ In the South, eradicating slavery was synonymous with rejection of constitutional rights. "The question is, Will the North remain content with the so-called Comprise Bills, or will her people persist in attempts to violate the Constitution?" a writer calling himself Georgia asked in a Northern magazine. He then said what would happen if efforts to end slavery continued. "But let the North refuse to abide by our rights, and the cry, which will go up from the hearts of the whole Southern people, will be, 'Let us go out from among them.'"³⁹

A new legislative compromise, the Kansas-Nebraska Act of 1854, would allow the Union to continue, but the press debate only intensified. A new political party calling itself Republican formed, and it increasingly made slavery its principal platform issue. By 1859, there was little, many felt, that could keep a division North and South from occurring. When John Brown led a raid on Harper's Ferry, Virginia, on October 16, 1859, newspaper commentators in the South saw his actions as a personification of the North's disdain of the South. Finally, the North was seen more as a threat and enemy than as a different region of the same country. Southerners wrote that Brown's raid was financed by Northerners. The Richmond *Enquirer* claimed the North "aided and abetted this treasonable invasion of a Southern State." South Carolina Governor William H. Gist stated that the South must "now unite for her defense."⁴⁰ Robert Barnwell Rhett, editor of the Charleston, South Carolina, *Mercury*, had long supported secession, and he said Brown's raid was part of "a wide-spread scheme [that] was maturing at the North for insurrections throughout the South." He claimed that there was proof that Horace Greeley and his *Herald* were the strongest advocates of this plan. "The experience of the last twenty-five years, of ignominious toleration and concession by the South," Rhett said in the conclusion to his fiery editorial, "show to the most bigoted Unionist that there is no peace for the South in the Union, from the forbearance or respect of the North. The South must control her own destinies or perish."⁴¹ Northerners tended to agree. "While millions of prayers went up for the old martyr [Brown] yesterday, so millions of curses were uttered against the hellish system which so mercilessly and ferociously cried out for his blood," Pittsburgh *Gazette* editor David White explained, while the abolitionist William Lloyd Garrison penned an editorial that declared "*John Brown was right*."⁴² The positions represented by the loudest and strongest press voices North and South meant the nation was at an impasse. The South, as Rhett said, believed that it had been under attack for decades. In the North, slavery had achieved anathema status among the majority. When the nation elected Lincoln president from the "black republicans," as Southerners referred to the party, the Richmond *Enquirer* announced, "This is a *declaration of war*." In South

Carolina, the Charleston *Mercury* explained the Union was no more. "Henceforth we are two peoples."[43]

The political and patronage press served as the sounding board for political issues. Ordinary citizens and politicians used it to discuss the issues of the day. Slavery increasingly became the prime motivator of nearly every act. By the end of November 1860, nearly all Americans knew troubling days lay ahead. It was obvious, no matter where one lived, from the news.

ADVOCACY AND ABOLITION

When the nation entered the Era of Good Feelings in the last years of the 1810s, an opening existed for the press to do more than concern itself with the critical political issues of the day. As we have seen, though, politics remained central to the information in the press, and slavery turned out to be at the core of politics. With such a powerful tool in their midst, Americans readily turned to the press as a tool for change beyond the press of patronage. Consequently, following the Second Great Awakening, many Americans saw it as their moral and religious duty to make life better for the less fortunate. As a result, an advocacy press developed. People realized that the most efficient and powerful way to spread a message was via the printed word. Because of the moral imperative that drove them, reformers often assumed a radical stance, and none turned out to be more radical than those who espoused the end of slavery. "I WILL BE HEARD," William Lloyd Garrison shouted from the first issue of his newspaper, the *Liberator*.[44] Abolition resonated throughout the nation. From humble beginnings at small newspapers printed in Ohio, Tennessee, and North Carolina, Americans who believed slavery violated the will of God turned to the press to call the nation to accountability. Benjamin Lundy, who operated one of the first abolitionist papers, believed that the press publications were the most powerful tools that anyone could employ, stronger even than the weapons of war. "Castles fall before them," he said, "cannons are silenced."[45] By 1835, more than one million antislavery publications were produced in New York City alone and disseminated to a national population of 15 million.[46]

While the moral and religious drove abolitionists initially, they also believed there were financial and patriotic ramifications that necessitated the end of slavery, and, according to Harriet Martineau, writing in 1839, abolitionists were comprised of "men and women of every shade and color, of every degree of education, of every variety of religious opinion, of every gradation of rank" bound together only in the conviction that slavery should no longer be interwoven with any element of American life.[47] While politicians looked for compromise concerning slavery and many in the North and the South agreed that slavery would eventually be abolished, abolitionists almost universally spurned compromise and believed manumission should occur immediately. Garrison, as the most vocal of abolitionist printers stated this non-compromising position: "On this subject, I do not wish to think, or to speak, or to write in moderation....I am in earnest—I will not equivocate—I will not excuse—I will not retreat a single inch...."[48]

The reaction that many Americans had to the most radical or outspoken abolitionist editors is proof that their message reached deep into the nation, even though individual newspapers such as Garrison's *Liberator* never had large circulation figures. The response to the abolitionist message also reveals just how divided the United States was concerning slavery, and that divisiveness was not limited to North versus South. People in both regions could be found on both sides of the slavery question. Southerners tended to be more accommodating of the practice, but many whites in the North and Midwest saw no problems with slavery, either. People often attacked abolitionist editors, almost all of whom operated their publications in free states after 1831. Garrison was jailed in Boston in 1835 for his own protection after being attacked by an angry mob. In 1836, an irate Cincinnati mob destroyed the printing press of James G. Birney, whose *Philanthropist* promoted manumission of slaves. The most radical of reactions to abolitionist editors occurred in Illinois in 1837 when a mob broke into the *Alton Observer* office of Elijah Lovejoy and killed him. Lovejoy had been run out of Missouri for attacking a judge who imposed a lenient sentence upon men convicted of burning a black man alive. Angry citizens destroyed his press twice before Lovejoy moved across the Mississippi River into Illinois. On the night of Lovejoy's death, the *Observer*'s shop was surrounded by people who fired shots into the building and tried to set it on fire with Lovejoy and his employees still at work. Lovejoy died trying to stop someone from torching the office's roof. "About the time the fire was communicated to the building, the Rev. E. P. Lovejoy," Alton's mayor would later explain, "received four balls in his breast, near the door of the warehouse." The building was then burned to the ground, but not before the assailants threw the press out the window, broke it into pieces, and then dumped them into the Mississippi. Lovejoy's body was not removed until the next day.[49] Abolitionist newspapers used Lovejoy's murder to ramp up hostility for slavery. "Our country seems fast verging on a revolution," one editor observed.[50] Another said, "We trust it will have the effect to satisfy every body in the free States, that there are only two sides in this conflict: Liberty or Slavery—liberty for all, or slavery for all....'He that is not with us is against us.'"[51]

Issues surrounding Lovejoy's death and the dissemination of abolitionist publications turned into an issue of free speech in the eyes of abolitionists. "Mr. Lovejoy died in defence of the clearest and most precious rights of American citizenship, and of man," a writer to the *Vermont Chronicle* said, "and it may be a question how far duty requires one to peril his life for the freedom of the press among barbarians or where law is prostrate."[52] Government reaction to the abolitionist press during the 1830s provides another way to gauge its ability to affect society. A writer in Georgia noted that Northern abolitionist publications were "ALL-SUFFICIENT TO DO DESTRUCTION UPON THE SOUTH."[53] As a result, Amos Kendall, the nation's postmaster and a native South Carolinian, allowed Southern states to purge antislavery writings from the materials arriving via the mail even though he admitted they met all postal regulations. "[A]nti-slavery publications," Kendall said in a letter printed in the *New York Times*, "tend directly to produce in the south, evils and horrors surpassing those usually resulting from foreign invasion or ordinary insurrection. From their revolting pictures and fervid appeals addressed to the senses and passions of the blacks they are calculated to fill every family with assassins and produce at no distant day an exterminating servile war." He then asked, "Are the officers of the United States compelled by the constitution and laws, to become the instruments and accomplices of those who design to baffle and make nugatory the constitutional laws of the states—to fill them with sedition, murder and insurrection—to overthrow those institutions which are recognised and guaranteed by the constitution itself?"[54] The answer, for Kendall and for thousands of Americans North and South, was no. Eventually, every slaveholding state except Kentucky passed laws that made it illegal to disseminate abolitionist publications within their borders.[55]

LIFE AMONG THE LOWLY

The fact that abolitionist publications created such a fervor in the United States is a testament to the way the press disseminated information in antebellum America. Most abolitionist newspapers never passed three thousand issues per printing, but they and their subject matter garnered attention tremendously disproportionate to their circulation figures. Early in the 1850s, however, a Washington, D.C.-based paper, The *National Era*, topped 28,000 per issue in sales. How? The paper's editor, Gamaliel Bailey began running a serial novel written by Harriet Beecher Stowe. American newspapers had used serialized fiction for years to gain readers for newspapers and magazines. Stowe's work, however, was different. Though it was a work of fiction, the author claimed it was based on true-life facts and accounts. Before it ended its ten-part run, *Uncle Tom's Cabin, or Life Among the Lowly* was a work known by most Americans. In 1852, the story of the gentle and kind slave, Tom, and his treatment and eventual murder by the evil slave overseer Simon Legree appeared in book form. It became the best-selling novel of its time: five thousand copies sold in two days, nearly fifty thousand copies sold in eight weeks, and 300,000 within a year. Eventually, Americans would buy about 3 million copies of the book with another 3.5 million bought worldwide.[56] The technology of the day spread *Uncle Tom's Cabin*, one Southern writer said, as "steam-driven presses, steam-ships, steam-carriages, iron roads, electric telegraphs, and universal peace among the reading nations of the earth" faster than any other piece of information had ever been disseminated.[57]

Stowe came from abolitionist lineage. Her father was New England minister Lyman Beecher, and her brother, Henry Ward Beecher, was also a minister and would become an abolitionist editor. The 1850 Fugitive Slave Act so infuriated Stowe that she promised to write something profound that exposed the evils of slavery, and she did. Bailey said Stowe's writing "has done more to diffuse real knowledge of the facts and workings of American Slavery, and to arouse the sluggish nation to shake off the curse, and abate the wrong, than has been accomplished by all the orations, and anniversaries, and arguments, and documents, which the last ten years have been the witness of."[58] A decade later, President Abraham Lincoln reportedly said to Stowe when she visited the White House, "So you're the little woman who wrote the book that started this Great War!"[59]

Stowe's work was probably the most powerful antislavery polemic ever written, and it elevated U.S. press focus on it and upon slavery, if such was possible, considering the political climate of the nation, which always was affected by the issue of slavery. Frederick Douglass believed that "nothing could have better suited the moral and humane requirements of the hour" than *Uncle Tom's Cabin*. "Its effect was amazing, instantaneous, and universal," he wrote in his autobiography.[60] Because of the way that Stowe wrote the narrative, those who attacked it first had to establish that the book was fiction. Stowe claimed in her 1853 publication *A Key to Uncle Tom's Cabin* that her work was based, in part, on articles gleaned from Southern newspapers.[61] "'Uncle Tom's Cabin' is a fiction in every sense of the word," one writer said. "Mrs. Stow has borne false witness; she has slandered hundreds of thousand of her own countrymen."[62] Jonathan Thompson of the *Southern Literary Messenger* said sensible Americans would see through Stowe's lies. "Mrs. Stowe will not find many readers weak enough to believe it, even in New England....We have no words to express our scorn of such an effort....'THOU SHALT NOT BEAR FALSE WITNESS AGAINST THY NEIGHBOUR.'"[63] Hundreds of people responded in the press to Stowe's work, but James De Bow may have reached the most people and summarized all of the attacks on *Uncle Tom's Cabin* in his New Orleans-based *De Bow's Review*, a monthly periodical with a circulation of three thousand, the largest in the South. De Bow wrote that it was useless for Southerners to attempt to convince Northerners that slaves were well-treated, had lower instances of birth defects, crime, disease, and insanity as compared to the rest of the world. There was no need pointing out that living standards of slaves in the South exceeded that of poor whites in the North.

Truth, Southern truth, De Bow noted, is useless "when those who should hear us deafen themselves with this eternal 'ding, dong' of superstitious prejudices and pharisaical cant" that is supplied by *Uncle Tom's Cabin*.64

Ultimately, the abolitionist press, of which *Uncle Tom's Cabin* represents the pinnacle of anti-slave rhetoric, created a rift that forced the nation to civil war. John C. Calhoun of South Carolina knew this would happen and said so in *Niles' Weekly Register* seventeen years before Stowe's novel was serialized in the *National Era*. "The incessant action of hundreds of societies, and a vast printing establishment, throwing out, daily, thousands of artful and inflammatory publications, must make, in time, a deep impression on the section of the Union, where they freely circulate," Calhoun predicted.65 The abolitionist press shaped the way the nation thought about slavery. Most in the South resisted the verbal onslaught; most in the North ultimately agreed the practice must end. One group in the North, however, viewed the press as a means to take the next step beyond slavery. Free blacks turned to the press to advocate, as Frederick Douglass would later say, for "equal rights for all."

WE WISH TO PLEAD OUR CAUSE

For black Americans, the antebellum press represented the first opportunity to express their opinions in a printed public form. The conditions of slavery were a concern to free blacks, but the black press, according to Frankie Hutton, had a broader mission. It sought to "uplift and vindicate people of color in the true spirit of American democracy."66 The prospectus of the nation's first black-run paper said, "We wish to plead our own cause. Too long have others spoken for us. Too long has the publick been deceived by misrepresentations, in things which concern us dearly...."67 Free blacks living in the North lacked many of the rights that white citizens enjoyed. White males in New York, for example, needed only to meet an age requirement in order to vote. Black males, however, were required to own at least $250 in property with all taxes paid before they could cast a ballot.68 Many free blacks were illiterate, and most suffered from prejudice and poor working conditions. Those who were able to rise above these circumstances joined in the struggle to better their lives. White Americans seeking to do the same turned to a multitude of papers that already existed, but black Americans were not permitted access to this avenue of debate. When Willis Hodges tried to protest New York's double standard for voting, he was told by the editor of the New York *Sun* that "the *Sun* shines for all white men, not black men." If Hodges or any other black man wanted to have his voice heard, the paper suggested, "You must get up a paper of your own if you want to tell your side of the story to the public."69 Approximately forty black newspapers were founded before the Civil War, along with the first black magazines, the *Mirror of Liberty* in New

York City and the *National Reformer* of Philadelphia. The publications reflected the diverse measures black Americans were willing to take to alter their social status. Although many considered emigration as a solution—the focus of the Back-to-Africa movement—others argued that remaining in the United States and disproving the charges of black inferiority was the only way to fight prejudice. In addition to a press sensitive to their needs, free black Americans also banded together and established their own institutions, such as churches, schools, libraries, and labor organizations, to combat various forms of legal and extralegal discriminatory practices.[70]

African Americans who wanted to create their own publications faced numerous obstacles. In most cases, the need for a press dedicated to serving the interests of the black community was not enough to garner the support of a race that had only recently begun to show progress in educating itself. Financial difficulties plagued editors of the majority of the early newspapers and magazines. Editors and publishers earned their livelihood in a variety of occupations. Many were ministers, while others were businessmen and laborers. Most had received only a minimal amount of education. Free blacks used the press to fight against slavery and other social injustices. Before the Civil War about forty black newspapers were begun. Often, they were started by African American ministers and educators, like Samuel Cornish and John Russwurm, who published the first black-run newspaper, *Freedom's Journal*, beginning in 1827. Black newspapers railed against slavery, but their principal purpose was to fight for civil rights, pride, education, and progress for African Americans. The papers did, however, after a number of short-lived titles, expand throughout the northeastern United States. By 1850 black-run newspapers could also be found in Cleveland and in San Francisco with circulations ranging from 1,500 to 3,000 per issue.[71]

As America's largest city, New York became the place where many of the country's first black-owned newspapers published. Their editors quickly realized that education was a central element in the fight for equality of the races, and editors made its discussion a top priority. The city's education system did not provide for African Americans, so the black press began its advocacy campaign to provide public schools for African American children and for parents to send their children to the few privately supported "free" schools that existed for blacks. Cornish and Russwurm, in the initial issue of *Freedom's Journal*, called education "the object of highest importance to the welfare of society." Later in the year, the pair chided readers for not taking advantage of a school provided by the Manumission Society of New York. "The subject of Education is so important that we feel assured that it must recommend itself to every one: in the present case, the object is so benevolent, that no man of colour can hesitate one moment about embracing the generous offer of the Manumission Society."[72] Others joined the fight through *Freedom's Journal*. Anonymous writer F.A. noted that education held the key to the "elevation and happiness to our coloured population ... as a race of human beings....On this one principle alone rests the whole cause of African *education*.[73]

The fight for education among blacks continued after *Freedom's Journal* folded. In the 1830s, Cornish and Charles Ray edited the *Colored American*, the name chosen because they believed it best described the heritage of African Americans than any other term.[74] They also wanted to demonstrate that New York, a place where one would assume that blacks would possess more rights than in other places in the United States, limited equality, especially in terms of schools. "There is scarcely a place in the free States where the people of color possess so few advantages for education, as in the boasted City of New York," Ray wrote in 1839. The black community wanted schools to give its children the abilities *"to follow any profession which requires a liberal education."*[75] The constant calls for tax-supported education for blacks, especially in terms of secondary schools, finally occurred in the 1850s. In the late 1840s, the Society for Education among Colored Children operated several secondary schools in New York. In 1850, New York City mandated local school taxes, which paved the way for universal public education. Three years later, the City Board of Education, using tax funds, absorbed the SECC schools into the city system.[76] Pleased, but not content, African Americans began to push for the integration of public schools, keeping the issue in front of its audience. Realistically, though, they continued to stress that something close to comparable for black schools needed to exist, which was highlighted in the abolitionist *National Era*. New York City had 1,896 schools at the beginning of 1850, the paper stated, but of those only thirty-five were for African Americans. Funds for black schools barely topped $5,000, while more than $1.1 million was being spent on white education yearly.[77]

New York, nor any other city in the United States in the middle of the nineteenth century, provided equally for African Americans. The tenuous nature of the publishing of black newspapers meant that many were always only an issue or so away from termination. But, the black press kept issues like education in front of their readership, and like white papers of the era, black newspapers were shared, read aloud, and their content discussed. Because they continually discussed education for African American children, they kept the issue of government-supported schools in the news, and, according to Lauren Kessler, sent a clear message to white society about the potential and ability of black Americans.[78] Or, as Frederick Douglass said, the black press "has demonstrated, in large measure, the mental and literary possibilities of the colored race."[79]

THE POWER OF A PENNY

While the issue of slavery underlined and defined the political direction of the

United States during the antebellum period, the introduction of penny papers in 1833 radically changed journalism. Initially, according to Jerry Knudson, the penny press conveyed news rather than views with an emphasis on information that would interest the average person. This news often focused on what would later be called "human interest" stories,[80] but regardless of what the stories were tabbed, the penny papers found a way to elevate the sensational nature of many of their features. The lack of partisanship that existed initially in the penny press was a radical departure from the traditional method of presenting information in the United States. Patronage, as we have seen, was the rule among editors and politicians, and it dominated the way the press operated throughout this period.

Penny papers, however, were able to disassociate themselves from political ties, if they chose, because they operated on the principle that quantity in sales, rather than the quality of readership, mattered more. The more papers sold each day, the more advertising revenue could be obtained. Within a few years, penny papers were selling tens of thousands of copies per issue, rather than several thousand at best per issue as most patronage papers sold. They were sold to anyone who wanted to buy that day's issue rather than requiring people to buy the paper with a yearly subscription. The large circulation numbers, combined with competition, especially in New York City, led the penny papers to hire staffs of reporters, editors, engravers, and printers. The papers were able to cover events in ways that had not been feasible a decade earlier, and were still difficult for partisan papers that lacked the revenue that the penny papers were producing. Eventually, some of the penny papers acquired national circulation. Papers such as Horace Greeley's New York *Tribune*, James Gordon Bennett's New York *Herald*, and Samuel Bowle's *Springfield (Mass.) Republican*, for example, were able to influence the nation beyond their cities of origin. The penny papers—because of their success in attracting readership—were able to accomplish exactly what Benjamin Day proposed in the first issue of the New York *Sun*: coverage of all events that mattered to the public. "The object of this paper is to lay before the public, at a price within the means of everyone, ALL THE NEWS OF THE DAY, and at the same time afford an advantageous medium for advertising," Day said.[81] Penny papers paid attention to local news of politics, government, the courts, society, and business. Private calamities and triumphs became public because of this new definition. The concept of news expanded, and it also took on the characteristic of timeliness because the penny press operated in a world where speed—speed of travel and speed of communication—was increasing rapidly.[82] In the world of competition that the penny papers created, information in the newspaper reached farther and deeper into the American population than ever before because more copies were available than ever before and because expanding mail and rail systems could move information better than ever imagined. The idea that the press could affect the thoughts of the nation became easier to accomplish through the high circulation, national reaching penny papers.

The penny papers eventually turned their news back to views because of the political and social climate of the nation. While many of the partisan patronage papers were influential, it would be hard to deny the sphere of influence that Greeley, Bennett, and Bowles obtained because of the circulation numbers of their papers and ability to take whatever political stance they felt morally obligated to follow—Greeley's nationally circulated *Weekly Tribune*, for example, had a circulation of 112,000 by 1854.[83] Patronage editors, though, were limited by the party line. Both styles of newspapers were effective, but the reach of the major penny editors was, no doubt, much broader. Greeley, for example, became one of the great voices in the antislavery movement and for rectifying maltreatment of the less fortunate. He was a principal agent in the push for the nation to expand westward. In order to elect William Henry Harrison in 1840, Greeley started another paper, the *Log Cabin*. In six months, it accomplished Greeley's goal, reaching a circulation of 85,000.[84] Greeley's singular ability to shape the direction of the nation was probably unmatched during the years before the Civil War, though the press in general—as seen—kept a multiplicity of issues constantly in public dialog. Historian Allan Nevins said that for an entire generation, Greeley and the *Tribune* "stood pre-eminent among the organs of opinion in the United States; it [the *Tribune*] was one of the great leaders of the nation...."[85] One reason that Greeley and other editors achieved such success can be found in the inventions that acted as either the means of producing or for the disseminating of information during this time period.

STEAM, DOTS AND DASHES

Until the 1820s, few changes had been made to the way in which publications were produced for public consumption. Johann Gutenberg's printing press design was still the basis for what print shops used; manual production was the means of creating any publication. Some press operations—especially the successful ones in New York City—did not use the hand-operated screw press but one operated by levers. In the 1820s, though, inventors began applying technology to the printing press. Cylinders were the first improvement. The cylinders revolved and pressed the paper against the type, which was held on a flatbed. The addition of steam power, however, truly revolutionized printing. Steam-driven presses produced thousands of pages per hour instead of several hundred as produced on manual presses. The Koenig press, invented by the German Friedrich Koenig, employed a moveable bed that allowed easy inking of the frame between impressions. Koenig also added an additional cylinder to the press. This made it possible to print both sides of the paper at once.

Richard Hoe built the new presses in America, and by 1830, the new steam cylinder presses could produce up to four thousand newspapers per hour. Hoe continued to find ways to improve the steam press. In 1840, R. Hoe & Company introduced a revolving press that printed eight pages at a time. Typeset pages were attached to the cylinders, not a flatbed, and an automated process applied ink to type. The press produced nineteen thousand papers per hour. Steam presses were able to print large numbers of newspapers per hour, too, because of the invention of an automatic papermaking machine. The cost of newsprint dropped by 25 percent.[86] The ability to print thousands of copies in the time it used to take to produce a couple of hundred, plus the reduced cost of newsprint, meant that more publications and more copies of them than ever were readily available. Technology combined with the concept of the penny paper to expand the reach of information.

But one other invention transformed the movement of information and made timeliness a new element in the value of news. That invention was the telegraph, taken from the word telegraphy, which meant writing at a distance. "Telegraph" became a popular word to use as the name of a newspaper, and more than forty had incorporated it into their nameplate by the beginning of the antebellum period.[87] Despite the use of the term, news moved slowly and in the ways it had for a century. Newspapers traveled by mail and on ships. It took weeks for news to circulate from New England to states like Alabama and Mississippi. The Post Office Act of 1792 allowed printers and editors to send each other newspapers gratis through the nation's mail system. The legislation allowed editors to gather news from throughout the nation to share with local patrons. Nearly 20 percent of newspapers in 1810, for example, were delivered by mail to regions outside of the area of publication.[88] But legislation could not move information faster than the physical resources of the day permitted.

Samuel F. B. Morse's invention sent a series of dots and dashes along wires, the first set up between the nation's capital and Baltimore in 1844. Carrier pigeons and news boats, which met incoming ships to acquire information before they docked, had been ways to speed up information dispersal, but the telegraph ensured that information could travel almost instantaneously throughout the United States—so long as there were telegraph lines in place. The telegraph, the *Sun* of Baltimore said after Morse sent his first message, represented a "complete annihilation of space." James Gordon Bennett believed the telegraph had the power to "cause a unity of thought and action throughout the whole republic, similar to that exhibited by a single community."[89] Other editors called for an immediate expansion of telegraph lines and government control of them, and soon lines connected not only Baltimore and Washington, but those cities with Philadelphia and New York. The *Philadelphia Public Ledger* predicted that when the lines were extended throughout the

United States, "its effect will be beyond calculation."[90] By the end of the Mexican War in 1848, telegraph lines stretched to New Orleans and became a vital link in obtaining information from the war front in Mexico. Within five years, approximately 23,000 miles of telegraph lines stretched across the United States.[91]

Access to the telegraph stimulated circulation wars among New York City newspapers and led to news associations, groups of newspapers that joined together to collect news. In 1848, papers in New York created what would become the New York Associated Press, the precursor to the Associated Press. Other press associations followed. By the 1850s, the telegraph reached all major American cities with one exception. Lines did not reach San Francisco until 1861. The companies that owned the telegraph wires—like Western Union—gave special rates to the press, but they did more: They provided the press with preferential access to the nation's rapidly expanding telegraph lines.[92] The tenuous times no doubt played a crucial role in this decision, but it also points out just how important obtaining information was to the nation. While the press may not have been the largest user of the telegraph, its preferred status helps explain how the media were able to shape the agenda of the nation by spreading information literally to all parts of it instantaneously. When the nation divided in 1861, the telegraph had helped make newspapers the center of attention. People gathered at newspaper offices, awaiting the latest news from the front lines. Journalists in the field telegraphed their stories to their papers' offices. There, a newspaperman would come outside and read the latest dispatch to the eager crowds.[93] By 1872, the telegraph was referred to as the "Star of the Empire" because it was able "to flash intelligence throughout the land" to be disseminated by media.[94]

ILLUSTRATIONS AND IMAGES

The content that the press distributed was certainly critical to its ability to define the direction of the nation, and the growth of visual components greatly enhanced this ability. Printers realized the value of the visual soon after Gutenberg's invention of the printing press and began creating visuals with woodcuts, images carved in relief in a block of wood, though using woodcuts predated the invention of the printing press by about one thousand years. In the fifteenth and sixteenth centuries, broadside woodcut images of major European events became commonplace in Europe. Because the block could be fastened into the frame of a printing press along with lines of text for a explanatory caption, these visual representations became a key way to disseminate information in Europe. The woodcut itself also became complex and highly detailed.

Printers of newspapers and magazines had long used woodcuts for advertisements, and their use for editorial cartoons became a part of the political struggles that accompanied the partisan patronage press of the era. When Andrew Jackson died in 1845, James Gordon Bennett filled the front and second pages of the New York *Herald* with woodcut images of the funeral, running the images not once but twice, once in his daily and once in his weekly.[95] Similarly, *Godey's Lady's Book*, which began in 1830, was filled with engravings that were colorized by hundreds of women. The monthly magazine reached a circulation of 150,000 by the Civil War and was a staple among women for fashion, cooking, entertainment, and taking care of activities within what was often referred to as the "domestic sphere."[96]

In the 1850s, a number of enterprising artists and editors created illustrated newspapers. *Gleason's Pictorial, Illustrated News, Frank Leslie's Illustrated Newspaper*, and *Harper's Weekly* provided Americans with illustrations and stories, often capturing events from around the globe. *Leslie's Illustrated Newspaper* and *Harper's Weekly*, however, assumed a critical role when the nation fell apart in 1860 and 1861. Their sketch artists provided readers with images of war that had never been available. The expansion of the press in conjunction with technological advancements from the 1820s forward led people to expect considerable amounts of information, but it was almost always devoid of accompanying images. The illustrated papers changed that. They showed readers battlefield action, prisoner of war camps, the preparation of munitions, the daily life of soldiers—anything that an artist could draw and then send back to New York to be enhanced and published was a topic for the illustrated papers. "[T]hey have written with their pencils in the field," *Harper's* said of the illustrators, "a history quivering with life, faithful, terrible, romantic, the value of which will grow with every year."[97]

Even though *Harper's Weekly* was self-serving in its comments, there was great truth in the paper's editorial on the value of the images created by sketch artists and to how their images shaped the nation's understanding of the war. "To them we owe the national familiarity with the features of our famous soldiers, sailors, and statesmen," *Harper's* proclaimed. "It is they who have enabled the poorest boy in the land to own a portrait of the bravest hero. All over the country thousands and thousands of the faces and events which the war has made illustrious are tacked and pinned and pasted upon the humblest walls.…They have shared the soldier's fare; they have ridden and waded, and climbed and floundered, always trusting in lead pencils and keeping their paper dry. When the battle began they were there. They drew the enemy's fire as well as our own. The fierce shock, the heaving tumult, the smoky sway of battle from side to side, the line, the assault, the victory—they were a part of all, and their faithful fingers, depicting the scene, have made us a part also."[98]

One other visual element greatly affected Americans' perceptions of the war, and that came from the camera of the photographers. Photography gradually evolved from its first photograph in 1827 in France by Joseph Niepce to its introduction in the United States in 1839, which followed the creation of the daguerreotype, a photographic method created by Louis Daguerre. With it, images were directly imprinted upon silver-plated copper. Their realism startled those who saw them. *The Knickerbocker* claimed, "Their exquisite perfection almost transcends the bounds of sober belief."[99] The most common method of capturing photographic images, though, used glass plates cleaned with a solution called collodion, which was comprised of nitric and sulfuric acid and ether, and then treated with light-sensitive silver nitrate. Images were captured on the glass surface, which was coated with a developer after having been exposed to the image. After going through developing, the glass plate was washed and varnished. It could either become a photographic image or could be made into a negative. If a negative, then it could be used to make copies.[100]

The ability to make copies of photographs was key to their influence on the nation during the Civil War. Although newspapers would not be able to reproduce photographs until 1880 when the halftone process—which turned photos into a series of dots—was perfected, images were able to be transferred from glass negatives to cardstock. This led to the development of stereoscopic vision, which placed identical images side by side on a card, but one had its image moved slightly off center of the other. The card, called a stereo view, was then placed in a viewer called a stereoscope. The imbalance between the two images created a three-dimensional effect when looking at the photograph. During the 1850s, these types of images became hugely popular in the United States as parlor entertainment. With the onset of war and the photography that accompanied it, stereo views of the Civil War photographers were quickly mass marketed and eagerly purchased by Americans.[101] *The Nation* noted that stereo views "bring up vividly some of the most prominent scenes in the drama of the late rebellion. Here are alike the instruments of ruin and the ruin itself—views from the cradle and views from the grave of secession."[102]

After Southern states began to secede and Lincoln called for 75,000 volunteers, photographer Mathew Brady received presidential permission to follow Union troops into Virginia to capture the putting down of what many considered an insurrection. Brady, however, only accompanied the troops to Bull Run, where his photographic wagon was destroyed. After that, his assistant, Alexander Gardner, organized photographers, and Gardner, Timothy Sullivan, George Barnard, and others provided the majority of the photographs of the war, though the name "Brady" was assigned initially to nearly all the photographs produced by those in Brady's employ, though Brady did shoot some images after that. As a result, Brady initially received the credit for the photographs that began appearing for public consumption.[103] When newspapers referred to photographs from the battlefront, they almost always associated them with Brady. "Mr. BRADY is rendering us all a real service, in divers ways, by this work of his, undertaken so courageously," an editorial writer explained. "Look upon this picture—or upon this! ... Let us, then, heartily acknowledge our obligations to such an 'abstract and brief chronicle of the times' as this which Mr. BRADY has been so earnestly and unobtrusively making up for us."[104]

War photographs, because photography required its subjects to remain still in order to capture the image, could not provide viewers with images of actual fighting. It is here where the illustrated newspapers and the photographers complemented each other. The sketch artists captured the fighting, but the photographers were able to capture death. Photographic images that New Yorkers viewed following the Battle of Antietam slapped them in the face when they were displayed in the city. "Mr. Brady has done something to bring home to us the terrible reality and earnestness of war. If he has not brought bodies and laid them in our dooryards and along the streets, he has done something very like it," the *New-York Times* stated after photographs from the battlefield near Sharpsburg, Maryland, went on display in the city. "You will see hushed, reverend groups standing around these weird copies of carnage, bending down to look in the pale faces of the dead, chained by the strange spell that dwells in dead men's eyes."[105] Ironically, the images were the work of Gardner, not Brady. Photographs, contemporaries said, have allowed us "to see just how the men live and look who are fighting the battles of the Republic" and "catch our armies 'living as they rise,' and, alas! to embalm our falling heroes ere they fall."[106]

In 1866, the New-York Historical Society unanimously resolved that the "photographs by Mr. M. B. Brady, of scenes, incidents, and portraits connected with the late rebellion, and other material of historic incident, [to be] a nucleus of a national historical museum."[107] While the photographs of the war did provide a visual museum of elements of the war, they also opened Americans' eyes to the grim truth of war. Photographers captured the fact that war was anything other than chivalrous. The images that people saw of soldiers after battle revealed men who were mired in dirt and filth. Bodies twisted by pain and then frozen in that position by rigor mortis challenged popular cultural conceptions long in place that those who fought were noble. Photographs allowed those not directly involved in warfare "to see just how the men live and look who are fighting the battles of the Republic."[108]

At the end of the Civil War, Alexander Gardner published his two-volume *Gardner's Photographic Sketch Book of the War*. With it, the visual images of the war were available in a way newspapers had only been able to illustrate and with a story line that stereo views did not provide. The *Sketch Book* included one hundred images taken by Gardner and other photographers. Gardner wrote in the *Sketch Book* preface that the images contained in his volumes "will be accepted by posterity with an undoubting faith."[109] Those who viewed it agreed. The *Daily National Intelligencer* said Gardner's photographic collection was a "magnificient work ... each portraying, with that graphic truthfulness which can be attained only by the photographer's art, some scene made memorable by the momentous struggle through which our country has recently passed."[110] The technology that allowed photography to penetrate the psyche of the nation would continue to be refined and perfected. A century later in another war, a photograph—one by Associated Press photographer Eddie Adams—repulsed the American public. Adams captured the execution of a Viet Cong prisoner by Colonel Nguyen Ngoc Loan. Later, Adams said, "Still photographs are the most powerful weapon in the world."[111] For Americans at the end of the Civil War, the power to deliver visual images was no less powerful and no less an instrument for shaping the nation.

✱✱✱

The explosion in reach and size of the press in the middle of the nineteenth century meant that media possessed tremendous influence. Technology played a crucial role in the expansion of the press. Advancements in production meant thousands of papers were available where once only hundreds had been possible. Transportation of news via steam ships and by rail meant that papers could be sent faster and farther than ever imagined. At the same time, the telegraph transformed the timeliness of news. What had taken weeks in the eighteenth century and early nineteenth now took days or only a day to share with readers. In addition, the visual component increasingly became integral to the presentation of information and to shaping ideas about the issues facing the nation. The mainstream press of partisan and penny papers presented a cornucopia of news, but much of what people read in these papers centered on the pivotal issues of the day, which included slavery and ultimately whether the Union would remain intact.

It is nearly incomprehensible to understand the reach and power that the press acquired in the antebellum and Civil War period of the nation. Being able to publish and disseminate information became a staple of a growing and diverse society. Every advocacy group turned to the press. Everyone who wanted to share her or his ideas knew the printed word was the way to shape opinion. In the 1820s, for example, a number of "philosophical" societies began to promote new discoveries in medicine, mathematics, and science. There was, in the United States, a "spirit of inquiry," as Dr. James DeKay pointed out in 1826, that was overtaking the nation.[112] When DeKay made his speech, twenty-six scientific societies operated in New York alone, and twenty-four different U.S. societies regularly printed a journal, producing more than fifteen hundred articles on scientific issues before 1835.[113] Many of these journal articles were reprinted in newspapers, and most were sold via newspaper advertisements. These scientific journals were able to influence American opinion on nearly every subject, including life on other planets.[114]

Religious publications shaped opinion, too. The religious press competed successfully with secular papers.[115] Religious editors looked like their secular counterparts, and they published the same news. They also discussed theology and morality, along with group-specific subject matter. The 1850 Census listed 191 religious newspapers or periodicals in America,[116] and religious journalism, according to one source, may have accounted for three-fourths of America's reading material in 1840.[117] Those who believed it was critical to share the Christian gospel looked upon the press just as abolitionists, politicians, scientific societies—in short, anyone who wanted to effect change in the nation did. They saw media as the agent of change. "In a world where thought is unbound, and inquiry unrestricted, THE PRESS becomes, perhaps, the most important of human agencies for good or evil," the religious newspaper the *American Messenger* declared in 1849. "The world has gone to reading, and read they will, for weal or woe."[118] By 1850, according to David Nord, religious groups were selling about a half-million books a year and giving away another 35 million books and tracts.[119] The power and influence of this volume of printing and its dissemination was no doubt tremendous, but it is also almost universally overlooked when considering the power of media to shape a nation.

The press soon afforded a voice for nearly every cause or occupation in America. The Industrial Revolution provided America with a new job market, and labor newspapers began to address the circumstances of factory workers. These publications soon expanded to cover the needs of various professions—from bankers to miners. Agricultural newspapers sought ways to help farmers. Other specialized publications addressed inventions and discoveries aimed at making life easier.[120] Magazines aimed at women were successful. *Godey's Lady's Book* had a readership that surpassed every other American publication by the Civil War. In 1848, following the women's rights convention at Seneca Falls, New York, the press became a tool of suffragists. As Joseph Medill said of the media at the outbreak of the Civil War, they "shape and give directions to public sentiment. They are the narrators of facts, the exponents of policy, the enemies of wrong."[121] As widely as the media were able to establish their reach during this time period and shape national agenda, though, they were not finished growing. In the years after the Civil War, the press was set to grow at a rate even greater than it had in the antebellum era. More magazines and three times the number of daily papers in fifty years meant that the press had the potential to become even more influential, and with powerful figures like

Charles Dana, Joseph Pulitizer, William Randolph Hearst, and Adolph Ochs on the horizon, influence by media on the direction of the nation would continue.

NOTES

1. *New York Herald*, May 6, 1835.
2. Numbers are based on Alfred McClung Lee, *The Daily Newspaper in America* (New York: Macmillan, 1937); Edward Connery Lathem, comp., *Chronological Tables of American Newspapers, 1690–1820* (Barre, Mass.: American Antiquarian Society & Barre Publishers, 1972); Carol Sue Humphrey, *The Press of the Young Republic, 1783–1833* (Westport, Conn.: Greenwood Press, 1996); Wm. David Sloan and James D. Startt, *The Media in America: A History,* 4th ed. (North Port, Ala.: Vision Press), 250; and William E. Huntzicker, *The Popular Press, 1833–1865* (Westport, Conn.: Greenwood Press, 1999), 170.
3. Alexis de Tocqueville, *Democracy in America*, 2 vols. (New York: Alfred A. Knopf, 1946), 1:187–88.
4. Michael Shudson, "Was There Ever a Public Sphere? If So, When? Reflections on the American Case," in *Habermas and the Public Sphere*, Craig Calhoun, ed. (Cambridge, Mass., 1992), 149.
5. William J. Gilmore, *Reading Becomes a Necessity of Life: Material and Cultural Life in Rural New Britain, 1780–1835* (Knoxville, 1989), 193–94; Walter Dean Burnham, *The Current Crisis in American Politics* (New York, 1982), 129.
6. *Aurora* (Philadelphia), February 2, 1825.
7. Andrew Jackson, *The Correspondence of Andrew Jackson*, John Spencer Bassett, ed., 7 vols. (Washington, D.C., 1926–35), 4:19.
8. Quoted in Culver H. Smith, *The Press, Politics, and Patronage: The American Government's Use of Newspapers 1789–1875* (Athens: University of Georgia Press, 1977), 56.
9. Gerald J. Baldasty, "The Press and Politics in the Age of Jackson," *Journalism Monographs* 89 (1984): 7.
10. Study 00003: "Historical Demographic, Economic, and Social Data: U.S., 1790–1970." Ann Arbor: Inter-University Consortium for Political and Social Research. You may access the census records at an number of on-line sources including http://www.icpsr.umich.edu/ and http://fisher.lib.virginia.edu/census/.
11. John M. Blum, et al., *The National Experience*, 6th ed., 2 vols. (San Diego: Harcourt Brace Jovanovich, 1985), I:310.
12. "Historical Demographic, Economic, and Social Data: U.S., 1790–1970," http://fisher.lib.virginia.edu/census/.
13. Gilmore, *Reading Becomes a Necessity of Life*, 193–94.
14. Prospectus for *Christian Journal* (Richmond), published in *Family Visitor* (Richmond), October 8, 1825.
15. Lorman Ratner and Dwight L. Teeter, Jr., *Fanatics and Fire-eaters: Newspapers and the Coming of the Civil War* (Urbana: University of Illinois Press, 2003), 9.
16. Tocqueville, *Democracy in America*, 2:111.

17. Jürgen Habermas, *The Structural Transformation of the Public Sphere: A Structural Transformation of Bourgeois Society*, trans. Thomas Burger and Frederick Lawrence (Cambridge: The MIT Press, 1989), 29.
18. Isaac Bates, "Speech of Mr. Bates," *Massachusetts Spy* (Worcester), July 21, 1841.
19. *Connecticut Courant* (Hartford), January 31, 1832.
20. "Annexation," *United States Magazine and Democratic Review* 17:85 (July and August 1845). John Louis O'Sullivan, editor of the *Review*, is generally given credit for coining the term, though it may have been penned by one of the magazine's writers instead. See Linda S. Hudson, *Mistress of Manifest Destiny: A Biography of Jane McManus Storm Cazneau, 1807–1878* (Austin: Texas State Historical Association, 2001).
21. *Columbian Centinel* (Boston), January 22, 1820.
22. *Enquirer* (Richmond), January 6, 1820.
23. *St. Louis Enquirer* (Missouri Territory), April 7, 1819.
24. *General Advertiser* (Philadelphia), December 23, 1820.
25. *Enquirer* (Richmond), March 7, 1820.
26. *Daily Advertiser* (New York), March 7, 1820.
27. Letter to John Holmes, April 22, 1820, in *The Writings of Thomas Jefferson*, ed. Paul L. Ford, 10 vols. (New York: G.P Putnam's Sons, 1892–1899), 10:157.
28. *National Intelligencer* (Washington, D.C.), March 3, 1820.
29. David A. Copeland, *The Antebellum Era* (Westport, Conn.: Greenwood Press, 2003), 179.
30. Timothy Dwight, *Connecticut Courant* (Hartford), October 13, 1828.
31. James Buchannan, "Mexico," *Union* (Washington, D.C.), December 24, 1847.
32. Weston Gales, *Weekly Raleigh Register, and North Carolina Gazette*, May 22, 1846.
33. See, for example, *Sun* (Baltimore), October 15, 1847.
34. *Congressional Globe*, 29 Cong. 1 sess., 1217, in Champlain W. Morrison, *Democratic Politics and Sectionalism: The Wilmot Proviso Controversy* (Chapel Hill: University of North Carolina Press, 1967), 18.
35. David M. Potter, *The Impending Crisis, 1848–1861* (New York: Harper & Row, 1976), 22.
36. *Whig* (Boston), August 15, 1846.
37. *Tribune* (New York), June 3, 1854.
38. *National Era* (Washington, D.C.), September 7, 1854.
39. *American Whig Review* (New York), December 1850.
40. The *Enquirer* and Gist quoted in Potter, *The Impending Crisis*, 383.
41. Richard Barnwell Rhett, *Mercury* (Charleston, S.C.), November 1, 1859.
42. *Gazette* (Pittsburgh, Pa.), December 3, 1859; *Liberator* (Boston), September 7, 1860.
43. *Enquirer* (Richmond), November 19, 1860; *Mercury* (Charleston), November 19, 1860.
44. *Liberator* (Boston), January 1, 1831.
45. Quoted in Ford Risley, *Abolition and the Press: The Moral Struggle Against Slavery* (Evanston, Ill.: Northwestern University Press, 2008), x.
46. Thomas Leonard, *News for All: America's Coming-of-Age with the Press* (New York: Oxford University Press, 1995), 67.
47. Harriet Martineau, *The Martyr Age of the United States* (Boston: Weeks, Jordan & Co., 1839), 3–4.
48. *Liberator* (Boston), January 1, 1831.

49. John M. Krum, "The Alton Tragedy," *Daily National Intelligencer* (Washington, D.C.), November 24, 1837.
50. *Philanthropist* (Cincinnati), November 7, 1837.
51. Joshua Leavitt, "THE FIRST MARTYR HAS FALLEN, in the Holy Cause of Abolition!" *The Emancipator* (New York City), November 23, 1837. *The Emancipator* was published by the American Anti-Slavery Society.
52. *Vermont Chronicle* (Bellows Falls), November 29, 1837.
53. Quoted in Leonard, *News for All*, 73.
54. *New York Times*, August 22, 1835.
55. Risley, *Abolition and the Press*, 44.
56. Potter, *the Impending Crisis*, 140.
57. Quoted in Hollis Robbins, "*Uncle Tom's Cabin* and the Matter of Influence," *History Now* 16 (June 2008), http://www.historynow.org/06_2008/historian2.html.
58. *National Era* (Washington, D.C.), April 15, 1852.
59. Charles Edward Stowe, *Harriet Beecher Stowe: The Story of Her Life* (Boston: Houghton Mifflin, 1911), 203.
60. Frederick Douglass, *Life and Times of Frederick Douglass Written by Himself* (New York: Citadel Press, 1983), 289.
61. Harriet Beecher Stowe, *A Key to Uncle Tom's Cabin; Presenting the Original Facts and Documents upon Which the Story Is Founded. Together with Corroborative Statements Verifying the Truth of the Work* (Boston: John P. Jewett, 1853), 5–8.
62. *Courier and Enquirer* (New York), October 21, 1852.
63. *Southern Literary Messenger* (Richmond), October 1852.
64. *De Bow's Review* (New Orleans), March 1853.
65. *Niles' Weekly Register* (Baltimore), September 5, 1835, quoted in Leonard, *News for All*, 74.
66. Frankie Hutton, *The Early Black Press in America, 1827–1860* (Westport, Conn.: Greenwood Press, 1993), xiv.
67. *Freedom's Journal* (New York), March 6, 1827.
68. Garland Penn, *The Afro-American Press and Its Editors* (Springfield, Mass.: Wiley and Co., 1891), 61. Bernell Tripp, *Origins of the Black Press, New York, 1827–1847* (Northport, Ala.: Vision Press, 1992), 44.
69. Willard B. Gatewood, Jr., ed., *Free Man of Color: The Autobiography of Willis Augustus Hodges* (Knoxville: University of Tennessee Press, 1982), 75–76.
70. See, "The Abolitionist Press, in Wm. David Sloan and David A. Copeland, *Mass Media: A Documentary History* (Northport, Ala.: Vision Press, 2009).
71. Hutton, *The Early Black Press in America*, x.
72. *Freedom's Journal* (New York), March 16, 1827; December 21, 1827.
73. *Freedom's Journal* (New York), February 15, 1828.
74. Tripp, *Origins of the Black Press*, 37.
75. *The Colored American* (New York), October 19, 1839.
76. Graham Russell Hodges, *Root and Branch: African Americans in New York and East Jersey, 1613–1863* (Chapel Hill: University of North Carolina Press, 1999), 244.
77. *The National Era* (New York), January 24, 1850.
78. Lauren Kessler, *The Dissident Press: Alternative Journalism in American History* (Beverly Hills, Calif.: Sage, 1984), 25.
79. Quoted in Penn, *The Afro-American Press and Its Editors*, 448.

80. Jerry W. Knudson, *In the News: American Journalists View Their Craft* (Wilmington, Del.: Scholarly Resources, 2000), 35.
81. New York *Sun*, September 3, 1833.
82. Sloan and Copeland, *Mass Media: A Documentary History*, 56.
83. James Ford Rhodes, *Historical Essays* (New York: Macmillan,1909), 90.
84. Frank Luther Mott, *American Journalism, A History: 1690–1960*, 3rd ed. (New York: Macmillan Co., 1962), 268.
85. Allan Nevins, *American Press Opinion, Washington to Coolidge: A Documentary Record of Editorial Leadership and Criticism, 1785–1927* (Boston: D.C. Heath, 1928), 112–13.
86. Information on printing comes from Mott, *American Journalism*, 203–04; Michael Emery and Edwin Emery, *The Press and America: An Interpretive History of the Mass Media*, 6th ed. (Englewood Cliffs, N.J.: Prentice Hall, 1988), 112–14; and Sidney Kobre, *Development of American Journalism* (Dubuque, Iowa: Wm. C. Brown, 1969), 216–17, 219.
87. Richard R. John, "Recasting the Information Infrastructure for the Industrial Age," in *A Nation Transformed by Information*, eds. Alfred D. Chandler, Jr. and James W. Cortada (Oxford and New York: Oxford University Press, 2000), 74.
88. William A. Dill, *Growth of Newspapers in the United States* (Lawrence: University of Kansas, 1928), 11.
89. Baltimore *Sun*, May 27, 1844; New York *Herald*, June 6, 1844, quoted in Susan Thompson, *The Penny Press: The Origins of the Modern News Media, 1833–1861* (Northport, Ala.: Vision Press, 2004), 126.
90. *Philadelphia Public Ledger*, June 11, 1846.
91. Luther Thompson, *Wiring a Continent: The History of the Telegraph Industry in the United States, 1832–1866* (Princeton, N.J.: Princeton University Press, 1947), 241.
92. John, "Recasting the Information Infrastructure for the Industrial Age," in *A Nation Transformed by Information*, 83.
93. Huntzicker, *The Popular Press*, 96–97.
94. In *Crofutt's New Overland Tourist and Pacific Coast Guide*, 2 vols. (Chicago: Overland Publishing, 1879–1880), 2:12–13, quoted in John, "Recasting the Information Infrastructure for the Industrial Age," in *A Nation Transformed by Information*, 84.
95. *New York Herald*, June 25, 1845, June 28, 1845.
96. See Fred Lewis Patee, *The First Century of American Literature:1770–1870* (New York: Cooper Square, 1966), 495.
97. "OUR ARTISTS DURING THE WAR," *Harper's Weekly* (New York), June 3, 1865, 339.
98. Ibid.
99. *The Knickerbocker, or New-York Monthly Magazine*, quoted in Robert Leggat, "The Daguerreotype," http://www.rleggat.com/photohistory/history/daguerro.htm (September 2008).
100. National Public Radio, "Wet-Plate Photography," *The Wizard of Photography*, http://www.pbs.org/wgbh/amex/eastman/sfeature/wetplate_step1.html (2000).
101. Bob Zeller, *The Blue and Gray in Black and White: A History of Civil War Photography* (Westport, Conn.: Praeger, 2005), 21, 23, xiv.
102. "Current Literature," *The Nation*, August 17, 1865, 219, quoted in Zeller, *The Blue and Gray in Black and White*, 162.

103. Department of Photographs, "Photography and the Civil War, 1861–1865," in *Heilbrunn Timeline of Art History* (New York: The Metropolitan Museum of Art, 2000–), http://www.metmuseum.org/toah/hd/phcw/hd_phcw.htm (October 2004).
104. "Brady's Photographs of the War," *New-York Times*, September 26, 1862.
105. "Brady's Photographs," *New-York Times*, Oct 20, 1862.
106. "Brady's Photographs of the War," *New-York Times*, September 26, 1862.
107. "Brady's Historic Exhibition," *New-York Times*, February 26, 1866.
108. "Brady's Photographs from the Seat of War," *New-York Times*, July 26, 1864.
109. Alexander Gardner, *Gardner's Photographic Sketch Book of the War*, 2 vols. (Washington, D.C.: Philip & Solomons, 1865–1866).
110. *Daily National Intelligencer* (Washington, D.C.), February 26, 1866.
111. Eddie Adams, "Eulogy," *Time Magazine*, July 27, 1998, http://www.time.com/time/magazine/article/0,9171,988783,00.html.
112. James E. DeKay, *Anniversary Address on the Progress of the Natural Sciences in the United States: Delivered before the Lyceum of Natural History of New York, February, 1826* (New York, 1826), 7–8.
113. George H. Daniels, *American Science in the Age of Jackson* (New York: Columbia University Press, 1968), 13–15, 23.
114. See, for example, David Copeland, "A Series of Fortunate Events: Why People Believed Richard Adams Locke's 'Moon Hoax,'" *Journalism History* 33 No. 3, (Fall 2007): 140–150.
115. Frank Luther Mott, *American Journalism. A History: 1690–1960*. 3rd ed. (New York: The Macmillan Company, 1962), 206.
116. Frank Luther Mott, *A History of American Magazines, 1741–1850* (Cambridge, M.A.: Harvard University Press, 1966), 342.
117. Benjamin P. Browne, *Christian Journalism for Today* (Philadelphia: The Judson Press, 1952), 14.
118. *American Messenger*, May 1849, 18, quoted in David Paul Nord, *Faith in Reading: Religious Publications and the Birth of Mass Media in America* (Oxford: Oxford University Press, 2004), 114.
119. Nord, *Faith in Reading*, 151.
120. Huntzicker, *The Popular Press*, 56–60.
121. *Chicago Tribune*, October 3, 1861.

5. *We Expect Great Results from This Work: Post-Civil War and Yellow Journalism*

William Randolph Hearst's tenure as a New York publisher was only two years old when the brash, sinfully rich California transplant announced a new direction for the *New York Journal*. Hearst declared that with his paper he would change journalism, and that his journalism would change the world. "Action—that is the distinguishing mark of the new journalism," Hearst stated. "It marks the final stage in the evolution of the modern newspaper....The new journal of to-day prints the news, too, but it does more. It does not wait for things to turn up. It turns them up."[1] Hearst went on to say that his *Journal* was "an Advocate of the People" with "a vigilant eye" peering at government or anyone who might be ready to take advantage of an unsuspecting public. "We expect to see great results flow from this work," Hearst concluded.[2]

Hearst referred to how he wanted his newspaper to operate within the public sphere as the journalism of action. "The Journal holds the theory that a newspaper may fitly render any public service within its power. Acting on this principle, it has fed the hungry, brought criminals to justice and enforced by legal methods the responsibility of public officials," Hearst said as he described what he meant by journalism of action and promised, "The journalism that does things has come to stay."[3] While Hearst may have coined the term journalism of action, he was still not creating a new type of journalism. Active involvement by the press in the lives of the community it served—participatory journalism—had always existed in America. More directly, though, Hearst was copying the journalist he most emulated, Joseph Pulitzer. The press was "An institution which should always fight for progress and reform; never tolerate injustice or corruption," Pulitzer had long believed. Newspapers, he said, should "always remain devoted to the public welfare; never be satisfied with merely printing news."[4] Shortly after beginning his journalistic career in St. Louis, Pulitzer had said that the goal of journalism was to "expose all fraud and sham, fight all public evils and abuses, and to battle for the people with earnest sincerity." Decades later, Pulitzer still believed this. "He [the journalist] is there to watch over the safety and welfare of the people who trust him"[5]

This idea that the press should "watch over" government and those with power to affect the people had been espoused by Thomas Jefferson as early as 1786 when he said that "our liberty cannot be guarded but by the freedom of the press."[6] This concept—which would be referred to as the watchdog function of the press decades after Pulitzer's death—requires journalists to investigate perceived wrong doings and improprieties in the name of public welfare in an effort to stop the injustice or malfeasance. It requires, according to Ralph Negrine, active reporting where the journalist puts together pieces of information.[7] For Pulitzer, Hearst, their reporters, the muckrakers that followed in the early twentieth century (see Chapter 6), and fearless press crusaders like Ida B. Wells, the active element meant that the journalist and the media outlet could be a part of the story. In the era known as the Gilded Age, this became the norm in a growing number of publications, and Pulitzer and Hearst were the kings of producing this type of journalism. They were not, however, the only ones to do so. The press in the years after the Civil War, operated actively everywhere, from the largest media outlets like those in the major metropolitan centers to the smallest of rural newssheets, in towns with populations of only a few hundred.

The last thirty-five years of the nineteenth century created great changes in America's press. Almost exclusively partisan prior to the Civil War, newspapers moved away from political alignment. Congress officially ended political patronage, which had been the financial life blood of publishers since the 1790s, in 1860, and the State Department stopped handing out printing contracts to newspaper editors in 1875.[8] Papers were still highly political, and some remained partisan, but most sought an independent course, the end of patronage forcing the move in some cases, but in many others it was a conscious decision by editors to do so. Independence allowed the press to support parties based on platform and to "do no man's bidding," as Horace Greeley described it.[9] To be independent meant that other sources of revenue had to be available. After the Civil War, the growth in the production of goods and the accompanying advertising provided the revenue. Advertisers eventually demanded to know circulation figures of publications to find out how far and to how many their ads were reaching. This drove newspaper circulation because increased numbers meant a paper could demonstrate its reach into the populace and often ensured increased advertising income. Financial security became the lynchpin of independence and of one element that fueled the massive growth of the press.

But the press also grew during this period at an unprecedented rate because people depended upon it for the latest news. While press outlets grew by the hundreds in the first half of the nineteenth century, they grew by the thousands in the last quarter. The telegraph made the movement of news instantaneous. The telephone created more changes in the late 1880s and in the 1890s. In large cities, especially, word of mouth no longer served as the fastest way to share information. In the cities above the nation's capital, newspapers—opened by unsuspecting readers—blared the news of the evening assassination of Abraham Lincoln. Those readers, however, had come to expect that the morning paper would provide them with information they could not acquire anywhere else as quickly as in their newspaper.[10]

News organs during the last half of the nineteenth century were considered as vital as food. "Only *bread and the newspaper* we must have, whatever else we do without," Oliver Wendell Holmes said.[11] The increase in the number of publications and in their circulation added credence to the value of papers as described by the future Supreme Court justice. By 1870, approximately 4,500 newspapers published on a regular schedule in the United States. In New York City, the *Sun*, the *Herald*, and the *Daily News* boasted circulations of 100,000 daily—the *Sun* reaching 220,000 in 1876. Others in the city sold 50,000 to 75,000 daily. In Boston, Philadelphia, and Chicago, people bought newspapers that registered similar circulation figures. The number of papers in the United States doubled from 1870 to 1880, and by 1890, according to Frank Luther Mott, more than twelve thousand newspapers filled the nation with information.[12] The number of dailies increased by more than 500 percent from 1870 to 1900, reaching a circulation of 15 million.[13] Nearly sixteen thousand papers were published at the end of the century, according to Alfred McClung Lee.[14]

Magazines were as prevalent as newspapers in the country, and according to Mott, "Newspapers flourished, books at low prices multiplied ... but of all the agencies of popular information, none experienced a more spectacular enlargement and increase in effectiveness than the magazines."[15] By 1885, approximately 3,300 periodicals on numerous subjects served the nation. "Every field of human thought has been entered," the *Graphic* noted, "and the magazine has become not only a school of literature but of science, art and politics as well."[16] An additional eleven hundred began publishing in the next five years, and by 1900, *N. W. Ayer & Son's American Newspaper Annual* estimated that more than 5,500 magazines existed in the United States, with an average of one hundred new titles per year added in each of the first five years of the twentieth century.[17]

Literacy surpassed 90 percent among white Americans in 1880 and continued to increase,[18] meaning that the press had the potential of being used by the majority of Americans individually rather than it being a commodity used by some individually while many others experienced its content through public readings. Literacy combined with the number of papers available and gave the media great reach and consequently power. Newspapers may have never had the ability to shape and define American issues with as much saturation as they did at the end of the nineteenth century. According to Ted Curtis Smythe, 2.61 copies of papers were sold per urban dwelling in 1900. This penetration of papers into the fabric of society, he says, was not attained before this time period nor afterward,[19] though Americans continued to buy newspapers in excess of 100 percent of the number of households through the 1960s.[20] Penny papers opened the world of information to hundreds of thousands of Americans in the 1830s through 1850s. In the last two decades of the nineteenth century, the cost of the raw materials needed to produce papers dropped by around three-fourths, which meant that by the end of the century, 100,000 papers could be produced cheaper than 10,000 could have been in the 1870s. The introduction of pulp newsprint superseded the use of rag paper, which, *New York Tribune* editor Whitelaw Reid said, meant that the resources to produce newspapers had "become almost unlimited."[21] Costs of most dailies remained at one or two cents in the 1890s. The desire to know, cheap production, cheap paper prices, and saturation of product, therefore, made the newspaper within reach for practically all Americans.

Another great change in circulation occurred in the late 1870s. Until then, most weeklies published on Friday or Saturday. The large New York dailies—the *Herald* and the *Tribune*, for example—produced weekly papers, which were compiled editions of their dailies that were sent throughout the nation with circulations in the hundreds of thousands. Dailies published six days a week with no paper on Sunday. Once the first papers added a Sunday edition, they caught on quickly. Mott says that by 1890, more than 250 daily papers produced a Sunday edition. Sunday papers immediately became popular because of content and because people had the time to read them. Pulitzer carried the Sunday edition to new heights with features and color comics. Content was expanded, as well. Sometimes, the *Sunday World* published up to fifty pages. The *Sunday World* was selling a quarter of a million copies within four years of Pulitzer's purchase of the *World* in 1883, and the cost of the paper was increased over the daily version, which provided Pulitzer with even more revenue.[22] By 1897, the *Sunday World*'s circulation topped 600,000 at five cents a copy. The weekday edition sold for a penny. Pulitzer's Sunday edition, which was not considered a part of the weekday subscription, however, could not compare to that of Hearst's, which often ran to eighty-pages and included three free magazine supplements.[23]

When all that occurred within the sphere of the media during the last half of the nineteenth century is considered—the revenue available, the saturation of the press into society, and the political but independent nature of most press organs—one can begin to understand the influence the media possessed in the

Gilded Age. Newspapermen were powerful. They were easily elected to political office, but as Joseph Pulitzer discovered, he wielded far more power writing editorials and sponsoring exposés with the *World* than serving as a single member of New York's congressional delegation. Papers could elect presidents, and they could send politicians to prison. Whitelaw Reid unabashedly claimed, "One form of government in this country is the government by newspaper."[24] On many levels, he was absolutely correct.

THE JOURNALISM OF "SMALL PUBLICS": COMMUNITY WEEKLIES

William Randolph Hearst called it "News for small publics"— the specialized news that he carried that attracted assorted groups to his newspaper and increased its overall circulation into the hundreds of thousands.[25] Small publics, however, existed everywhere in the United States. The nation might have been moving to a more urban lifestyle in post-Civil War United States, but millions of Americans lived in small towns and communities. Newspaper editors by the thousands reached and influenced people in the years after the Civil War through weekly newspapers. These papers and their editors often became the moral guideposts, the cultural calendar, and principal source of information within their sphere. As a result, the local media became the shapers of public opinion. In an era that began with no standardized time for the nation and every community operating with noon based on when the sun appeared directly overhead there,[26] the editor became perhaps the most powerful individual in rural America. Most communities of three hundred or more had a newspaper to serve them.

As the nation's dailies continued their growth in numbers and circulation, editors on another level with significantly smaller readership were shaping ideas in a much smaller sphere. Still, these media were able to influence perhaps in a way the large dailies could not because of the unique connection between editor and the community served by his publication. The influence of the community publication continued into the twentieth century, and, in fact, it continues today. In many areas of the United States, the small-community weekly is the only media outlet that covers all elements of life for those living there. Only large events—a murder, some other horrific tragedy, or some serendipitous event that qualifies as unique from a human-interest standpoint—earn coverage from media outlets that cover a region or the nation. In this media-to-patron environment that existed in rural America and among the "small publics" that comprised so much of the nation's cities in the last half of the nineteenth century, media became the way to share commonalities. More importantly, those creating the media message almost always knew the general beliefs of a community. As community leaders, editors were able to shape opinion because their opinions were grounded in those basic ideas. Communication, as John Dewey said in 1916, created commonality in communities because communication was the only way to reach consensus.[27] "Your paper tells you when to go to church, to county, circuit and probate court, and when to send your children to school and anywhere else you want to go," as one resident of a rural Southern town said in a letter to the local editor. "It calls attention to business enterprise, advocates the best of schools, of law and order in town."[28] In other words, the community press covered all aspects of life, and the majority of citizens subscribed to the agenda. As one community newspaper patron stated, "If the *Gazette* said it, it was right."[29]

The media that served rural America were able to direct the agendas of their respective communities so well because they were gatekeepers of information. Following a long-established trend in American journalism of clipping articles from other newspapers and reprinting them, rural newspapers selected the information from sources that supported their local agendas and ran them. Perhaps none did this better than William Allen White of Emporia, Kansas.

White assumed control of his newspaper, the *Emporia Gazette* in the 1890s. Though he worked late in the time period of this chapter and continued his work well into the twentieth century, his tenure with the *Gazette* provides a perfect example of the way small-town editors shaped their communities, and, in White's case, became a voice on the national stage by consulting presidents. Rural editors were not generally concerned with providing the latest national intelligence, though, unless national issues could affect local ones. White explained the way rural newspapers affected the communities they served in an article titled "The Country Newspaper." "It is the country newspaper bringing together daily the threads of the town's life," White said, "that reveals us to ourselves."[30]

Rural communities, however, were not comprised of thoughtless followers. Editors had to convince readership when controversial issues arose. That is exactly what White did in 1896 with his famous editorial, "What's the Matter with Kansas?" The 1896 national election was playing out on the local level, but the problems that made this election so important were rooted in the financial crisis that the entire nation faced. In 1893, bankruptcies among railroad companies and uncontrolled selling on the New York Stock Exchange sent the nation's economy into a downward spiral. Before year's end, more than five hundred banks and fifteen thousand businesses had failed. Approximately 18 percent of the nation's work force was soon to be unemployed, and the administration of Grover Cleveland was powerless to stop depression. In 1894, unemployed workers led by Ohioan Jacob Coxey organized out-of-work men into what the press called "Coxey's Army." Twelve hundred marched to Washington seeking assistance where they were met by federal troops and at least twenty thousand sympathetic spectators. Similar protests occurred throughout the nation along with strikes. Economic woes continued, and in 1896 the issue of how to ensure the value of the nation's currency took center stage.[31]

In Emporia, citizens were divided over a proposed policy by Democrats and Populists that would use silver as the basis of the nation's money supply instead of gold. After being accosted on the streets of Emporia by a group of Populists who demanded he explain why the *Gazette* opposed them, White returned to his office and wrote his classic editorial.[32] Though the issues of the monetary standard and of economic depression were national in scope, White localized them in "What's the Matter with Kansas?" White pointed out (though not exactly accurately) that "the nation has grown rich" while "Kansas has gone downhill." Because White felt that the national election of William Jennings Bryan would be detrimental to Emporia and Kansas, he attacked Bryan's platform, showing that it would be detrimental for all Kansans. The only way to defeat economic depression, White said, was for Emporia and Kansas to focus upon themselves. If they were able to do so, he figured, capital would flow into the state, something that had not happened during the decade.[33] White's editorial struck a chord not only in Kansas but among likeminded Americans everywhere. Not only did he solidify support for his cause, he was seen as the man who spoke for the people. While Kansas ultimately gave its electoral votes to Bryan, White's effort and those that would follow at the *Gazette* proved that an editor at a small paper in rural Kansas could shape the direction of a local community and, to a certain extent, the nation.

CREATING THE "NEW SOUTH"

Another group of journalists grew out of the rural editors, but they went to work at city newspapers. This was especially the case in the South's rapidly growing urban and industrial centers. Whether they worked for small publications—like that of William Allen White—or larger dailies, both sets of editors had one thing in common: They were interested in setting the political agenda for the areas they served. They voiced opinions on local and national issues—when those issues affected the local sphere. In most cases the press directed, and readership followed. Georgia journalist Henry Grady was one of a cadre of Southern big-city editors who accomplished this. After having bought several newspapers only to have them fold, Grady purchased a percentage of the *Atlanta Constitution* and became the paper's editor. Grady had proclaimed in 1874 in one of his failed papers, the *Atlanta Herald* that a "New South" existed. He would use that term, as did other Southern editors, for many years to work to change the South and opinions of the region throughout the nation. The *Constitution* followed the pattern that had been established by the large Northern dailies and created a weekly edition that was filled with the best of the paper's daily issues. Circulation of the weekly soon exceeded 140,000,[34] and each issue spread the concepts of the New South throughout the region.

Grady was not ashamed of his Southern heritage, but he was certain that if the South continued to pin its economic success solely on agriculture that the region was doomed. He and others like Henry Watterson, the editor of the Louisville *Courier-Journal*, felt that the South had to adopt the industrial model that had made the North so prosperous. He talked of the resources that the South possessed in terms of people and minerals. He talked of the region's potential in terms of climate and growing transportation. After trumpeting his message repeatedly to a Southern audience, Grady took it to the North in an effort to secure financial input to create the New South. "There was a South of slavery and secession—that South is dead," Grady told the New England Society in words they no doubt wanted to hear. "There is a South of union and freedom—that South, thank God, is living, breathing, growing every hour."[35] *The North American* of Philadelphia noted "that if there is to be a 'New South,' it will be mainly due to the wise influence of such journalists as the editor of the Atlanta Constitution," adding, "His words will reach to every corner of the south."[36]

While Grady's *Constitution* served as the principal mouthpiece of the concept of the New South, other writers and editors bought into Grady's vision and potential of the New South. "The dawning of the *New South* is already visible," John H. Seals said. "Our people are learning slowly, though surely from hard necessity ... and an earnest and industrious population will cover the entire area, which the music of machinery will sound along our streams."[37] "There are better times ahead," the *Memphis Appeal* promised. "From Maryland to Texas the press is teeming with valuable information as to the best means of promoting the great material interests of the South....A new era is dawning. Everything calculated to retard the progress of the South has been removed, and if we are not prosperous, it will be because we are as indolent as our enemies have represented us."[38]

As proof that a New South indeed existed, the *Constitution* hosted General William T. Sherman on a post-Reconstruction visit to Georgia. Afterward, the general wrote to the *Atlanta Constitution* with observation on the South, twenty-five years after his infamous march from Atlanta to the sea. Sherman acknowledged that he was considered "the bête noir" for Southerners because of his duties as commander of the Northern forces during the war, but he also noted that he had been warmly received in his latest travels through the South. Sherman said Atlanta and north Georgia were ready to assume a place of prominence in the United States. Though he praised much of the South, Sherman was particularly impressed with Atlanta. "I am satisfied, from my recent visit, that Northern professional men, manufacturers, mechanics and farmers may come to Atlanta ... with a certainty of fair dealing and fair encouragement."39 While much about the South would not change for decades—such as in terms of race relations—and some discounted that the South had changed much since the pre-Civil War period, Grady and his disciples effectively shaped the understanding of the South among a majority of its residents and throughout the nation. "The South has been rebuilt by Southern brains and energy," he said.40 Even two weeks before his untimely death from pneumonia at age 39, Grady was expounding the New South through a series of editorials in the New York *Ledger*. Voices in the North and the West soon began to use Grady's New South rhetoric to describe the changes that were taking place below the Mason-Dixon line. "Much as there is in the South that is discouraging, there is much also that is full of hope and promise for the future," the *Milwaukee Daily Sentinal* observed. "Every year diminishes the number of the old irreconcilables, and fills their places with young men who care little for the past and everything for the future ... the New South which is growing out of the ruins of the Old, and which will ere long be everywhere the controlling force."41

EXPOSE AND ATTACK: PULITZER AND "NEW JOURNALISM"

America's population grew tremendously during the last half of the nineteenth century. Immigrants accounted for approximately 14 million of the increase in population from nearly 40 million people in 1870 to the nearly 76 million Americans in 1900.42 One immigrant who vowed to change his adopted country and did was Joseph Pulitzer. Pulitzer emigrated to the United States as a mercenary soldier from Hungary. After the Civil War, he moved to St. Louis after failing to find a way to make a living in New York City. St. Louis proved to be the right spot for Pulitzer. After several dead-end jobs—including taking care of U.S. Army mule teams—he secured a job writing for one of the city's German-language papers, the *Westliche Post*. Pulitzer's English was poor, so learning the ropes of journalism on a paper that operated in a language that he had learned as a boy was perfect. As Pulitzer worked, he sharpened his English in the library and by covering politics in the state capital. Soon, he had saved enough money and stock from the *Post* to buy another German-language paper. He immediately sold that paper, making enough money to buy two English-language papers, the *St. Louis Post* and the *St. Louis Dispatch*. He merged them and began his career as a newspaper publisher and editor at the age of 32.43

Pulitzer had definite ideas of what a newspaper should do. As a *Westliche Post* reporter, he had made few friends with his hard-hitting style of writing. The same would be true of Pulitzer and his English-language newssheet. "Entirely independent and free from ties and trammels of any sort, the *Dispatch* will be in an honest and sincere opposition to that fraudulent, dangerous and usurping minority which rules over the nation," Pulitzer said of the *Post-Dispatch* in an interview with a rival paper shortly after the purchase. "To wage war against all frauds and sham, whether in the city, the State or the republic, shall be the first political duty of the *Dispatch*."[44] His plan was to create a politically active newspaper that worried less about the political party and more about what was best for the people. He would go after wrongdoing on any level. Or, as he would say later in the *New York World*, "The newspaper that is true to its highest mission will concern itself with the things that *ought to happen tomorrow*, or next month, or next year, and will seek to make what ought to come to pass," adding, "The highest mission of the press is to *render public service*."[45] To Pulitzer, that meant that he would "always fight for progress and reform; never tolerate injustice or corruption; always fight demagogues of all parties; never belong to any party; always oppose privileged classes and public plunder; never lack sympathy for the poor; ... never be satisfied with merely printing the news; always be drastically independent; never be afraid to attack wrong, whether by predatory plutocracy or predatory poverty."[46]

Pulitzer's mission plan combined with political exposé came at a price early in his career, however. Pulitzer's managing editor, John Cockerill, shot and killed a disgruntled lawyer who stormed the *Post-Dispatch* offices in 1882 after the paper printed an article attacking the man. A rival St. Louis editor immediately blamed the shooting on Pulitzer's "aggressive and sensational journalism" and the dead man "a victim of its venom."[47] That sparked a trip to Europe by Pulitzer who bought one of New York City's struggling newspapers, the *World*, before he returned to St. Louis. In 1883, Pulitzer assumed leadership of the *World* and left St. Louis, bringing Cockerill with him. Pulitzer immediately reiterated what his purpose was as a newspaper owner. Like the *Post-Dispatch*, the *World* would be "truly democratic—dedicated to the cause of the people rather than that of purse-potentates—devoted more to the news of the New than the Old World—that will expose all fraud and sham, fight all public evils and abuses—that will serve and battle for the people with earnest sincerity."[48]

Pulitzer's New York newspaper achieved success almost immediately. He doubled the twenty thousand circulation in four months. Within a year, circulation topped 100,000, and by 1886 a quarter-million copies were sold each day, including the Sunday edition. In 1887, Pulitzer decided to publish two papers daily, and within five years, the combined circulation of the two neared 375,000 copies per day.[49] His business model, exposés, sensational stories, independent political stance, and targeting of audiences—like that of New York's immigrant population—made his style of journalism one that other editors emulated and dubbed "new journalism." Other editors, like Henry Grady, Henry Watterson, and those who supported the concept of the "New South" also practiced a form of new, independent journalism, but none did it on the level and in the style of Pulitzer. The platform of the press provided editors and publishers with a way to effect change that all realized, but none more so than Pulitzer. "The *World* should be more powerful than the President," he said. "He is fettered by partisanship and politics and has only a four-years term. The paper goes on year after year and is absolutely free to tell the truth and perform every service that should be performed in the public interest."[50] Pulitzer and other journalists exposed corruption in government and elected presidents.

MAKING AND BREAKING POLITICIANS

The ability of the media to take on corruption in government may have been made easier in the Gilded Age because of the amount of vice that existed, but it also speaks to the strength of voice with which the nation's media spoke. "The press ... is the most magnificently repressive moral agent in the world today. More crime, immorality and rascality is prevented by the fear of exposure in the newspapers than by all the laws, moral and statute ever devised," Pulitzer said.[51] In New York, before Pulitzer ever entered the world of journalism, two publications, the *New York Times* and *Harper's Weekly*, brought down the political machine of William M. "Boss" Tweed. A state legislator who became head of New York's Democratic party, Tweed and his ring of conspirators effectively ran politics in the city by controlling the

Rigged elections were one of the chief charges leveled against William Tweed and his cronies.

mayor's office and almost all of the activities of government through bribes and payoffs. According to the *New York Herald*, Tweed and his cronies bilked approximately $200 million from the city from 1868 until his Tweed Ring was stopped in 1873.⁵² The *Times* attacked Tweed with thousands of investigative stories that turned up how Tweed's ring had bribed contractors and anyone that did business with the city. The *Times* said that they were all "living joyously on the drippings from Boss TWEED's punch bowls and pastries."⁵³ The *Times* named all the politicians that it could who were in cahoots with Tweed and stated: "These men are in the position of a gang of burglars, who, having stolen all your silver-ware and jewelry and placed them under lock and key, turn round and challenge you to identify your property....[T]he Tammany Ring are a law unto themselves; they are the Government; they control the machinery of justice, and own a large share of the Judges. Hence they defy your efforts to detect their villainy, and laugh at the idea of restitution." The paper then proudly pointed out that "the TIMES has succeeded in exposing" the corruption.⁵⁴

Thomas Nast often portrayed money as what drove the Tweed.

The *New York Times*' stories were made even more powerful in the eye of the public, because of the attacks on the Tweed Ring by *Harper's Weekly*, especially its editorial cartoonist Thomas Nast. Nast provided caricatures of Tweed with money and corrupt activities as their central elements. Even the *Times* admitted that Nast's cartoons were critical in the demise of Tammany Hall's political stranglehold on the city. "A man who can appeal powerfully to millions of people with a few strokes of the pencil must be admitted to be a great power in the land. No writer can possibly possess a tenth part of the influence which Mr. NAST exercises," the paper observed. "It is this power of accurately *describing* men in a picture which renders NAST so formidable." ⁵⁵ Tweed said the same about the attacks on him: "I don't care a straw for your newspaper articles; my constituents don't know how to read, but they can't help seeing them damned pictures."⁵⁶ Tweed was arrested in 1871, but his political ties with the judicial system ensured a hung jury. He was arrested again in 1873 on 204 counts of fraud. He was fined more than $120,000 and sentenced to twelve years in prison. Again, his political ties got that sentence reduced to one year. Arrested a third time, Tweed was able to visit Cuba and

This cover of Harper's *led to the arrest of Tweed in Spain after it was seen and the politician, who had fled the United States, was extradited and imprisoned.*

Tweed's image dragging two boys along the street helped identify him after he went to Spain to avoid prison time.

there absconded to Spain. Nast's cartoons, however, were well known in Europe, and Tweed was recognized, extradited to the United States, and returned to prison where he died at age 55.[57]

The undoing of the Tweed Ring by the *New York Times* and *Harper's Weekly* had another effect. It shamed other city papers that were reticent to print articles against such a powerful public figure as Tweed or take on political corruption. It fostered the independent attitude that other media figures were demanding in other parts of the country. Journalist W. L. Stone, editor of the New York *Commercial Advertiser*, pointed this out, saying that "the Press has become in this country an engine of such great importance in the daily affairs of life," and that "its energies of such tremendous power, either for good or evil."[58]

In many ways, the bringing down of Tweed opened the way for Pulitzer's style of journalism in New York City, and Pultizer, along with Nast, helped change the course of the nation in 1884 by influencing—and in reality—changing the outcome of the presidential election.

In 1884, Joseph Pulitzer was the new publisher in New York City, but he was certainly not an unknown. His journalistic ferocity was well known. When Republican James Blaine and Democrat Grover Cleveland began their run for the White House, Pulitzer backed the Democrat, calling him an honest man, but the lessons of the Tweed scandal had not been forgotten, and the press peered closely into both candidates' backgrounds. Blaine, by some, was seen as a continuation of corrupt politicians, and Thomas Nast even linked Blaine to Tweed in his *Harper's* cartoons.[59] Blaine had been suspected of corrupt activities in dealing with railroads, and the press uncovered his efforts to keep these exchanges secret. Pulitzer ran a letter in *World* on September 15, 1884, that confirmed this, the last line saying "Burn this letter," a request that the cover-up document be burned after completion of the deed. Papers and Republicans took the content of that letter and created a poem as a campaign tool: Blaine, Blaine, James G. Blaine / The continental liar from the State of Maine, / *Burn this letter*!

Cleveland was the victim of doggerel, as well, when it was revealed that he had fathered a child out of wedlock with a woman from Buffalo. His opponents announced: Ma! Ma! Where's my Pa? / Where's my Pa, Ma? / Where's my Pa?[60]

To the American public, immorality trumped corruption. With less than a week to go before the election, though, a simple statement by a clergyman at a meeting with Blaine in his presence provided Pulitzer the chance to change the course of the election with the *World*. He knew that the rising popularity of his newspaper, especially among immigrants in the nation's largest city, might be able to win the election for Cleveland. On October 30, the *World*'s front page described the scene with a huge editorial cartoon titled "The Royal Feast of Belshazzar Blaine and the Money Kings." Blaine was surrounded by some of the nation's wealthiest men, all with large jewels shining from their chests. They fed on "lobby pudding" and "monopoly soup," and other foods that represented their fortunes coming at the expense of "ordinary" Americans. In front of the table stood an out-of-work man with his wife and child, asking for assistance but being ignored by the gluttonous Republicans, financiers, and industrialists. Behind the table the Old Testament warning that God had numbered the days of the kingdom of Belshazzar and the Persians appeared, written with a hand just as mysterious as the one that had written it in the book of Daniel.[61] The inference for the days of Republican control of the nation was not missed by the *World*'s readers.

Accompanying the editorial depiction, Pulitzer added a series of stories that included headlines that said, "Republican Corruption Fund" and "Starving While Blaine Feasts." In the text, the *World* compared Blaine's closing of Maine textiles mills, putting hundreds out of work, with the extravagances pictured in the cartoon. "Delmonico's was filled with millionaires last night. The object of the banquet was two-fold—nominally to honor Mr. James G. Blaine but really to raise a corruption fund of $500,000 with which to defeat the will of the people," the article explained. Inside the *World*, Pulitzer ran an editorial that provided his readership with the incentive to elect Cleveland. Quoting from the minister at the gathering, Pulitzer pointed out that Blaine and the Republicans disdained "RUM, ROMANISM AND REBELLION." The attack intimated that Blaine supported prohibition, was strongly anti-Catholic, and felt that the Civil War had been a mistake. All three were issues with which the readers of the *World*, especially Irish-Catholic immigrants who were estimated to number one-half million in New York, took issue. New York, which had been considered a sure bet for Blaine on October 29, ended up in the Cleveland column on November 4 by 1,149 votes, each of which Pulitzer claimed—correctly—to have won for Cleveland.[62]

STUNT JOURNALISM

Pulitzer's ability to shape the direction of New York and other parts of the nation expanded following the 1884 election. Through the pages of the *World*, he raised the money to build the pedestal for the Statue of Liberty when public funds could not be obtained for the base for the gift from France. But one of the ways that the *World*—followed by other media outlets that copied it—was able to effect societal changes came to Pulitzer purely by luck. Elizabeth "Pink" Cochrane had been a reporter for the *Pittsburgh Dispatch*. As with many reporters of the age, she adopted a pen name. As Nellie Bly, she had written stories of bull fighting in Mexico and of the plight of women in Pittsburgh who worked in the steel mills. Bly, however, felt that she was not being used to her potential in Pittsburgh and moved to New York City in 1887. There, she called on editors but was unable to land a job until she proposed to *World* editor John Cockerill a story that would make New York readers pay attention. Bly would go undercover as a mentally ill immigrant, get herself locked up in the women's mental institution on Blackwell's Island, and then write an exposé of the ill treatment that reportedly occurred there. If it worked, the *World* would give Bly a job. If not, the paper owed her nothing.

Bly spent several days preparing for her story while living in a boarding house. Acting disoriented and as if she might hurt others, she was arrested, taken before a magistrate, and then sent to Blackwell's asylum. Once in confinement, Bly said, "I made no attempt to keep up the assumed *role* of insanity. I talked and acted just as I do in ordinary life. Yet strange to say, the more sanely I talked and acted the crazier I was thought to be by all except one physician, whose kindness and gentle ways I shall not soon forget."[63] Bly remained in confinement for ten days, until Pulitzer, now in on the undercover assignment, sent a lawyer to free his new reporter. "The insane asylum on Blackwell's Island is a human rat-trap," Bly said. "It is easy to get in, but once there it is impossible to get out."[64] When Bly's stories began appearing in the *World*, they caused an immediate sensation. She told of how inmates were treated—of rotten food, of ice-cold baths, of beatings, of incompetent physicians, of cruel nurses, of deaths.

The public outcry about conditions on Blackwell's Island created questions for the city and for the staff at the asylum. Bly wrote about it. "Since my experiences in Blackwell's Island Insane Asylum were published in the *World* I have received hundreds of letters in regard to it," she said in the introduction to her collected *World* articles on her confinement. "I am happy to be able to state as a result of my visit to the asylum and the exposures consequent thereon, that the City of New York has appropriated $1,000,000 more per annum than ever before for the care of the insane. So I have at least the satisfaction of knowing that the poor unfortunates will be the better cared for because of my work."65

Being able to effect change was something Pulitzer had come to enjoy from his role as a publisher and journalist. The fact that his reporter became an integral part of the story was of no concern. In fact, it made the *World* and its reporters even more important because they were working tirelessly, the stories said, to make things better for the ever-increasing numbers of people who subscribed to Pulitzer's paper. The centrality of the media outlet and the reporter would also become one of the basic tenets in the journalism of action. Public reaction to Nellie Bly was so positive that Pulitzer had her doing an undercover or investigative story each week while he was hiring new reporters to copy her style, some calling these reporters stunt journalists. Stunt journalists did all sorts of things—such as going undercover in sweat shops or posing as prostitutes to catch community leaders soliciting for sex. Eventually, many of the "stunts" served only to increase circulation because of their oddity and lack of news value, Bly's competition to beat the eighty days it took Jules Verne's fictitious Phileas Fogg to make a trip around the world being one example.66 Stunt journalism opened the door of the newsroom to women, though, but some female reporters changed society through hard-hitting, investigative reporting.

IDA B. WELLS

Born to slaves in Mississippi who sent her to college after the Civil War, Ida B. Wells became a voice of social change in the United States when few African Americans had the power to change anything. Wells' attacks on injustice and on the lynching of blacks in particular spread through the United States via newspapers, books, and pamphlets. Wells' work was impossible to ignore because she used facts and figures gleaned from the white press to make her points. She mixed this information with interviews with African Americans to tell powerful stories. Wells was thrown off a railroad car in 1884 when she refused to give her seat to a white man. She successfully sued the railroad, but the Tennessee Supreme Court overruled the decision. Soon after, she gave up her job as a teacher and joined the staff of the Memphis *Free Speech* and used its pages to organize blacks in the city in a series of protests against the city's transit system and against white-owned businesses, the latter occurring after three black businessmen were executed by a mob in 1892.67 Wells advised black citizens of Memphis to leave the city. "There is therefore only one thing left that we can do," she said in the *Free Speech*, "save our money and leave a town which will neither protect our lives and property, nor give us a fair trial in the courts, but takes us out and murders us in cold blood when accused by white persons."68

Wells championed numerous causes with her journalism, including suffrage for women and equal rights for blacks, but the deaths of the grocers in Memphis caused Wells to focus her commentary on lynching, which reached an all-time high that same year. Wells' advocacy journalism called for action, which she espoused in two articles in the *New York Freeman*. "There is no dodging the issue; we have got to take hold of this problem ourselves, and make so much noise that all the world shall know the wrongs we suffer and our determination to right those wrongs," she wrote, adding two months later, "We have reached a stage in the world's history where we can no longer be passive onlookers, but must join in the fray for our recognition, or be stigmatized forever as a race of cowards."[69] After a mob destroyed the offices of the *Free Speech* and left notice that Wells would be killed if she returned, Wells, who had traveled north, remained there. On June 25, 1892, using the pseudonym "Exiled," she wrote an article titled "The Truth About Lynching," which filled nearly the entire front page of the *New York Age*, the principal African-American newspaper in the United States. There, she rebutted the white rationale for the destruction of the *Free Speech* office and provided her first systematic attack on lynching. "Somebody must show that the Afro-American race is more sinned against than sinning, and it seems to have fallen upon me to do so," she said. "The awful death-roll that Judge Lynch is calling every week is appalling, not only because of the lives it takes, the rank cruelty and outrage to the victims, but because of the prejudice it fosters and the stain it places against the good name of a weak race."[70]

The *New Age* had a readership comprised of blacks and whites, but the New York paper did not have the reach nationally that was required. Therefore, T. Thomas Fortune, editor of the *New Age*, published Wells' editorial in pamphlet form. *Southern Horrors: Lynch Law in All Its Phases* could now be easily distributed throughout the country and abroad, and it became, according to Patricia Schechter, the point of origin in the United States for all battles to expose the evils and immorality of lynching and racism and to put an end to them.[71] Wells continued to write for Fortune's paper on lynching as she lectured on it throughout New England. In an effort to create international pressure on the United States to stop the practice, she traveled throughout the United Kingdom in the spring of 1893 speaking on the content of her pamphlet. Once she returned, she expanded her efforts to the Midwest and Chicago.[72] Wells' writings were powerful and systematic in their refutation of lynching. What made them impossible for white America to ignore, however was the way she used what appeared in the white press to support her contentions.

Wells was soon joined by other African-American leaders in the fight to end lynching. "Alone Miss Ida Wells may not be able to accomplish much, but hers is a move in the right direction, and she will undoubtedly be followed by others of her color working in the same cause," one periodical said.[73] She was. Dozens of African Americans took up the fight to end lynching and toward equal rights. The *Milwaukee Sentinel* reported how African Methodist Episcopal Zion ministers listened to Wells speak and then urged the forming of anti-lynching leagues.[74] Charles H. Williams put the onus of ending lynching where it originated, with white America. For lynching to end, he said, it "MUST ORIGINATE WITH THE PEOPLE OF THE NORTH."[75] A number of white

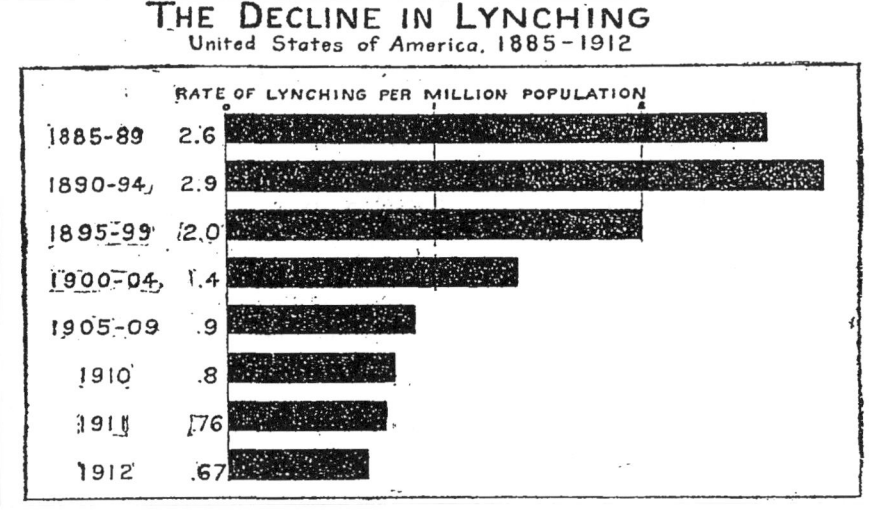

The New York Times *ran this graph of the decline in lynchings of African Americans in the United States in March 1913.*

publications began to support Wells and report on her efforts. In 1894, the Chicago *Daily Inter Ocean* hired Wells to report on a lynching in Kentucky. The story ran with a graphic illustration of the hanging.[76] The *Rocky Mountain News* in Denver ran a story about Wells and her anti-lynching crusade.[77] In 1895, Wells published *The Red Record*, another pamphlet on lynching. This one, however, included photographs, illustrations, and statistics on lynchings that were tabulated from news accounts that ran in the *Chicago Tribune*.[78]

Wells continued her fight against lynching after she married Ferdinand Barnett in 1895 and into the twentieth century. "If we only had men with the backbone of Mrs. Barnett, lynching would soon come to a halt," the *Chicago Defender* noted as Wells successfully lobbied to have a sheriff removed from office who tacitly allowed hangings in Cairo, Illinois.[79] This occurred after she had led the battle that ended in the passage of anti-lynching legislation in Illinois.[80] Lynchings did not end in the United States because of the efforts of Wells and others, but they did create changes. Including Illinois, forty states from the 1890s to the early 1930s adopted codes to deal with lynching and race riots even though they sometimes proved ineffectual.[81] No national anti-lynching law was ever adopted, however.[82] In 1913, the *New York Times* reported on a steep decline in lynchings. Though the article never referred to the efforts of African Americans—or Ida B. Wells specifically—Wells' writings and public appearances raised awareness throughout the United States. "It is no cause of satisfaction that there should have been sixty-four lynchings during 1912 in the United States, chiefly in the South; but the country may well be satisfied with the fact that, with a single exception, this was actually the lowest number of lynchings during the last twenty-eight years," the *Times* article said.[83]

CREATING YELLOW JOURNALISM

Yellow journalism, according to Joseph Campbell, has been equated with a lurid and sensational treatment of the news, accomplished through journalistic misconduct—a story at any cost, obtained in any manner. The principal practitioners of yellow journalism transcended the walls of the newsroom and became men Americans disliked because of their wealth and because they wielded so much power.[84] Though yellow journalism may have been sensational in some cases, it surely changed the press at the end of the nineteenth century as it changed the nation. With large headlines and innovative layouts that drew in readers by the millions to front pages filled with news on every imaginable subject that was accompanied by photographs, illustrations, and graphs, the yellow press took on causes and wrote stories to end corruption on any level, as seen with Nellie Bly and the Blackwell insane asylum and with the election of Grover Cleveland in 1884.[85] If Bly was glorified in the process or if the *World* took credit for Cleveland's election, so much the better for the yellow journals, according to their reasoning. Joseph Pulitzer and William Randolph Hearst became synonymous with the yellow press, though they were certainly not the only practitioners of this new style of journalism. Some considered the pair to be "entirely irresponsible to the public and utterly unscrupulous as to what they print or how they obtain it."[86] Others said that their "news has been shown to be so reliable, that it has deserved the immense circulation and great influence it has attained."[87]

The journalism of action that Pulitzer, Hearst, and other journalists practiced required personal involvement on many levels. Reporters and their publications assumed roles in stories, and the press outlet often demanded action. Though Campbell argues that the yellow press did not cause the Spanish-American War because elements beyond the control of even Pulitzer and Hearst were at play on the international stage, the yellow press—as well as more traditional press outlets—certainly led Americans to support the war. "The shameful influences of yellow journalism upon the masses is bad enough," the *Macon Telegraph* observed in February 1897 following the death of insurgent Ricardo Ruiz in Spanish custody, "but when its malignant and far-reaching ingenuity encompasses the senate chamber and brings the country to the very threshold of war, it is time for the strong men of the country to cry out against it."[88] The coverage of the troubles in Cuba by the yellow press may not have caused the war, but many—like Charles Rittenhouse Pendleton of the *Macon Telegraph*—believed that their highly charged stories were affecting the populace as well as politicians in Washington, a belief that seems justified given the reach of the yellow press and the commentary about Cuba that appeared in papers throughout the nation.

Hearst, more so than Pulitzer, elevated the rhetoric of war and the number of sensational stories coming out of Cuba. Pulitzer, though, worked to compete because he, too, believed that Cuba should be independent. By the time the insurrection that turned into a war between Spain and the United States ended, American media had sent more than five hundred reporters, illustrators, and photographers to cover it with Hearst reportedly spending up to $50,000 a week to gather and disseminate the news.[89] In 1896, Hearst initiated *Cuba Libre*, the *Journal*'s campaign for a free Cuba. It dubbed the Spanish captain-general of Cuba, General Valeriano Weyler, the "Butcher of Cuba," claiming that he was responsible for unimaginable atrocities including burning alive insurrectionists and the rape of women.[90] Pulitzer's *World* cried that Weyler's tactics in Cuba produced "Blood on the roadsides, blood in the fields, blood on the doorsteps, blood, blood, blood! The old, the young, the weak, the crippled, all are butchered without mercy."[91]

In 1897, the stories and the horror of what was happening in Cuba increased. And, the reporters and the yellow journals played an increasing role in the stories. Hearst sent reporter Richard Davis and artist Frederic Remington to Cuba. On January 19, they witnessed the execution of Adolfo Rodríguez, a Cuban freedom fighter. The *Journal* contained a full, front-page illustration of Rodríguez' calmly standing in front of his firing squad. The headline proclaimed, "DAVIS AND REMINGTON TELL OF SPANISH CRUELTY."[92] Davis then described how Rodríguez had been forced to march to the site of his execution, of the wait for the guns to fire, of the prisoner's bravery, of how his head snapped back as he fell motionless to the ground. Later in the year, Hearst inflamed the public with the story of the rescue of Evangelina Cosío y Cisneros, who had been accused of treason. Her story was the epitome of the journalism of action because Hearst sent in a reporter, Karl Decker, who brokered Cisneros out of jail and smuggled her to the United States.[93] "An American Newspaper Accomplishes at a Single Stroke What the Red Tape of Diplomacy Failed Utterly to Bring About in Many Months," the *Journal* boasted.⁹⁴ Hearst then arranged a New York reception for her, and nearly 75,000 people attended, and later Cisneros met with President William McKinley. "The American people," one newspaper observed after the Cisneros

The New York Journal *published numerous editions on February 17, each updated as news was obtained from Cuba. The* Evening Journal *had at least nine different editions, the last claiming that naval officers, the president, and even England believed "foul play" was involved in the Maine's sinking. The edition above is the first of the day for the morning* Journal.

rescue, were "ready at any and all times to take sides with the oppressed and especially ready now to show that they are not selfish in the freedom they enjoy."⁹⁵ The sensational stories by yellow journal reporters were buoyed by reports at the end of 1897 that more than 300,000 Cubans were locked away and that at least 100,000 had died there of hunger and disease.⁹⁶

As the yellow journals continued a crusade against Spanish cruelty in Spain, President McKinley sought diplomatic means to end the abuses. In January 1898, with Spain's permission, the U.S.S. *Maine* sailed into Havana harbor with three hundred soldiers on board. Greeted politely, the *Journal* published a letter from Spain's U.S. Ambassador that called the president "weak" and a "low politician."[97] Six days later, the *Maine* exploded, killing 266. Hearst and Pulitzer responded with banner headlines that blamed Spain. "DESTRUCTION OF THE WAR SHIP MAINE WAS THE WORK OF AN ENEMY," the *Journal*'s banner headline said on February 17. Hearst then offered $50,000 for proof of who sank the ship. The *World* said, "MAINE EXPLOSION CAUSED BY BOMB OR TOPEDO?" On February 20, the *Journal* claimed to have "proof of a submarine mine" having sunk the *Maine* as he and Pulitzer sent large numbers of reporters to Cuba.

President McKinley tried to avoid war, but many voices in the press clamored for it, some giving the yellow press credit for uncovering what the government couldn't. "The days immediately following the Maine disaster, the Journal showed that it could not have been an accident or an internal explosion, but that the Maine was blown up by external causes," one editor said. "The bold and manly course of the Journal, backed by the other 'yellow journals' soon made it clear that the American people were in no humor to accept 'blood money,' and that the independence of Cuba must be exacted by the United States. One by one the papers have come around to that position until now it is not suggested in any quarter that the payment of money by Spain would make the murder of the Maine officers and the Cuban horrors 'a closed incident.'"[98] Others noted the war spirit that swept the nation. "Not since the day when the surrender of Lee was announced has Washington been so wildly and patriotically excited as to-day," the *Denver Evening Post* reported. "The people are war mad, as the saying goes, and nothing can restrain them."[99] On April 23, 1898, the United States declared war on Spain two days after Spain declared war on the United States.

After the sinking of the *Maine* in Havana harbor and the outcry from the *World* and *Journal*, other papers reacted. "The newspapers that are giving the public a splendid idea of what is meant by yellow journalism are clamoring for war, and unless those who are in charge of national affairs are calm and deliberate some overt act may be committed that will make war inevitable," the *Arkansas Gazette* predicted. "The newspapers do not always tell the truth concerning developments in great questions, and certainly a fine illustration of the disposition to inflame the public mind with sensational news has been had in the investigation of the Maine disaster. If our congressmen and senators at Washington believe all the newspapers are saying they will very soon get the idea that it would be unsafe for them to return home unless they use all efforts to precipitate a war no matter what course Spain may pursue."[100]

Another paper declared that the yellow journals should be silenced because there was no constitutional protection for what they were saying: "the freedom of speech and of the press gives no person or paper license to utter falsehood or to incite a revolt against the law of its administrators."[101]

The reaction to what the newspapers of Hearst and Pulitzer were producing in cities around the country—be it positive or negative—surely attests to the reach of the yellow journals and their ability to set the public agenda. Even if editors disagreed with what the yellow journals said, they kept it alive in public by reacting to the bold headlines and accusations made by Hearst, Pulitzer, and others. And, many editors praised their work for "opening the

Many Americans believed that the way in which the yellow press covered the sinking of the Maine and the accusations that Spain had intentionally sunk the ship were intended to start a war. In this cartoon from the Minneapolis Journal *of February 25, 1898, the cartoonist shows the yellow journals, headed by their cartoon mascot, the Yellow Kid, ratcheting up war rhetoric.*

eyes" of Americans to the cruelties and injustices taking place in Cuba. "[T]he events of the past six weeks prove conclusively that the *New York Journal* is America's greatest newspaper," the Raleigh *News and Observer* declared. "It pleads guilty to a sensationalism that prints all the news from freshest sources, and often before its contemporaries have realized the importance of the news which its correspondents have given in detail....It is not too much to say that but for the faithful pen pictures in the *Journal*, the American people would not to-day know of the cruelty and barbarity which moved the Senate as that august body has not been moved in a generation."

Hearst bragged, "How do you like the Journal's War?"[102] The *Journal's* circulation reached more than 1.6 million daily. The *World* topped 1 million in circulation, too. Though the yellow journals could not force a president to declare war, they could take credit for one. Hearst did; others agreed, and even if many believed that the stories in the yellow press were "threatening to place the nation in a false position,"[103] their war hype caused reaction in America and helped steer the nation toward war.

✳✳✳

From the end of the Civil War until the end of the nineteenth century and then into the twentieth, media's rapid expansion in the number of outlets and the number of copies per issue meant that media saturated the nation. Powerful individuals—like Pulitzer, Hearst, Grady, Watterson, and White—made their names synonymous with the media, but they were only examples of what was happening collectively with the media in the United States. What may have given the media such power was its ability to reach the subgroups within the mass of Americans. As one writer said of Adolph Ochs of the *New York Times*, he was able to give "to each reader the things in which he was personally interested, printing the news in such volume as to attract a great variety of interests."[104]

The press was able to expand size and content, in part, because of the growth of advertising revenue. Industrialization created more income and more products. The combination led to other new jobs—those of advertising pitch men. In turn, men like Volney Palmer, George Rowell, and Francis Ayer started ad agencies and represented companies. They filled the pages of newspapers and periodicals with advertising. The combination of expendable income, increased products, and venues for their promotion—media advertising—supported the increase in the number of publications through advertising dollars and through increased subscriptions. Even though depression rocked the nation in the mid-1890s, the desire for information allowed the number of media outlets to continue to grow, coupled with new technology that made it possible to print publications for considerably less money than it cost to do so in the 1870s as well as gather information. Many believed that the press began to exist "ultimately to make money for its owner ... and to draw business through its circulation for advertisers."[105] Though true, this did not mean that the media that created the new journalism of the Gilded Age had lost their bent for political issues.

Media had been the focal point of discussion for the public for more than a century by the end of the Civil War, but the growth of the press in the last half of the nineteenth century gave it new power. While Pulitzer may have run articles in the *World* that helped immigrants learn the proper etiquette of being American, it was the politics of the era that drove the desire for information. From 1872 through 1900, turnout for presidential elections never dropped below 70 percent. Corruption in a gilded age could be checked by powerful media reporting. Wrongdoings could be exposed when intrepid reporters were willing to go undercover to obtain stories at any cost to themselves or their publications.

The "small publics" that had developed throughout the nation were concerned with anything that affected their livelihoods, and so were editors. Specialized audiences for media had already begun to develop. They would mean that the press could continue to specialize content in the years ahead. More media outlets both large and small would mean even more ways to set direction in the United States. And, at the beginning of the twentieth century, magazines would become the home of the new journalism that fought corruption, even as newspapers continued to set agendas.

The explosion of magazines and their saturation into society meant that weekly and monthly periodicals were a staple for consumers by the beginning of the twentieth century. Their writers, adapting the styles of the stunt journalists like Nellie Bly and reporters for the black press like Ida B. Wells, were set to take on government and business in America that some felt had grown too large, powerful, and nonresponsive to the people. These reporters and their magazines would work to muck out corruption. Newspapers would continue to grow in number and influence, but the progressive bent of magazines, which sought to make life better for all—especially those who had no voice in society—captured the day and defined the next twenty years for media and the nation.

NOTES

1. *New York Journal*, October 13, 1897.
2. *New York Journal*, December 3, 1897.
3. *New York Journal*, October 5, 1897.
4. Quoted in Wm. David Sloan, Cheryl S. Wray, and C. Joanne Sloan, *Great Editorials*, 2nd ed. (North Port, Ala.: Vision Press, 1997), 157.

5. Quoted in Denis Brian, *Pulitzer: A Life* (New York: John Wiley & Sons, 2001), 390.
6. Thomas Jefferson to John Jay, 1786, in Paul Leicester Ford, ed. *The Writings of Thomas Jefferson*, 10 vols. (New York: G.P. Putnam's Sons, 1892–99), 4:186.
7. Ralph Negrine, *The Communication of Politics* (London: Sage, 1996).
8. Culver H. Smith, *The Press, Politics, and Patronage: The American Government's Use of Newspapers 1789–1875* (Athens: University of Georgia Press, 1977), 245.
9. Horace Greeley to Thurlow Weed, in Henry Luther Stoddard, *Horace Greeley: Printer, Editor, Crusader* (New York: G.P. Putnam's Sons, 1946), 321.
10. Ted Curtis Smythe, *The Gilded Age Press, 1865–1900* (Westport, Conn.: Greenwood Press, 2003), 1.
11. Oliver Wendell Holmes, "Bread and the Newspaper," *Atlantic Monthly*, September 1861: 346, reprinted in Oliver Wendell Holmes, *The Works of Oliver Wendell Holmes*, 13 vols., *Pages from an Old Volume of Life: A Collection of Essays 1857–1881* (Boston and New York: Houghton, Mifflin & Co., 1892), 8:1.
12. Frank Luther Mott, *American Journalism, A History: 1690–1960*, 3rd ed. (New York: Macmillan, 1962), 404, 421, 403, 411.
13. Sidney Kobre, *Development of American Journalism*, (Dubuque, Iowa: Wm. C. Brown, 1969), 350.
14. Alfred McClung Lee, *The Daily Newspaper in America: The Evolution of a Social Instrument* (New York: Macmillan, 1937), 748–49.
15. Frank Luther Mott, *A History of Magazines, 1885–1905* (Cambridge, Mass.: Harvard University Press, 1957), 10.
16. *Graphic* 6 (February 6, 1892): 107, quoted in Mott, *A History of American Magazines*, 12.
17. Mott, *A History of American Magazines*, 11.
18. *Historical Statistics of the United States, Colonial Times to 1970*, Part 1 (Washington, D.C., U.S. Bureau of Census, 1975), 382.
19. Ted Curtis Smythe, "The Diffusion of the Urban Daily, 1850–1900," *Journalism History* 28, 2 (2002): 73.
20. Wilson Dizard, Jr., *Old Media New Media: Mass Communications in the Information Age*, 3rd ed. (New York: Addison Wesley Longman, 2000), 157.
21. Whitelaw Reid, "Recent Changes in the Press," *American and English Studies*, 2 vols. (New York: Charles Scribner's Sons, 1913; reprint, London: Smith, Elder & Co., 1914), 2:297.
22. Mott, *American Journalism*, 481.
23. Kobre, *Development of American Journalism*, 383, 392.
24. Whitelaw Reid, "Journalistic Duties and Opportunities," *American and English Studies*, 2 vols. (New York: Charles Scribner's Sons, 1913; reprint, London: Smith, Elder & Co., 1914), 2:317.
25. Quoted in Smythe, *The Gilded Age Press*, 149.
26. Late in 1883 the General Time Convention sponsored by the railroads created a standardized time schedule for trains. Adopting standardized time, however, was voluntary. Not until 1918 did Congress make standard time in the United States official and mandatory. Robert E. Riegel, "Standard Time in the United States," *American Historical Review* 33, 1 (October 1927): 84–89.
27. John Dewey, *Democracy and Education* (New York: Henry Holt, 1916), 6.

28. Quoted in Thomas D. Clark, *The Southern Country Editor* (Indianapolis, Ind.: Bobbs-Merrill, 1948; reprint, Columbia: University of South Carolina Press, 1991), 33.
29. Quoted in Sally Foreman Griffith, *Home Town News: William Allen White and the* Emporia Gazette (New York and Oxford: Oxford University Press, 1989), 180.
30. William Allen White, "The Country Newspaper," *Harper's Monthly Magazine* 132 (May 1916): 887–91, quoted in Griffith, *Home Town News*, 159.
31. See Elizabeth Burt, *The Progressive Era* (Westport, Conn.: Greenwood Press, 2004); David Whitten, "Depression of 1893," *EH.Net Encyclopedia*, ed. Robert Whaples. (August 15, 2001) http://eh.net/encyclopedia/article/whitten.panic.1893; Douglas Steeples and David Whitten, *Democracy in Desperation: The Depression of 1893* (Westport, Conn.: Greenwood Press, 1998).
32. William Allen White, *The Autobiography of William Allen White* (New York: Macmillan, 1946), 280.
33. *Emporia (Kansas) Gazette*, August 15, 1896.
34. Mott, *American Journalism*, 456.
35. Grady delivered the speech to the New England Society at its annual dinner in New York City, December 12, 1886. It was reprinted in a number of newspapers. See, for example, "The New South," *St. Louis Globe-Democrat*, December 26, 1886.
36. "A Suggestive Speech," *The North American* (Philadelphia), December 27, 1886.
37. "A Great Speech," *Daily Arkansas Gazette* (Little Rock), August 10, 1877. The speech originally ran in the *Atlanta Daily Constitution*, July 14, 1877.
38. *Memphis Appeal,* in "The New South," *Daily Arkansas Gazette*, (Little Rock), May 11, 1877.
39. W. T. Sherman, "Letter from General Sherman," *St. Louis Globe-Democrat*, (St. Louis), February 11, 1879. The letter first appeared in the *Atlanta Daily Constitution*, February 1, 1879.
40. Grady's New York *Ledger* editorials were compiled and printed in 1890 after Grady's death on December 23, 1889. Henry W. Grady, *The New South*, (New York: Robert Bonner's Sons, 1890), 184.
41. "The New South," *Milwaukee Daily Sentinel*, May 23, 1881.
42. James West Davidson, et al., *Nation of Nations* (New York: McGraw-Hill, 1990), Appendixes A-33, A-37.
43. Information on Joseph Pulitzer comes from Denis Bryan, *Pulitzer: A Life* (New York: John Wiley, 2001) and W. A. Swanberg, *Pulitzer* (New York: Charles Scribner's Sons, 1967).
44. "The Dispatch Deal," *St. Louis Globe-Democrat*, December 11, 1878.
45. *New York World*, October 2, 1886.
46. *St. Louis Post-Dispatch,* April 11, 1907. From Pulitzer's retirement speech of April 10.
47. Quoted in Bryan, *Pulitzer: A Life*, 59.
48. *New York World*, May 11, 1883.
49. Mott, *American Journalism*, 435.
50. Joseph Pulitzer to Frank Cobb, August 6, 1910, in Swanberg, *Pulitzer*, 386.
51. *St. Louis Post-Dispatch*, May 18, 1882. (Swanberg, 51)
52. *New York Herald*, January 13, 1901.
53. "Largess," *New York Times*, January 3, 1871.
54. "More Ring Villainy," *New York Times*, July 8, 1871.

55. *New-York Times*, March 30, 1872. Tammany refers to an organization originally formed following the Revolutionary War for charitable purposes. By 1820, the organization, which operated out of Tammany Hall, had turned into a political operation that sought to control various activities in New York City. Tweed became the head of the organization in 1868 and led it to new levels of political corruption.
56. Quoted in John Durham Peters and Peter Simonson, *Mass Communication and American Social Thought* (Lanham, Md.: Rowman & Littlefield, 2004), 234.
57. Donna L. Dickerson, *The Reconstruction Era* (Westport, Conn.: Greenwood Press, 2003), 320. See, also, Alexander B. Callow, Jr., *The Tweed Ring* (New York: Oxford University Press, 1966) and Seymour J. Mandelbaum, *Boss Tweed's New York* (New York: John Wiley, 1965).
58. W. L. Stone, "Old Time Journalism," *New York Times*, February 7, 1871.
59. See, for example, "Very Democratic," *Harper's Weekly*, June 28, 1884, and "Grave Regrets," *Harper's Weekly*, September 27, 1884.
60. Poems quoted in Smythe, *The Gilded Age Press*, 111.
61. Daniel 5:25–28. "And this is the writing that was inscribed: MENE, MENE, TEKEL, UPHARSIN. This is the interpretation of the thing: MENE; God hath numbered thy kingdom, and brought it to an end; TEKEL; thou art weighed in the balances, and art found wanting."
62. Smythe, *The Gilded Age*, 112–13; Swanberg, *Pulitzer*, 92–93. See, also, Mark Wahlgren Summers, *Rum, Romanism, & Rebellion: Making of a President* (Chapel Hill: University of North Carolina Press, 2000).
63. Nellie, Bly, *Ten Days In a Mad-House* (New York: Ian L. Munro, 1887), Chapter 1.
64. Bly, *Ten Days In a Mad-House*, Chapter 16.
65. Bly, *Ten Days In a Mad-House*, introduction.
66. Nellie Bly, *Nellie Bly's Book. Around the World in Seventy-Two Days* (New York: The Pictorial Weeklies Co., 1890). For a thorough understanding of Nellie Bly, see Brooke Kroeger, *Nellie Bly: Daredevil, Reporter, Feminist* (New York: Crown, 1994).
67. Miriam Decosta-Willis, ed., *The Memphis Diary of Ida B. Wells: An Intimate Portrait of the Activist as a Young Woman* (Boston: Beacon Press, 1995), 2.
68. *Free Speech* (Memphis), quoted in Ida B. Wells, *Crusade for Justice: The Autobiography of Ida B. Wells*, ed. Alfreda M. Duster (Chicago and London: University of Chicago Press, 1970), 52.
69. *New York Freeman*, May 28, 1887 and July 9, 1887, quoted in Linda O. McMurry, *To Keep the Waters Troubled: The Life of Ida B. Wells* (Oxford: Oxford University Press, 1998), 121.
70. Ida B. Wells, *Southern Horrors. Lynch Law in All Its Phases* (New York: New York Age Print, 1892), preface.
71. Patricia A. Schechter, *Ida B. Wells-Barnett and American Reform, 1880–1930* (Chapel Hill: University of North Carolina Press, 2001), 85.
72. Wells, *Crusade for Justice*, xix.
73. "Ida B. Wells's Crusade Against Lynching," *Public Opinion* (August 9, 1894): 440, quoted in Schechter, *Ida B. Wells-Barnett and American Reform*, 109.
74. "Endorse Ida Wells' Work," *Milwaukee Sentinel*, September 15, 1894.
75. C. H. Williams, *The Race Problem*, quoted in Schechter, *Ida B. Wells-Barnett and American Reform*, 110.
76. Ida B. Wells, "The Brutal Truth," *The Daily Inter Ocean*, (Chicago), July 19, 1893.
77. "She is against Lynching," *Rocky Mountain News* (Denver), July 30, 1894.

78. Ida Wells-Barnett, *The Red Record: Tabulated Statistics and Alleged Causes of Lynching in the United States* (1895). Wells married Chicago attorney and newspaper editor Ferdinand Barnett in 1895. She hyphenated her name because of the recognition of Ida Wells in the United States.
79. *Chicago Defender*, January 1, 1910, quoted in Paula J. Giddings, *Ida: A Sword Among Lions* (New York: HarperCollins, 2008), 487.
80. Giddings, *Ida: A Sword Among Lions*, 483.
81. James Harmon Chadbourn, *Lynching and the Law* (Chapel Hill: University of North Carolina Press, 1933).
82. Robert Siegel, "Anti-Lynching Law in U.S. History, *All Things Considered*, NPR (June 13, 2005), http://www.npr.org/templates/story/story.php?storyId=4701576.
83. Frederick L. Hoffman, "Fewer Lynchings," *New York Times*, March 4, 1913.
84. W. Joseph Campbell, *Yellow Journalism: Puncturing the Myths, Defining the Legacies* (Westport, Conn.: Praeger, 2001), 4–5.
85. Campbell, *Yellow Journalism*, 7–8.
86. "Yellow Journalism," *Indiana State Journal* (Indianapolis), March 16, 1898.
87. "So Called Yellow Journalism Vindicated," *News and Observer* (Raleigh, N.C.), March 27, 1898.
88. "Yellow Journalism and War," *Macon (Ga.) Telegraph*, February 28, 1897.
89. "The Year's Record," *Fourth Estate*, January 12, 1899, 2, in Burt, *The Progressive Era*, 107.
90. Ben Proctor, *William Randolph Hearst: The Early Years, 1863–1910* (New York: Oxford University Press, 1998), 102.
91. *New York World*, May 17, 1896.
92. *New York Journal*, February 2, 1897.
93. For a full account of Cisneros, see W. Joseph Campbell, *The Year That Defined American Journalism: 1897 and the Clash of Paradigms* (New York: Routledge, 2006), 161–94.
94. *New York Journal*, October 10, 1897.
95. "Reception of Miss Cisneros," *Philadelphia Press*, October 17, 1897, quoted in Campbell, *The Year That Defined American Journalism*, 163.
96. David F. Trask, *The War with Spain in 1898* (New York: The Free Press, 1981; reprint, Lincoln: University of Nebraska Press, 1996), 9.
97. "The Worst Insult to the United States in Its History," *New York Journal*, February 9, 1898.
98. *News and Observer* (Raleigh, N.C.), March 27, 1898.
99. "'REMEMBER THE MAINE!' 'VIVA CUBA LIBRE!'" *Denver Evening Post*, April 20, 1898.
100. *Arkansas Gazette* (Little Rock), February 28, 1898, 2.
101. *Morning Oregonian* (Portland), March 3, 1898.
102. *New York Journal*, May 8, 9, 10, 1898.
103. "Government and Yellow Journalism," *Morning Oregonian* (Portland), March 3, 1898.
104. Chester S. Lloyd, *The Young Man and Journalism* (New York: Macmillan, 1922), 91, quoted in Smythe, *The Gilded Age Press*, 177.
105. "Newspaper Evolution," *Newspaperdom*, 7 November 24, 1898, quoted in Smythe, *The Gilded Age Press*, 213.

6. *There Is Filth on the Floor and It Must Be Scraped Up: The Muckrakers and Press of the Early 20th Century*

By 1906, the exposés on corruption, manipulation, and malfeasance targeting the nation's businesses and governmental entities seemed to spring daily from the pages of powerful and ever-growing mass media. The nation's most visible social reformer, President Theodore Roosevelt, championed the work of the growing number of investigative journalists as they informed the nation of greed, arrogance, and the mistreatment of Americans. But in 1906, *Cosmopolitan*, one of the best-read crusading magazines of the era, ran David Graham Phillips' stinging indictment of corruption on Capitol Hill, "The Treason of the Senate." Phillips stated that "bribery and party prejudice is potent everywhere."[1] He then supported his charge by naming names and giving examples. Embarrassingly for Roosevelt, many of those "exposed" in Phillips' nine-part series were tied strongly to the president.

Even though Roosevelt had lauded the press earlier for its work to end corruption in the nation, he was furious with *Cosmopolitan* and responded. Referencing John Bunyan's seventeenth-century novel *Pilgrim's Progress*, the president compared the crusading journalists to "the Man with the Muckrake" who removed filth. "There is filth on the floor, and it must be scraped up with the muckrake," Roosevelt said. "But the man who never does anything else, who never thinks or speaks or writes save of his feats with the muckrake, speedily becomes, not a help to society, not an incitement to good, but one of the most potent forces of evil." Roosevelt could not in good conscience completely discredit what journalists had been doing to make America a better place in which to live, though. "I hail as a benefactor every writer or speaker, every man who, on the platform or in book, magazine or newspaper, with merciless severity makes such attack provided always that he in his turn remembers that the attack is of use only if it is absolutely truthful."[2]

Roosevelt had used the term "muckraker" derisively, but the crusading journalists accepted the moniker as a badge of honor. They quickly began to refer to themselves as muckrakers, thinking of it only in a positive way. Reform became a chief component of the first two decades of the twentieth century, and those who sought to create a more egalitarian nation considered themselves progressive, a name soon applied to the period. The Progressive movement grew out of the Populism of the 1890s that believed that the rights of the common people were being trampled. Politicians such as Robert La Follette and William Jennings Bryan led the fight for change. Bryan delivered fiery rhetoric in support of "the laboring interests and all the toiling masses" and promised at the 1896 Democratic National Convention that the powerbrokers of the nation "shall not press down upon the brow of labor this crown of thorns. You shall not crucify mankind upon a cross of gold."[3] Progressivism expanded this theme by advocating political reform, legislative protection for workers, and a distrust of the ever-growing monopolies in American business.

Theodore Roosevelt assumed control of the political element of the Progressive Movement when a bullet from assassin Leon Czolgosz ended the presidency of William McKinley almost as quickly as it had begun in 1901. Roosevelt asked for legislation to fight any number of ills that was called the Square Deal. But the real champions of progressivism were the journalists—the muckrakers—who were willing to take up the cause of any group that felt it was being maltreated by the powerbrokers of society, and that is how many Americans viewed them. "No progress was ever made without the difficult and thankless work of the muck-rakers," a writer to the *New York Times* noted. "Without them the world would not be a fit place to live in.... The pens of these men who are now so prominently before the public have won for the people more than all the swords that were ever made. Those pens hold the balance of power in these United States, and our hope of the future is in them."[4] Even Roosevelt, who derided the muckraking journalists in 1906, admitted that they were essential to the nation. "In our country I am inclined to think that almost, if not quite, the most important profession is that of the newspaper man, including the man of the magazines," Roosevelt said in 1910.[5]

The journalists of the first decades of the twentieth century were able to shape and influence the nation so powerfully because of the scope and reach of media. Newspapers, which had been the backbone of the nation's information system, continued to grow in number and circulation. By the beginning of World War I, more than fifteen thousand newspapers were printed in the United States. Most of them were weeklies and semi-weeklies, but 2,250 of them were dailies, according to the *American Newspaper Annual Directory*, marking an increase of nearly 20 percent in the number of daily newspapers printed in 1892.[6] American dailies passed the 22.4 million circulation figure in 1910,[7] the year after daily newspapers reached their zenith of 2,600 in America.[8] The sheer number of newspapers and the number of total issues of each ensured that the media at the beginning of the twentieth century could reach nearly any target audience it desired. The number of dailies with circulations in excess of 100,000 continued to grow in number, joining several in the nation's largest cities that reached gargantuan size—in excess of 1 million sales per day—at the end of the nineteenth century. Publications for the nation's pluralistic society expanded, too, making it much easier to target groups with more focused agendas. Publications aimed at the nation's farmers, for example, topped 450 with a circulation of 15 million by 1910,[9] while labor publications promoted workers' issues and produced 2 million copies of 323 publications annually by 1913, which helps explain the 1 million votes that socialist Eugene V. Debs received in his run for the presidency in 1912.[10]

The explosion in the number of magazines, which began in the last quarter of the nineteenth century, continued, and magazine journalism truly came into its own at the beginning of the twentieth century. At the turn of the century, more than 5,500 periodicals published in the United States, an increase of more than 66 percent in the number of magazines that had been published just fifteen years before. The number increased by another five hundred by 1905, with estimates of at least another fifteen hundred titles having begun but failed from 1885 to 1905.[11]

In the 1890s, a group of magazines initiated a new business plan. They dropped their price per issue to ten or fifteen cents by printing on cheaper wood pulp paper. With magazines previously selling for two to three times that much, *McClure's, Cosmopolitan, Munsey's, Collier's, Saturday Evening Post*, and others gradually increased their circulation into the hundreds of thousands. *Saturday Evening Post* reached the 1 million circulation figure in 1904, and another fourteen magazines were selling a million copies of each issue by 1920.[12] The weekly format—along with an extended writing style—allowed the magazines to tell stories in ways that was sometimes impossible in daily newspapers, and they provided the only means of collectively reaching Americans from coast to coast with those stories or with advertising for the exploding number of products now available.[13] These magazines slowly began to expose corruption, sometimes following the lead of social reformers and sometimes with their own undercover reporters. The muckrakers rarely offered a fix for the problems they uncovered; they simply exposed it for society. They left it to citizens to demand change and to politicians to make it happen. Even though many magazines were less expensive at the beginning of the twentieth century than they had been in the early 1890s, most were still for the most part a class-driven medium appealing to the more affluent.[14] The muckraker magazines, though, expanded the circulation base to a much broader demographic both economically and culturally, and their stories were often reprinted in the weekly newspapers that flooded much of the United States. As a result, their reach and their effect on society was much greater than circulation numbers reveal much like the newspapers of the eighteenth and early to mid-nineteenth centuries.

Consequently, magazines and newspapers gave media the ability to cover events on all levels of society in depth. Because there were so many publications and reporters vying to break news or to uncover scandals, malfeasance, or any type of corruption, the media truly set the agenda for the nation. As one writer observed, the press "determines the questions of popular interest, sets the trend of public speech, public thought and general customs. The community is dependent on it for information and guidance concerning commercial, social and political matters."[15]

AMERICA AT THE BEGINNING OF A NEW CENTURY

By the beginning of the twentieth century, more than 76 million people lived in the United States. That number grew by 16 million by 1910. In 1920, the nation's population reached 106 million and then 123 million by 1930.[16] The demographics of the age help explain why the work of the progressives was essential. In 1900, a mother of four children knew she had a fifty-fifty chance of having one of them die before the age of 5. On average at least one parent died before their children turned 21. White Americans could expect to live an average of forty-eight years, African Americans just thirty-three years. In 1900, the average family's annual income was around $450–$3,000 if tabulated in twenty-first-century values. The majority of families had no indoor plumbing, no phone, and no car. About half of all American children lived in poverty. Most teens did not attend school; instead, they labored in factories or fields.[17] But this last fact was changing. An estimated 15.5 million young people were now attending high school, compared with less than half that number during the Reconstruction era. As a result, approximately 11.3 percent of all Americans were illiterate at the beginning of the twentieth century, according to U.S. Census data, a number that represented a drastic drop from 1880, when about 17 percent of Americans were considered illiterate.[18]

Wealth was concentrated. One-tenth of the nation's population controlled nine-tenths of its wealth, and a nation that knew only three millionaires before the Civil War had at least 3,800 by 1900[19] with approximately 40 percent of all Americans living in poverty. Industrialization led a migration from rural America to the cities. By 1920, more Americans lived in towns or cities than in the country. Another 14.5 million people immigrated to the United States from 1900 to 1915, and almost all of them remained in the nation's urban centers. Immigrants and the traditional American worker averaged about fifty-two hours each week on the job.[20] Job-related injuries and illnesses were common, and there was little to no help for workers if they could no longer work as a result of job-caused reasons. Workers tried to unionize, but they had little success against powerful companies that were often aligned with government officials. As a result, a tremendous gap existed between the haves and have-nots in the United States. The work of progressives was an attempt to close this gap in society and in a belief that democracy, if it worked properly, could and would cure the ills that disparity had created in the nation. Media gave the movement its way of reaching the people and revealing what needed to be changed.

RAKING THE MUCK

When the 10-cent magazines turned to exposé journalism at the beginning of the twentieth century, they were not doing something that was new. Joseph Pulitzer had uncovered and reported on corruption in St. Louis early in the 1880s. Stunt journalists like Nellie Bly had been going undercover for years to expose wrongdoing to force changes in policy and law as had intrepid writers like Ida Wells. All types of publications in the 1890s produced exposés, too, like Jacob Riis's photo-filled book *How the Other Half Lives*, which chronicled life in the slums of New York. Hearst's mantra that "The journalism that does things has come to stay"[21] might not have been accepted by all journalists, but it seemed to many that the times demanded such work by media. It was in this journalistic environment that S. S. McClure directed his triumvirate of college-educated reporters, Ray Stannard Baker, Lincoln Steffens, and Ida Tarbell, to tackle what McClure called "some of the problems that were beginning to interest the people,"[22] things his reporters believed to be major ills in society. *McClure's* magazine was soon subscribed to by a half-million Americans, and the nation's total magazine circulation reached 5.5 million in 1905, thanks in large part to the crusading of the muckrakers.[23]

McClure's journalists exposed any number of wrongdoings. Josiah Flynt, in 1901, wrote about the criminal underworld that seemed to run rampant in some of the nation's cities.[24] Francis Nichols revealed the abuse of children by the coal mines in 1903.[25] But it was the work of Ida Tarbell, Lincoln Steffens, and Ray Stannard Baker for the magazine that helped make muckraking standard practice among so many journalists of the era and helped rewrite American legislation. Tarbell's eighteen-part series, published between November 1902 and October 1904, "The History of Standard Oil Company," was the epitome of investigative journalism. She spent four years gathering her information, and in the final installment concluded, "The truth is, blackmail and every other business vice is the natural result of the peculiar business practices of the Standard."[26] Tarbell followed the Standard Oil series with a two-part story on the driving force behind Standard Oil, John D. Rockefeller.

Focusing on Rockefeller's physical features, she called him "the oldest man in the world—a living mummy," and accused him of being "money-mad" and "a hypocrite." And on his effects on the United States, she believed that "Our national life is on every side distinctly poorer, uglier, meaner, for the kind of influence he exercises."[27]

Steffens' exposé on municipal corruption, which would appear in a collective volume titled *The Shame of the Cities* in 1904, began with "The Shame of Minneapolis." McClure, in an editorial comment, said that Steffens' report could just as easily have been called "the American Contempt of Law." McClure said that in "'The Shame of Minneapolis,' we see the administration of a city employing criminals to commit crimes for the profit of elected officials, while the citizens ... stood by complacent and unalarmed."[28] Baker tackled the nation's railroads with a series titled "The Railroads on Trial," which appeared in five installments covering about nine months. His work, especially the last article titled "How Railroads Make Public Opinion," suggested that the nation's large, privately owned public-service sector was employing promoters, high-pressure tactics, and the press to make what many considered corrupt business practices palatable to the public via the media. Just as Thomas Jefferson said that the way to unseat the Federalists was via the engine of the press, Baker said that the railroads have turned to newspapers as "the fountainhead of public information." By feeding the press with stories, Baker said of the railroads, they were able to convince the public that railroad legislation—which had been designed to protect people—was really not in the best interests of the public. Such a misuse of the media was immoral and coercive, Baker explained.[29]

THE PARTING OF A FOOL AND HIS MONEY

The muckrakers also attacked one of the foundations of media—its advertisers. At the beginning of the twentieth century, America's 76 million people were spending as much as $100 million a year on medicine by some estimates.[30] Patent medicines, as they were called, had been advertised and sold via media publications since the eighteenth century. Made of anything that could be put in a bottle—but almost always with a high-percentage base of alcohol—patent medicine manufacturers claimed their products could cure almost anything, and Americans were always in search of a cure for their ailments. Since the patent medicines rarely accomplished what they claimed and sometimes actually did harm, they became a natural target for muckrakers. They wanted the public protected from the potential harms of patent medicines, and they wanted their makers to be held accountable in their advertising for the product—they wanted truth in advertising. Some Americans had been trying to establish federal standards for both medicine and food for decades. As early as 1879, the first of 190 pieces of legislation was introduced in Congress to protect the consumer of food and drugs, but none ever received approval by both the Senate and the House of Representatives.[31] When the exposés of the patent medicine industry appeared in *Ladies' Home Journal* and *Collier's* beginning in 1904, however, Congress was forced into action and approved the Pure Food and Drug Act in 1906.

In 1904, Edward Bok shocked the nation when he announced, "A mother who would hold up her hands in holy horror at the thought of her child drinking a glass of beer, which contains from two to five per cent of alcohol, gives to that child with her own hands a patent medicine that contains from seventeen to forty-four percent of alcohol—to say nothing of opium and cocaine!" Bok then went on to take on one of the nation's most trusted elixirs, Lydia Pinkham's Vegetable Compound. Promoted as an all-natural remedy for any number of female problems, "The 'Patent-Medicine' Curse" told its readers that the chief "natural ingredient" in the compound was corn liquor. Bok finished his article by playing to proponents of the temperance movement, saying that taking patent medicines was "planting the seed of a future drunkard!"[32]

In October 1905, Samuel Hopkins Adams published the first of ten installments in "The Great American Fraud" series in *Collier's*. He promised "a full explanation and exposure of patent-medicine methods, and the harm done to the public by this industry, founded mainly on fraud and poison." Adams also attacked the most common method of advertising patent medicines, the fake testimonial, which he said ensured "the parting of a fool and his money."[33] In another article, *Collier's* asked other publications to follow its example and no longer accept advertisements for patent medicines, saying that after its series had exposed the misrepresentation of both contents and efficacy of patent medicines that any publisher who agreed to run patent medicine ads "knows what it is doing," helping to defraud and poison the public.[34] The articles on patent medicines focused the public's attention on their dangers, and Adams, especially, drove home the value of regulation for the patent medicine industry, telling the public that it had to lobby Congress for federal protection. As a result, thousands of consumer groups—especially women—petitioned Congress for protection. The challenge to mothers to protect their children obviously was a successful tactic.[35]

FOR PURITY IN FOOD AND SAFETY FOR WORKERS

Closely tied to the move for accountability in medicine and its advertising was the muckrakers' attacks on the nation's food stuffs, and nothing shocked the nation more than Upton Sinclair's revelations about the meatpacking industry. Sinclair's work, combined with *Collier's* revelations about how the patent medicine industry used economic power to coerce editors to run advertisements that were known to be misleading and inaccurate, pushed people to action and Congress to reaction.³⁶ Sinclair accepted the assignment to uncover the workings of the meatpacking industry from the publishers of *The Appeal to Reason*, which had strong socialist and worker leanings. *The Appeal* began running Sinclair's novel-exposé in serial form in February 1905.

The description of what took place inside a meatpacking plant revolted Americans. Sinclair talked of unsanitary food production, but his descriptions of working conditions and what happened to workers truly sickened readers. He described the hands of butchers as so often cut that they were "a mere lump of flesh," and noted that among the wool-pluckers that "acid had eaten their fingers off." In what Sinclair called the tank-rooms, men had to work around large, open vats. Sometimes, he said, workers fell in, "and when they were fished out, there was never enough of them left to be worth exhibiting,— sometimes they would be overlooked for days, till all but the bones of them had gone out to the world as Durham's Pure Leaf Lard!"³⁷ *The Jungle* was the tipping point in terms of forcing Congress' hand toward legislation because of the public outcry in the aftermath of its revelations, which the *Washington Post* described as "public interest to the highest pitch."³⁸ Where neither branch of the legislature had been able to agree on laws to protect people for thirty-five years, the Senate voted 63 to 4 for the Pure Food and Drug Act, which became law in June 1906.³⁹ Similarly, a meat inspection bill sailed through the Senate, the *New York Times* said, "the direct consequence of the disclosures made in Upton Sinclair's novel, 'The Jungle.'"⁴⁰

Not everyone found Sinclair's exposé—or that of the other muckrakers—of positive benefit for the nation. Horace C. Duval, a former Senate staffer, was especially displeased with Sinclair. "One year ago to-day we were the most prosperous nation in the world," he said after returning from Europe. "Upton Sinclair's attack upon the beef trust cut down exports to Europe and Africa from $6,500,000 to $1,000,000."⁴¹ Despite the fact that some found the muckrakers' methods objectionable, they were able to force the passage of a number of regulations that changed the United States. The *laissez-faire* approach that business and industry had assumed and their "let the buyer beware" mentality did not square with the ideas of Progressives or the muckrakers.

In addition to the Food and Drug Act, muckraker exposés helped spur multiple pieces of legislation. The Elkins Act of 1903 and the Hepburn Act of 1906 increased legislation on the railroads and provided the Interstate Commerce Commission with more regulatory oversight. The Mann Act, in 1909, also dealt with interstate commerce, but its chief purpose was to stop the transportation of women across state lines for prostitution. Barton J. Hendrick's *Collier's* article, "Daughters of the Poor," brought the issue to light. William Hard wrote "Making Steel and Killing Men" for *Everybody's*

Magazine, reporting on the fact that forty-six workers were killed on the job with 368 disabled permanently at the United States Steel South Chicago plant in 1906 alone.[42] Hard's exposé on what happened to workers helped pave the way for the Workmen's Compensation Act, which provided job-related injury and death benefits. Reporter Edwin Markham produced a series in 1906 and 1907 for *Cosmopolitan* titled "The Hoe-Man in the Making," which included articles like "Children of the Looms," "Child-Wrecking in the Glass Factories," and "Little Slaves of the Coal Mines." The series was a key component in the controversy that produced the Child Labor Law of 1916. Ironically, the law was later ruled unconstitutional, based in part on *The Child That Toileth Not* by Thomas Robinson Dawley, who employed the same muckraking investigative techniques—and extensive photographs—to argue that child labor was beneficial, not detrimental for young people.[43]

The Progressives, however, did not let setbacks deter them. Late in 1904, the muckraking magazines took on the insurance industry. *Collier's*, *The Era*, *Everybody's Magazine*, and others wrote on mismanagement in the industry. Their stories revealed that insurance companies were paying off politicians, that insurance executives were skimming funds for personal profit, and that companies were defrauding customers—all leading to insurance reform.[44] David Graham Phillips' "Treason of the Senate" helped change the Constitution, serving as the catalyst for the seventeenth amendment, which upon ratification in 1913 established that senators would be elected directly just like representatives. The amendment rewrote Article 1, Section 3 of the Constitution, which had given selection of Senators to state legislatures. And Ida Tarbell's "History of Standard Oil" led to a 1911 Supreme Court ruling that dissolved the giant Standard Oil into smaller companies, a ruling that the *Los Angeles Times* said brought to an end "a menace to the industrial and economic advancement of the entire country."[45]

The principal activities of the muckrakers took place in the first decade of the twentieth century. Though muckraking would continue, it would not be so widespread among journalists. Perhaps reporters tired of the exposés; perhaps the American people did. World war surely redirected the nation's attention, but the muckrakers truly changed the national mindset, and in the process, many of the nation's laws.

LOCKED DOORS AND DEATH

American magazines may have been the best way to effect change nationally in the United States in the Progressive era, but newspapers continued to be advocates for change, too, especially when events occurred that required immediate response. When a garment factory in Washington Place in New York City erupted in fire in March 25, 1911, newspapers, led by the nation's chief proponents of the journalism of action, the *New York Journal* and the *New York World*, joined by the *New York Times* and others, quickly exposed a massive list of wrongdoings and spurred New York state to institute safety codes and reduce the hours in the work week for laborers. The Triangle Shirtwaist fire was one of America's worst industrial accidents. The events surrounding it encapsulated much of what muckrakers and other Progressives believed to be wrong with the nation in terms of the lack of rights of workers who were powerless against rich business owners. In the case of the Triangle Shirtwaist fire, almost all of the 146 people who either jumped to their deaths or who died in the fire were female, adding to the argument of a helpless workforce. The tragedy followed the nation's first large-scale strikes by female workers from 1909 through 1910, called the "women's movement strike" and "the uprising of the thirty thousand," among others, which targeted primarily the New York garment district.[46]

When workers first proposed to strike in 1909, American Federation of Labor president Samuel Gompers told them that "there comes a time when to refuse to strike is to be as slaves."[47] The strikers adopted the slogan "Starve to win, or you'll starve anyway" in an effort to keep workers on picket lines. In the process, other women—some from the more affluent part of New York, joined the shirtwaist workers in the strike.[48] As the protests continued, numbers of strikers were arrested, and the boycott threatened to spread to garment manufacturing centers in Pennsylvania and Connecticut.[49] As the 1909 strike continued, one writer said, "In the history of the world no such scenes have been witnessed as those which for nearly two months past have characterized the strike of the shirt waist makers in New York." The article explained that the nearly 35,000 girls and women, who "were engaged at first in this greatest strike of women workers ever known," were "compelled to work several hours' overtime four nights a week, with no time off for supper....Most of the labor is paid for by the piece. The girls are not asking higher pay for piecework, but merely a readjustment as to working hours so that they will not be worked beyond their endurance." The story, complete with photographs of picketers, explained that the shirtwaist workers only made between $3.50 and $4 per week. "Thousands of them support not only themselves, but sick or disabled parents and several little brothers and sisters," it told readers.[50] In February 1910, the strike ended after garment-factory owners agreed to several minor concessions, which included a promise of fire escapes to the operations in the upper stories of buildings. In reality, though, almost all of the women who went out on strike gained nothing from their nearly four months on the picket lines, especially fire escapes or safety features for their workplaces.[51]

The makers of the popular-styled shirt had nearly completed their day of work on Saturday, March 11, 1911, at the Triangle Shirtwaist Company when one of the five hundred workers smelled smoke. Within seconds, fire erupted in the lint and cloth that filled the building, and the panicked employees rushed the two doors in an effort to escape. One of them was locked, and both opened into the workspace. As women rushed the doors, it became impossible to open them against the crush of workers rushing to the exits. Those who had managed to escape from the rooms discovered that the stairwell was impassible and that the elevators were either blocked or the passageway to them too filled with smoke to get through. Because the owners of the Triangle Shirtwaist Company did not comply with any safety regulations, there were no sprinklers in the building. In addition, the fire escape that was on the building led to a courtyard that was enclosed by the building. Plus, the fire escape had been constructed of inferior metal and soon was a twisted, heated mass outside floors eight through ten, where the trapped workers were. When the fire department arrived, firemen soon discovered that their ladders would only reach to the sixth floor, making it impossible to help any who might have reached the safety of the windows. The Triangle Shirtwaist Factory workers faced two possibilities: burn to death or jump.[52]

The shocking stories of the Shirtwaist fire filled New York's papers on March 26, and the graphic descriptions left nothing to the imagination. "They jumped," the *New York Times* said. "They crashed through broken glass, they crushed themselves to death on the sidewalk. Of those who stayed behind it is better to say nothing."[53] "Thud–dead; thud–dead; thud–dead; thud–dead. Sixty-two thud–deads. I call them that, because the sound and the thought of death came to me each time, at the same instant," United Press reporter William Shephard said in even more visual terms. "I even watched one girl falling. Waving her arms, trying to keep her body upright until the very instant she struck the sidewalk, she was trying to balance herself. Then came the thud—then a silent, unmoving pile of clothing and twisted, broken limbs."[54]

Immediately, newspapers began to call for accountability and for changes. "Is any one to be punished for this?" the *New York Journal* asked on Page 1 of its March 27 edition. The *New York World* revealed that "Reporters for the World" discovered the doors in the factory were locked, that the stairway was not usable, and that only one elevator in the building even worked.[55] "If all those girls were locked in, as they were, and deprived, as they were, of all chances to save their lives, the owners of the building, the employer of the girls, and the city that did not give them protection, should be made responsible," the *Journal* announced.[56] The *Times* revealed that city inspectors had known such a disaster was possible in the city. They knew that fire escapes were inadequate if they existed at all. They knew that nearly 98 percent of all doors in the garment shops opened in instead of out. They knew that many of those doors remained locked during the day. "In a word, the investigation has shown that even with the low standards for fire protection as demanded at the present by the labor law there are hundreds and thousands of violations in one industry alone." Yet, they did nothing to correct the problems.[57] The *Journal* called for the city to "DRAFT NEW LAW TO SAVE SHOP WORKERS" in a banner above the nameplate on March 28. With the press raising questions about every aspect of the Triangle Shirtwaist fire, the city was forced to act, and the coroner's jury in April leveled the first official blame in the fire against the business owners, Max Blanck and Isaac Harris, and against the shabby inspection system in the city.[58] The pair were found not guilty of murder at their trial, though, but a subsequent civil suit required them to pay $75 to the families of those who died. Only twenty-three cases were actually filed, though.[59]

The tragedy at the Shirtwaist Factory added to the explosive debate that had been taking place about workers' rights and safety, adding support to the *Everybody's Magazine* series "Making Steel and Killing Men" by William Hard and *Cosmopolitan*'s series "The Hoe-Man in the Making" by Edwin Markham to name just two of the muckraker exposés on worker conditions. Immediately after the Triangle Shirtwaist Factory fire, New York created the Factory Investigating Commission, which led to factory safety legislation and new set of safety codes. By 1913, twenty-five different bills that provided greater protection for workers had become law in New York state. The state now required all high rises to be equipped with sprinklers and fire escapes. Large operations were required to hold fire drills for workers. Doors could no longer be locked, and they had to open outward, not inward,[60] one of the main reasons that the Triangle workers could not escape the fire. Garment workers pushed the Joint Board of Sanitary Control to monitor more closely the health and safety conditions in the garment district to ensure that the new laws were put into practice. They also monitored whether factories were adhering to the newly mandated shortened work week.[61]

Newspapers in other cities followed the Triangle Shirtwaist tragedy closely, too, and used it as a platform to question factory safety in their cities. The *Chicago Tribune*, for example, asked, "Is any employer gambling with death to save a little money? Is the public inspectorship strict, or is it winking at law evasion, or is it inadequate and overworked, or is it incompetent? ... How is it with Chicago?"[62] The tragedy of the fire, the public and press reaction afterward, and the exploits of muckrakers for workers ultimately created better and safer work conditions.

SUFFRAGE

Universal voting rights for women followed a long and controversial path in the United States. Women had voted in the United States shortly after the Declaration of Independence was signed because New Jersey's 1776 constitution gave all the state's residents the right to vote if they met property and residence requirements.[63] Votes for both sexes remained law there until the words "free white male citizen" were introduced into the governing document in 1807.[64] Women in Utah voted from 1870 to 1887, when it was a U.S. territory and before Congress stripped Utah females of the right as a precondition for statehood and as part of the process for having the Latter-day Saints reject polygamy. The *Salt Lake Herald* pointed this out in 1895 when the state was set to vote on a new constitution. To anyone who believed that women voting would lead to social turmoil, the paper asked, "Where is the degradation of woman as its effect? Where are the discordant and wifeless and motherless homes as a result?"[65] The constitution was adopted along with women's suffrage. Utah joined the Union the next year.

The battle for universal suffrage for women, however, reached its climax in the early twentieth century, capped by the nineteenth amendment, which became law in 1920. The issue, however, had been debated, supported, and scoffed at in the press since the first women's rights convention was held in Seneca Falls, New York, in 1848. There, led by Elizabeth Cady Stanton, Lucretia Mott, and others, the convention declared, "We hold these truths to be self-evident: that all men and women are created equal."[66] With the Seneca Falls Convention, women began a battle that would require more than seventy years of protests, parades, lobbying, and thousands of press stories to gain the right to vote. The battle in the press was often nasty. Horace Greeley, after the second women's convention was held in Worcester, Massachusetts, said he doubted the vote would greatly improve the lot of women in America, but that he believed the nation had "to give the Democratic theory a full and fair trial," which meant women should be included in the process.[67] His rival at the *Herald*, James Gordon Bennett, however, called the suffragists "fanatical mongrels" and their efforts "balderdash, clap-trap, moonshine, rant, cant, fanaticism, and blasphemy."[68] Most of America's males subscribed to Bennett's views for decades.

In 1872, Susan B. Anthony and seven other women voted in the presidential election in Rochester, New York. Anthony was arrested a week later for violating the 1870 Enforcement Act, which made it a felony to vote without having a lawful right to do so. She was fined $100, though she could have been imprisoned for three years.[69] One editor, E. L. Godkin of *The Nation*, chastised Anthony for what she did but also provided the women's movement its greatest bit of advice. If women want the right to vote, he said, they need to "devote themselves to working on public opinion."[70] And, that is exactly what women and supporters of suffrage did. They formed advocacy groups such as the American Equal Rights Association and the National American Woman Suffrage Association. They took any cause that involved women and made it a reason for suffrage. When the shirtwaist workers went out on strike, suffragists joined them. "The Women that need help now, in the future, and always are the women workers, the women alone, the widow and the fatherless, the weak, sadly unable to fight the good fight. Suffrage is going to care for them," a writer to the *New-York Tribune* said, a direct reference to the thousands of women who were employed in the garment industry who had gone on strike in November 1909.[71]

Not having the ability to vote, however, did not keep women out of the political process. In the 1890s, they began forming political clubs to support candidates on all levels, from mayors to presidents. Women even ran for office, some being elected. The presidential election of 1900 was sometimes referred to as the "petticoat campaign" because of the massive number of women who worked to register male voters and to solicit support for William McKinley.[72] The *New-York Tribune* acknowledged that "the Republican women of New York City have been most effective helpers to the Republican party, whose leaders have been glad to utilize the women's services," while the

New York Evening Post explained that in the process of registering new voters, women "do all the tedious preliminary work."[73] As women became politically active, they met, marched, and published. All became subject matter for media. Getting out the male vote to support candidates of which women's groups did not necessarily coincide with the effort for suffrage, but in the minds of many, they were closely related.

In 1913, the National American Woman Suffrage Association organized a march on the nation's capital. On the day before Woodrow Wilson's inauguration, women took to the streets in Washington, but angry and sometimes drunken males disrupted the event, injured some of the women, all under the watching but non-interfering eyes of the District's police force.[74] "Can you imagine it? Ten thousand women in a line," Nellie Bly wrote. "I did not hear anyone speak about the new President....This was not his day with the public. It was out to see the women, and it meant to see them."[75] Other papers, however, belittled the event. "The ability of a woman to walk a long distance hardly established her fitness to vote," the *San Francisco Chronicle*, said. The paper was more astute, though, in what the entire event—including the violence that occurred—did for the cause of suffrage. "But as a publicity measure—a means of keeping alive the cause of woman suffrage—it certainly was successful."[76] By 1915, twelve states allowed women to vote either in local, state, or even national elections, but twelve of forty-eight meant that three-fourths of the states in the nation still did not provide women suffrage. The public relations' efforts, however, were paying off. Some of the nation's most important figures were coming out in support of suffrage. William Jennings Bryan, the voice for workers, Christian fundamentalists, and the secretary of state, endorsed suffrage that year when he said, "If the women have sense enough to keep out of the penitentiary and morals enough to go to church, who will say that they are not fit to go to the polls?"[77]

Other leaders, including President Woodrow Wilson, still opposed suffrage, saying that "changes of this sort ought to be brought about State by State."[78] In 1917, with the United States prepared to enter the Great War, suffragists staged protests outside the White House in freezing January temperatures, garnering massive press attention.[79] As the protest continued, angry men spit on the women, slapped them, threw cigar butts at them, and at least 168 protesters were arrested. Suffrage groups, however, kept up their campaign to sway the president. "Mr. President," suffragists continually asked the president when given the public opportunity, "if you sincerely desire to forward the interests of all the people, why do you oppose the national enfranchisement of women?"[80] A year after the White House protests and hundreds of stories and editorials in the press, Wilson and Congress capitulated to the ever-growing number of Americans who supported suffrage. "President Wilson this evening came out squarely for the proposed amendment to the Federal Constitution giving women the right to vote throughout the Union," the news report said. The president, however, did not waiver from his states' rights view on suffrage in his statement. Instead, he used the selfless contributions of women to the war effort as his rationale for the amendment: "The President told committeemen that he felt that in view of the fact that Great Britain had granted the franchise to women and the general disposition among the Allies to recognize the patriotic services of women in the war against Germany, this country could do no less than follow that example."[81] "After all these years of propaganda, picketing and incarceration, the suffrage forces will go 'over the top' in their great drive for liberty," the *Washington Post* predicted.[82] The House voted 274-136, providing exactly a two-thirds majority to support the amendment in January 1910, but the Senate required another fifteen months before it approved the amendment and sent it to the states.[83]

From this point forward, the president served as a leading proponent of suffrage, often sending personal messages to state legislatures as they prepared to vote on the amendment.[84] On August 18, 1920, Tennessee became the thirty-sixth state to ratify the nineteenth amendment; women had the right to vote.[85] The mainstream American press and an advocacy press that championed suffrage kept the issue as a topic constantly in front of the public for decades because, from Seneca Falls forward, suffragists adhered to the *Declaration of Sentiments*' strategy that called for "every candid person to hear every proposal for the elevation of our race."[86] By keeping the issue in front of people and constantly in debate, media helped women in America to think of themselves as competent and qualified. It also convinced males that females were deserving of the vote.[87] Especially in the 1910s, the constant barrage of information that increasingly portrayed suffrage as the right move for the nation ensured that state after state would approve the nineteenth amendment.

THE LUDLOW DISASTER AND FRAMING NEWS

The realization that media power could affect issues on any level from local to national introduced yet another element to media—public relations. Even though the press had been used as a PR tool for more than a century, enterprising individuals began to offer their services to help individuals and corporations change public opinion about them. None did a better job of this in the Progressive era than Ivy Lee, who, as one writer said, believed public relations "was entirely consistent with the aspect of the Progressive ethos that sought reform through exposure."[88] The difference in what the muckrakers did and what the PR agents did, then, was not the means of effecting change, but the end result. Both turned to media as the principal conduit of change. Muckrakers, however, sought to better society; public relations men worked with the truth to produce results for clients. Self-interest, not societal well-being, was the desired outcome. "We aim to supply news," Lee said in 1906. "In brief, our plan is, frankly and openly, on behalf of the business concerns and public institutions, to supply to the press and public of the United States prompt and accurate information concerning subjects which it is of value and interest to the public to know about."[89] So, Lee promised to do the same as the muckrakers, only he would be working for those elements that were the target of muckraking journalism.

Lee did just that with two pivotal issues. Both, however, demonstrate the power of controlling public opinion and how media may be used to do so through carefully constructed media messages. Lee had worked as a journalist before he opened a publicity bureau in Boston in 1904. Railroads had been one of the muckrakers' targets, especially for Ray Stannard Baker. Baker's work was central to congressional passage of the Elkins and Hepburn acts, which increased legislation on the railroads. Later, the rails wanted a rate increase, and Lee initiated a campaign that explained why the increase was justified. He targeted various "publics" with his message—the general public through media press releases and specialized correspondence to opinion leaders in an effort to have them speak in public and be quoted by the press in support of the rate increase. It worked, and the government allowed a 5 percent increase for hauling freight.[90]

Perhaps Lee's greatest achievement with public relations was in diffusing national disgust at the "massacre" that occurred at the Ludlow mine in Colorado where an estimated seventy-five people died, many of them women and children reportedly gunned down in tents and trenches following a clash between miners and the Colorado National Guard at the Colorado Fuel and Iron Company mines, which were owned by Standard Oil.[91] The massacre only increased public dislike of John D. Rockefeller Jr., who still suffered under Ida Tarbell's attack on his father and Standard Oil. Rockefeller turned to Lee who mounted what might be considered the first major crisis communications campaign. He provided media numerous fact sheets about what happened, mailing them out to at least forty thousand different outlets and influential people.[92] He talked about paid union agitators who caused problems; he played up the mine owner's good-faith efforts, going so far as to have them signed by prominent individuals; he provided press releases that made it appear that most newspaper editors supported the CF&I; he ghostwrote letters from Colorado officials, including the governor, supporting Rockefeller and the mining company; he even had Rockefeller photographed in worker's clothing as he held a miner's child.[93] "The fight at Ludlow was the direct outgrowth of a 'scab war,'" is the way many of the articles written using Lee's releases termed the event.[94]

Lee's work to paint Rockefeller and CF&I as the innocent parties in the Ludlow massacre proved that the PR practitioner was correct in his assessment of the power of media. Lee's promise to provide "prompt and accurate information concerning subjects which it is of value and interest to the public to know about"[95] only angered contemporaries who saw what Lee was doing as "utterly unscrupulous and mendacious." Muckraker Upton Sinclair dubbed him "Poison Ivy" and poet Carl Sandburg, "Ivy L. Lee—Paid Liar," [96] because Lee was able to manipulate the truth, have people believe the information that became a part of media presentations, and ultimately affect public opinion and policy.

CATCHING CHARLES PONZI

The laissez faire approach to business that had been the hallmark in America before Progressives and muckrakers opened the door to any number of con artists who sought to take advantage of people. That was the case with Italian immigrant Carlo Ponzi, who anglicized his first name to Charles after coming to America in 1903 at the age of twenty-one. By the time the *Boston Post* unraveled his investment scheme in 1920, he had bilked thousands of their financial security, made himself a wealthy man, and infamously given his name to any scandal that involved get-quick proposals via investments.[97]

Ponzi promised his investors that an investment would yield 50 percent returns within ninety days as he used their funds to purchase something called "International Reply Coupons," which Ponzi said, could be bought in a European country using that country's currency. The rate on the coupon meant that it could be redeemed for whatever its value in any country. Through his Securities and Exchange Company, Ponzi explained that he could buy the IRP in devalued Italian lira, for example, and then redeem in U.S. dollars for a massive profit. To do this, however, Ponzi needed large sums of cash, something he didn't have. Once investors bought into his proposal, however, large sums rolled in. Ponzi, however, banked most of it himself, while using some of the money to pay initial investors so that all appeared to be legitimate. To make his swindle even more attractive, he set up a series of foundations. He even used some of the funds to support the Boston Police Relief Association, which no doubt endeared him to law enforcement officials.[98]

By July, Ponzi was pulling in $1 million a week and had collected $10 million. By the end of the month, an estimated twenty thousand investors would give their savings to Ponzi, who, all the while, promised that, ultimately, they would probably receive a 400-percent yield on their monies.[99] Initially, his deceit was aided by media. The *Boston Traveler*—one of Boston's ten dailies—promoted Ponzi's scheme, telling its readers on July 9 that "WE GUARANTEE" investments with Ponzi's Securities and Exchange Company. "We haven't figured out how they make their enormous profit," the paper said, "but they seem confident in their ability to do so."[100] The *Post*, too, believed Ponzi initially, writing about how his company could double an investment in three months and about the man behind the company.[101] The Associated Press reported the federal government was the largest user in the world of the International Reply Coupons that Ponzi said were the basis of his investment bonanza, which provided even more credence to what Ponzi promised as a return on an investment with him.[102]

Newspaper reports began to refer to Ponzi as "the wizard of finance."[103] But even though the *Boston Post* had written positively about the Securities and Exchange Company, it was also looking into whether Ponzi might be pulling a scam. First, it interviewed C.W. Barron, the Boston financial expert who owned controlling interest in the *Wall Street Journal*. Barron questioned the efficacy of Ponzi's financial promises. "No man of wide financial or investment expereience," Barron said, "would look twice at a proposition to take his money upon a simple promise to pay it back with a 50 percent increase in three months....But it is unreasonable to ask anybody to believe that any large amount of money can be so invested."[104] Barron's paper waged in on Ponzi, too. "It is not within the range of human probability that an Italian immigrant, arriving in this country several years ago, and confessedly having nothing seven months ago, has discovered a hole in international postal arrangements by which he can pull out practically from treasuries millions a month," Barron was quoted in the *Wall Street Journal* directly from his Boston News Bureau financial newsletter. He then explained exactly what Ponzi was doing, even though he may not have known for sure: "Right under the nose of Government officials he is paying United States money to one line of depositors from deposits made by a succeeding line."[105]

If Barron smelled a rat, then the *Post* needed to prove that a scam was being run. With two reporters, P. A. Santosuosso and Herbert Baldwin, assigned to uncover exactly what was going on with Ponzi, the meticulous work of digging out the truth began.[106] The *Boston Post* printed at least one story that discussed Ponzi every day in early August, and another Boston paper called him "the most talked of man in America."[107] The *Post* established a contact on the inside of Ponzi's company, its publicity man, William McMasters. "After this edition of *The Boston Post* is on the street there will be no further mystery about Charles Ponzi. He is unbalanced on one subject: his financial operations," the *Post* declared as it introduced McMaster's revelations. "He is hopelessly insolvent. Nobody will deny it after reading this story."[108] The same day, the *Wall Street Journal* told its readers that with Ponzi an investor could realize a 2,463 percent profit in a single year.[109] The next day, hundreds of Boston depositors lined up outside the Securities and Exchange Company ready to cash in their Ponzi accounts, and Ponzi promised a libel suit against the *Post*.[110]

Meanwhile, Santosuosso interviewed Ponzi, claiming he wanted to do a story on the immigrant's life. What he discovered was a tale that had a series of unexplained gaps. One of them, the *Post* learned, might be because, in 1907, Ponzi had served time in prison in Canada. Indeed, the *Post* confirmed, someone named Charles Ponsi had worked for a Montreal bank and served time in prison for forgery. Baldwin chased down the lead and confirmed that Charles Ponsi was, indeed, Boston's Charles Ponzi.[111] On August 11, the *Post* ran a pair of front-page stories exposing Ponzi's past and that he'd served time in Canada for forgery.[112] Though he would deny the arrest and continued to claim his business was legitimate, the *Post*'s stories effectively ended Ponzi's scam. The *Wall Street Journal* put the *Post*'s investigative work into perspective: "The Boston Post uncovered Ponzi and arrested his further loot from savings banks and wage earners. It was the first not only to expose him, but to show the immorality of his scheme, the impossibility of its success and to declare him a bankrupt and then to trace out his criminal record in Canada." The paper then attacked all the other Boston papers that supported Ponzi blindly. "And all the while there were papers catering to people of financial intelligence that pussyfooted on Ponzi, declaring nothing yet proven, that Ponzi might yet turn up all right and this to the disgrace of journals that boast of intelligence of their readers."[113]

Ponzi was arrested two days after the *Post* broke the story of his incarceration in Canada. The *Post* declared, "Of all the get-rich-quick magnates that have operated, Ponzi is the king."[114] The Associated Press explained that Ponzi had been charged with mail fraud and larceny and that he owed his investors more than $7 million. Also, the AP said, investigations into illegal activity in Europe by Ponzi were being opened and that the Hanover Trust Company, the bank that Ponzi had used for investment and for deposits "is probably wiped out completely."[115] Hanover failed shortly afterward. Ponzi spent the next four years in jail, was convicted on more charges related to his scheme in 1927, and was finally deported to Italy in 1934. He emigrated to Brazil and died there in 1949.[116] Ironically, when the United States established a federal agency to oversee and regulate the stock market and to prevent abuses in security sales as Ponzi had done, it named the new entity the Security and Exchange Commission, a name hauntingly similar to Ponzi's pyramid-scheme organization's title, the Security and Exchange Company.

***

At the beginning of the twentieth century, one observer noted that "the power of the American press for good or evil is too great to be estimated."[117] Few Americans would have disagreed with that assessment during the next two decades. The number of media outlets available to the public had never been greater. Advocacy for causes, never stronger. Muckrakers exposed corruption, despicable working conditions, and unfair business practices, which led to governmental oversight, new regulations, and increased public awareness. Personal perspective determined whether this journalism was good or evil.

The African-American press continued the fight begun in the 1890s against lynching. In Chicago, Robert S. Abbott published the *Chicago*

Defender. A close observer of the way media were operating at the time, Abbott made his newspaper a combination of the journalism of action practiced by Pulitzer and Hearst and of the exposé style of the muckrakers. He took on all issues that concerned black Americans and shipped his paper everywhere. The fact that the *Defender* was available in the South meant that it had an audience hundreds of times larger than it had in Chicago alone. Abbott could make any issue into one that promoted the cause of African Americans, something he did in the June 20, 1925, issue of the *Defender* when he used the Scopes trial that was taking place in Dayton, Tennessee, as the basis for an editorial cartoon that attacked lynchings.

Other African Americans turned to the mainstream white press. The National Association for the Advancement of Colored People was organized in 1909, and its leaders immediately turned to the press for advocacy, and its meetings quickly became news. W.E.B. Du Bois, the *Washington Post* said, spoke of the way that race prejudice "is spoiling our ideas of democracy" to a gathering in New York City.[118] While many whites would have disagreed, the fact that the mainstream press even reported on what was happening under the auspices of the NAACP meant that the organization's message would reach beyond African-American society in the nation. The organization kept racial equality in the public sphere during World War I, too, as it pondered whether "America could make the Negro 'welcome to the bullet and deny him the ballot.'"[119] While no great changes occurred during this time period in terms of equality for African Americans, the issue was still part of what media presented to the populace.

Advocates of an alcohol-free America had more luck, however, though they had been working for a century to outlaw alcohol for consumption. Groups such as the Women's Christian Temperance Union and the Anti-Saloon League focused on establishing prohibition on local and state levels. "We will organize so that every town, village, and ward in the State will have its quotas of antis," an Anti-Saloon League member announced in January 1900.[120] The group used all media, creating fliers, pamphlets, and books to go along with stories generated for newspapers and magazines. Women formed the backbone of the movement, but they were joined by ministers and other men who were convinced that alcohol was the bane of the nation. As the movement progressed, pressure by dry forces on the press was limiting the number of outlets that would run liquor advertisements. At least forty magazines including *Collier's, Ladies Home Journal*, and *McClure's*, and 250 newspapers, according to the ASL, had agreed to the ban by 1912. That number continued to grow.[121] By 1913, prohibitionists had achieved enough success that the Anti-Saloon League proposed "Nation-wide prohibition, to be attained by means of a federal constitutional amendment, and a plank urging states to pass prohibition shipping laws."[122] In three years, twenty-three states had passed anti-saloon legislation, and temperance organizations had successfully put enough candidates into office to ensure that a constitutional amendment was truly possible. World War I interrupted progress toward prohibition, but after the war, tens of thousands of articles about prohibition, the temperance organizations, and the constitutional amendment for prohibition filled papers. The states ratified the Eighteenth Amendment in January 1919, and it went into effect on January 29, 1920.[123]

Media helped change many aspects of life in the United States in the early twentieth century, but media made news, too. As the nineteenth century closed, new inventions were being introduced that would revolutionize the dispersal of information. News about those inventions would also become a part of the information that people craved. Those inventions—motion pictures, the phonograph, radio, and ultimately television—would quickly become an everyday part of American life in the twentieth century.

NOTES

1. David Graham Phillips, "The Treason of the Senate," *Cosmopolitan* (April 1906): 628–38.
2. Theodore Roosevelt, At Washington, April 14, 1906, in Albert Bushnell Hart and Herbert Ronald Ferleger, eds. *Theodore Roosevelt* Cyclopedia, 2nd. ed. (Westport, Conn.: Meckler Corp., 1989), 357.
3. *Official Proceedings of the Democratic National Convention Held in Chicago, Illinois, July 7, 8, 9, 10, and 11, 1896*, (Logansport, Indiana, 1896), 234. Reprinted in *The Annals of America*, Vol. 12, *1895–1904: Populism, Imperialism, and Reform* (Chicago: Encyclopedia Britannica, Inc., 1968), 105.
4. One of Your Readers, "Value of Muck-Raking," *New York Times*, May 17, 1906, 8.
5. Theodore Roosevelt, "The Public Press," Address to the Press Club, Milwaukee, Wisconsin, September 7, 1910.
6. Frank Luther Mott, *American Journalism, A History: 1690–1960*, 3rd ed. (New York: Macmillan, 1962), 359.
7. Elizabeth V. Burt, *The Progressive Era* (Westport, Conn.: Greenwood Press, 2004), 7.
8. Alfred McClung Lee, *The Daily Newspaper in America: The Evolution of a Social Instrument* (New York: Macmillan, 1937), 65.
9. W. S. Crowe, *Batten's Agricultural Directory* (New York: George Batten, 1908), 15, in Leonard Ray Teel, *The Public Press, 1900–1945* (Westport, Conn.: Greenwood Press, 2006), 47.
10. Michael Emery and Edwin Emery, *The Press and American: An Interpretive History of the Mass Media*, 6th ed. (Englewood Cliffs, N.J.: Prentice Hall, 1988), 250.
11. *N. W. Ayer & Son's American Newspaper Annual*, in Frank Luther Mott, *A History of American Magazines, Volume IV: 1885–1905* (Cambridge, Mass.: Harvard University Press, 1957), 11.
12. Carl F. Kaestle, "Literacy and Diversity: Themes from a Social History of the American Reading Public," *History of Education Quarterly* 28,4 (1988): 528.

13. Matthew Schneirov, *The Dream of a New Social Order: Popular Magazines in America, 1893–1914* (New York: Columbia University Press, 1994), 5.
14. See Donald L. Shaw, Bradley J. Hamm and Diana L. Knott, "Technological Change, Agenda Challenge and Social Melding: Mass media studies and the four ages of place, class, mass and space," *Journalism Studies* 1, 1 (2000): 57–79.
15. "Need a Conscience: Growing Public Demand for Newspaper Guided by Right Principles," *Editor & Publisher*, August 13, 1910, 6.
16. *Measuring America: The Decennial Censuses From 1790 to 2000*, U.S. Bureau of the Census (Washington, D.C.: U.S. Census Bureau, 2000), Appendix A.
17. Steven Mintz, "The Twentieth Century," *Digital History*, February 17, 2009, http://www.digitalhistory.uh.edu/database/article_display.cfm?HHID=205 (accessed Feb. 17, 2009).
18. "Current Population Report," U.S. Census Bureau, February 12, 1963 (Washington, D.C.: Bureau of the Census, 1963), 1.
19. Arthur Weinberg and Lila Weinberg, eds., *The Muckrakers* (New York: Simon and Schuster, 1961; reprint, Urbana: University of Illinois Press, 2001), xvii.
20. John M. Blum, et al., *The National Experience*, 6th ed., 2 vols. (San Diego: Harcourt Brace Jovanovick, 1984), 246, 248.
21. *New York Journal*, October 5, 1897.
22. S. S. McClure, *My Autobiography* (New York: Frederick A. Stokes, 1914), 253.
23. John Tebbel, *The Media in America* (New York: Thomas Y. Crowell, 1974), 280.
24. Josiah Flynt, "In the World of Graft," *McClure's* 16 (February 1901).
25. Francis H. Nichols, "Children of the Coal Shadow," *McClure's* 20 (February 1903).
26. Ida M. Tarbell, *The History of the Standard Oil Company*, 2 vols., ed. David M. Chalmers (New York: McClure, Phillips & Co., 1904; reprint, Portland, Ore.: Dover, 2003), 225.
27. Quoted in "People & Events: Ida Tarbell, 1857–1944," *The American Experience: The Rockefellers*, 2000, http://www.pbs.org/wgbh/amex/rockefellers/peopleevents/p_tarbell.html.
28. S. S. McClure, "Editorial Announcement," *McClure's* 20 (January 1903): 336.
29. Ray S. Baker, "How Railroads Make Public Opinion," *McClure's* 26 (March 1906): 535–49 in J. Michael Sproule, *Propaganda and Democracy: The American Experience of Media and Mass Persuasion* (Cambridge: Cambridge University Press, 1997), 23.
30. Judith Serrin and William Serrin, eds. *The Muckrakers: The Journalism That Changed America* (New York: The New Press, 2002), 310.
31. C. C. Regier, "The Struggle for Federal Food and Drugs Legislation," *Law and Contemporary Problems* 1, 1 (December 1933): 3–4.
32. Edward Bok, "The 'Patent Medicine' Curse," *Ladies' Home Journal* 21 (May 1904):18. Bok wrote a number of articles on the dangers of patent medicine. See, for example, Edward Bok, "For the Safety of Yourself and Your Child," *Ladies' Home Journal* (February l906): l.
33. Samuel Hopkins Adams, "The Nostrum of Evil," *Collier's Weekly*, October 7, 1905 reprinted in Samuel Hopkins Adams, *The Great American Fraud* (Chicago: Journal of the American Medical Association, 1905), 3, 5.
34. "The Patent Medicine Conspiracy Against Freedom of the Press, *Collier's Weekly*, November 4, 1905.
35. Marc T. Law and Gary D. Libecap, "The Determinants of Progressive Era Reform: The Pure Food and Drugs Act of 1906," in *Corruption and Reform: Lessons from*

America's Economic History, eds. Edward L Glaeser and Claudia Goldin (Chicago: University of Chicago Press, 2006), 332.

36. See *The Nation*, June 28, 1906. To understand congressional opposition to food and drug regulation, see Thomas A. Bailey, "Congressional Opposition to Pure Food Legislation, 1879–1906," *American Journal of Sociology*, 36, 1 (July 1930): 52–64.
37. Upton Sinclair, *The Jungle* (New York: Doubleday,1905; reprint by the author, 1920), 116–17.
38. "MEAT TRUST IN A PICKLE," *Washington Post*, May 31, 1906, 1.
39. Robert Cherny, "*The Jungle* and the Progressive Era," *History Now* 16 (June 2008), http://www.historynow.org/06_2008/historian2.html.
40. "Meat Inspection Bill Passes the Senate," *New York Times*, May 26, 1906, 1.
41. "Setback by Muckrakers," *Washington Post*, September 28, 1907, 3.
42. William Hard, "Making Steel and Killing Men" *Everybody's Magazine* (November 1907): 580–584.
43. Thomas Robinson Dawley Jr., *The Child That Toileth Not: The Story of a Government Investigation* (New York, Gracia, 1912).
44. Louis Filler, *The Muckrakers*, 2nd ed. (University Park: Pennsylvania State University Press, 1976), 190–95.
45. "Standard Oil Company Must Dissolve Within Six Months," *Los Angeles Times*, May 16, 1911, 1. The Supreme Court case was *Standard Oil Co. of New Jersey v. United States*, 221 U.S. 1 (1911).
46. Leonard Ray Teel, *The Public Press, 1900–1945* (Westport, Conn.: Greenwood Press, 2006), 44.
47. "40,000 Called Out in Women's Strike," *New York Times*, November 23, 1909, 16.
48. "Suffragists to Aid Girls Waist Strikers, *New York Times*, December 2, 1909, 3.
49. "Suffragettes Fail," "More Pickets Arrested," "To Strike in Philadelphia," *New-York Tribune*, December 5, 1909, 12.
50. Robertus Love, "Shirt Waist Girls' Strike the Greatest Struggle of Women In History of Labor," *Washington Herald* (Washington, D.C), January 9, 1910.
51. "Garment Workers Return to Work, *New York Call*, February 15, 1910, 1.
52. Burt, *The Progressive Era*, 206; "Fire," *The Triangle Factory Fire* (January 20, 2004) http://www.ilr.cornell.edu/trianglefire/narrative3.html. For an indepth review of the Triangle Shirtwaist fire, see Leon Stein, *The Triangle Fire* (New York: J. B. Lippincott, 1962) and David Von Drehle, *Triangle: The Fire That Changed America* (New York: Grove/Atlantic, 2003).
53. "141 Men and Girls Die in Waist Factory Fire," *New York Times*, March 26, 1911, 1.
54. William G. Shepherd, "Eyewitness at the Triangle," *Milwaukee Journal*, March 27, 1911. The story appeared in numerous papers that subscribed to the United Press.
55. "World Reporter Finds Indications that Locked Doors Caused Big Loss of Life in Fire," *New York World*, March 27, 1911.
56. "The Murdering of Those Unhappy Girls on Saturday," *New York Evening Journal*, March 28, 1911.
57. "Fire Traps There Are Here," *New York Times*, March 27, 1911, 1.
58. "Placing the Responsibility," *The Outlook*, April 29, 1911, 949.
59. "Investigation, Trial and Reform," *The Triangle Factory Fire* (January 23, 2004) http://www.ilr.cornell.edu/trianglefire/narrative6.html.
60. Von Drehle, *Triangle*, 215.

61. "Investigation, Trial and Reform, *The Triangle Factory Fire* (January 23, 2004) http://www.ilr.cornell.edu/trianglefire/narrative6.html.
62. "Again," *Chicago Tribune*, March 27, 1911, 8.
63. Irwin N. Gertzog, "Female Suffrage in New Jersey, 1790–1807," in Naomi B. Lynn, ed., *Women, Politics, and the Constitution* (New York: Haworth Press, 1990), 48–49.
64. "New Election Law," *Trenton Federalist*, November 16, 1807, 3.
65. "No Just Cause for Fear," *Salt Lake Herald* (Salt Lake City), April 7, 1895, 4.
66. Quoted in Elizabeth Cady Stanton, *A History of Woman Suffrage*, 2 vols. (Rochester, N.Y.: Fowler and Wells, 1889), 1:70–71.
67. Horace Greeley, "Remarks to 'A,'" *New York Daily Tribune*, November 2, 1850.
68. James Gordon Bennett, "Woman's Rights Convention," *New York Herald*, October 28, 1850.
69. Donna L. Dickerson, *The Reconstruction Era* (Westport, Conn.: Greenwood Press, 2003), 347.
70. E. L. Godkin, *The Nation* (New York), June 26, 1873.
71. "The Woman Handicap," *New-York Tribune*, May 8, 1910.
72. Jo Freedman, *We Will Be Heard: Women's Struggles for Political Power in the United States* (Lanham, Md.: Rowman & Littlefield, 2008), 23, 36.
73. *New-York Tribune*, September 8, 1900; *New York Evening Post*, October 18, 1900.
74. Burt, *The Progressive Era*, 313.
75. Nellie Bly, "'Marching Women Have Made History,' Declares Nellie Bly," *New York Evening Journal*, March 4, 1913.
76. "Suffrage Publicity," *San Francisco Chronicle*, February 21, 1913, quoted in Burt, *The Progressive Era*, 314.
77. *Greensboro (N.C.) Daily News*, January 31, 1915.
78. "Suffrage Plea Vain," *Washington Post*, January 7, 1915, 1.
79. See, for example, "Silent Pickets of Suffragists at White House," *Christian Science Monitor* (Boston), January 10, 1917, 1; *New York Times*, January 11, 1917, 13; "Freezing Suffrage 'Sentinels' Ignore Invitation by Wilson to Come Inside and Get Warm," *Washington Post*, January 12, 1917, 13.
80. Quoted in "Feminism: The Fight for Suffrage," *American Decades*, ed. Vincent Tompkins, 10 vols. *Gale Virtual Reference Library* (Detroit: Gale, 2001), 2: 1910–1919.
81. "Wilson Backs Amendment for Woman Suffrage," *New York Times*, January 10, 1918, 1, 3.
82. "Suffrage at Issue," *Washington Post*, January 10, 1918, 6.
83. "House for Suffrage," *New York Times*, January 10, 1918, 6; "Equal Suffrage Won," *Washington Post*, June 5, 1919, 1.
84. See, for example, "Let the Women Vote, Is Wilson's Pleas: Wires Louisiana Legislature to Adopt Suffrage," *Washington Post*, June 6, 1918, 2; "President Writes 2 Suffrage Pleas," *Washington Post*, July 30, 1918, 5; "President Asks Vote for Suffrage," *New York Times*, July 31, 1918, 4.
85. "Ratification of Federal Suffrage Amendment Won," *Christian Science Monitor* (Boston), August 19, 1920, 1.
86. George C. Cooper, "Woman's Rights Convention," *National Reformer* (Rochester, N.Y.), August 10, 1848.

87. See "Suffrage Newspapers," in Maurine H. Beasley and Sheila J. Gibbons, *Taking Their Place: A Documentary History of Women and Journalism*, 2nd ed. (State College, Pa.: Strata Press, 2003).
88. H. M. Gitelman, *Legacy of the Ludlow Massacre: A Chapter in American Industrial Relations* (Philadelphia: University of Pennsylvania Press, 1988), 35.
89. Ivy Lee, "Declaration of Principles," quoted in Sherman Morse, "An Awakening in Wall Street: How the Trusts, After Years of Silence, Now Speak Through Authorized and Acknowledged Press Agents," *The American Magazine* 62 (September 1906): 460. Full article, 457–63.
90. "Thirty-Eight Roads Will Get Full Rate Increase," *Wall Street Journal* (New York), August 4, 1914, 5.
91. See Scott Martelle, *Blood Passion: The Ludlow Massacre and Class War in the American West* (New Brunswick: Rutgers University Press, 2007).
92. Gitelman, *Legacy of the Ludlow Massacre*, 35.
93. Priscilla Long, *Where the Sun Never Shines: A History of America's Bloody Coal Industry* (St. Paul, Minn.: Paragon House, 1989).
94. See, for example, "Defender of Militia," *Washington Post,* June 24, 1914, 13.
95. Ivy Lee, "Declaration of Principles," in Morse, "An Awakening in Wall Street," 460.
96. Samuel Gompers, Upton Sinclair, and Carl Sandburg quoted in Gitelman, *Legacy of the Ludlow Massacre*, 59.
97. See, Mitchell Zuckoff, *Ponzi's Scheme: The Story of a Financial Legend* (New York: Random House, 2005).
98. Roy J. Harris, Jr., *Pulitzer's Gold: Behind the Prize for Public-Service Journalism* (Columbia: University of Missouri Press, 2007), 125–26.
99. *West's Encyclopedia of American Law*, online edition, s.v. "Ponzi Scheme"; "How the Bubble Grew," *Boston Evening Transcript*, November 6, 1922, 24.
100. Quoted in Harris, *Pulitzer's Gold*, 126.
101. "Doubles the Money Within Three Months," *Boston Post*, July 24, 1920, 1.
102. "'50-Per-Cent-Profit' Magnate Says He Has Legitimate Get-Rich-Quick Plan Through Foreign Exchange," *Washington Post*, July 27, 1920, 2.
103. See, for example, "Trace Ponzi Funds," *New York Times*, August 17, 1920, 20; and "Ponzi's Scheme Was to Expose 'High Finance,'" *Wall Street Journal* (New York), August 28, 1920, 1.
104. "Questions the Motive Behind Ponzi Scheme," *Boston Post*, July 26, 1920, 1, quoted in Zuckoff, *Ponzi's Scheme*, 185–86.
105. "C.W. Barron Skeptical About 'Exchange Wizard,'" *Wall Street Journal* (New York), July 30, 1920, 2.
106. Unless otherwise cited, information on the fall of Ponzi is adapted from Zuckoff, *Ponzi's Scheme*, 227–85, including quotes from the *Boston Post*.
107. "Crisis for Ponzi," *Boston American*, August 4, 1920, 1.
108. "Declares Ponzi Is Now Hopelessly Insolvent," *Boston Post*, August 2, 1920, 1.
109. "Ponzi Plan Means 2,463% Profit In Year," *Wall Street Journal* (New York), August 2, 1920, 10.
110. "Hundreds of Ponzi Depositors in Line," *Wall Street Journal* (New York), August 3, 1920, 12.
111. Patrick Hinton, *A Biography of Charles Ponzi and his Scheme* (January 9, 2009) http://investment.suite101.com/article.cfm/a_biography_of_charles_ponzi_and_his

_scheme; Herbert Kenny, *Newspaper Row: Journalism in the Pre-Television Era* (Boston: Globe Pequot Press, 1987), 197.
112. Herbert Baldwin, "Candian 'Ponsi' Served Jail Term," *Boston Post*, August 11, 1920, 1; "That's Ponsi," *Boston Post*, August 11, 1920, 1.
113. "A Public Service," *Wall Street Journal* (New York), August 23, 1920, 1.
114. "Ponzi Arrested," *Boston Post*, August 13, 1920, 1.
115. Associated Press, "Ponzi Under Arrest; Debts Mounting Up," *Washington Post*, August 13, 1920, 1.
116. Mark C. Knutson, *The Remarkable Criminal Financial Career of Charles K. Ponzi* (1996), http://www.mark-knutson.com.
117. "The Press Club Dinner," *New York Times*, February 22, 1900, 3.
118. "Bane of Race Prejudice," *Washington Post*, December 8, 1910, 11.
119. "Full Equality Of Races Demanded," *Christian Science Monitor* (Boston), January 5, 1918, 13.
120. "Anti-Saloon Crusade Begun," *New York Times*, January 29, 1900, 5.
121. Burt, *The Progressive Era*, 289, 293.
122. "League Demands Prohibition Law Be Nation-Wide," *Christian Science Monitor* (Boston), November 14, 1913, 7.
123. "Nation Voted Dry; 38 States Adopt the Amendment," *New York Times*, January 17, 1920, 1.

7. *My Medium Is Everywhere: New Media and the New Century*

As the nineteenth century ended, new media that differed from the world of print began to capture the attention of people. Motion pictures, recorded music, and radio were in various stages of implementation, but all held the potential for changing the consumption of information as well as the entertaining of society. In less than thirty years, people would start to talk about another new medium—television. In the 1890s, Nathan Stubblefield began to experiment with a new invention in his hometown of Murray, Kentucky. In 1902, he tested publicly his wireless telephone there, on the Potomac River outside the nation's capital, and in a number of other locations.[1] He boldly proclaimed, "My medium is everywhere,"[2] in reference to his assertion that his wireless could travel through air, water, and even earth.

But Stubblefield's pronouncement had much greater implications. He, along with others whose names had already or would soon become known in households throughout the nation, were in a race to develop media that would transform the way society received information. In Stubblefield's case, the new "wireless" medium was broadcast through the air. Its content could move almost instantaneously between points. All one needed to listen to the transmission was a receiver. As radio waves filled the air, this new, wireless medium, indeed, had the potential to be everywhere. It and other new means of sharing with the masses became as much a part of what would shape the nation in the twentieth century as any event, including world wars. Or, as one historian explained, "New electric media were sources of endless fascination and fear, and provided constant fodder for social experimentation."[3]

The development of these new media was a part of the growing desire to transmit information as quickly as possible. The introduction of the telegraph in the 1840s had been the first step. By 1890, approximately 1 million miles of telegraph wire created a web of interconnectivity across the United States.[4] In 1856, workers began laying the first transatlantic telegraph cable between the United States and England. Completed in 1858, it failed almost immediately, but in 1866, a second, successful cable was laid. Now, communication between America and the United Kingdom could occur almost instantaneously. British colonialism ensured that telegraph wire stretched around the globe to any location within the empire. As a result, information could be obtained and shared in only a matter of days where it had taken weeks and months before.

Consider, for example, the massive, catastrophic eruption of the volcano on Krakatau in the Sundra Strait of Indonesia on August 27, 1883. The eruption, which destroyed two-thirds of the island and killed tens of thousands, was reported in the *Boston Daily Advertiser* on August 28. Under the dateline "BATAVIA, Java, Aug. 28," the *Advertiser* explained that areas had been swept away because of the rush of sea inland; that stones fell as showers, that total night persisted, and that a terrible calamity had occurred.[5] The same day, the *Daily Evening Bulletin* in San Francisco revealed that towns had been wiped away; that a tidal wave had engulfed Batavia; and that ash, mud, and stones covered much of Java.[6] The same stories appeared in papers around the nation on that Tuesday.[7] Even though these initial reports did not mention anything about the loss of life (because no one yet knew the extent of the volcano's destruction), the fact that news of the eruption took only one day to travel from Java in the Indian Ocean to people on the streets of Boston, Philadelphia, St. Louis, Milwaukee, San Francisco, and other U.S. cities reveals how quickly technological advancements pushed communications and why the creation of new means of sharing and informing became so newsworthy and an integral component in the shaping of the nation.

The telephone, invented by Alexander Graham Bell in the late 1870s, became the second technological advancement that would provide the basis for new media at the turn of the twentieth century. One writer even said the telephone was the "greatest industrial and commercial achievement of the American people."[8] Bell, in 1878, predicted that the telephone would become the medium of national communication, even though no service yet existed for any community and messages could only travel twenty miles over experimental lines set up in Connecticut.[9] By 1890, though, Americans were sharing more than a million telephone messages a day, and by 1920, Americans spent more per year for telephone service than they paid to use the U.S. Postal Service.[10]

The telegraph and the telephone became foundational tools for those who would go on to develop radio, though it was called "the wireless" well into the twentieth century and radiotelegraphy initially because it was a means of using radio waves—as discovered by Heinrich Hertz in 1887 but theorized by James Clerk Maxwell about twenty years earlier—to send Morse Code messages and then voice transmissions without the telegraph wire.

The photograph served as the foundational technology, initially, for motion pictures and for television, though the latter also transmitted wirelessly like radio. Mass news print media did not use photographs extensively even after the halftone process, which turned images into a series of dots so that they could be printed, allowed the *New York Graphic* to publish the first newspaper photo in 1880. Illustrations, not photographs, would dominate the images in newspapers and magazines into the twentieth century. Photographs, however, were available for public consumption in the last half of the nineteenth century. They could be seen in public exhibits; with stereoscopes, a viewer that held cards with images affixed to them; and in books. By the middle of the nineteenth century, photographs could be printed on paper rather than on glass or metal. The paper images were called calotypes by their British inventor, Henry Fox Talbot. This new technology created a means of producing thousands of photographic books, like *Gardner's Photographic Sketch Book of the War*, which gave Americans everywhere the chance to view the often-disturbing images of the Civil War, taken by Alexander Gardner and other photographers.[11]

Moving images had been a fascination of people, though, for years. Early in the nineteenth century, children played with a thaumatrope, a disk with images on both sides and string attached so that the disk could be spun. The resulting movement made the images on either side appear to be one and to move. But it was the work of another Briton, Edweard Muybridge, that revealed that a series of photographic images taken sequentially of action could produce the illusion of motion in pictures. In 1878, using wires to trip the shutter mechanism of cameras, Muybridge photographed a horse running on a California track. The result revealed, first, that horses had all four hooves off the ground at once when in a gallop. But, Muybridge discovered, if the photographs were then flipped through that the horse appeared to move. He began experimenting with a number of subjects, producing near motion pictures of people and animals running, working, and playing.[12] By the last decade of the nineteenth century, the elements were in place for Thomas Edison, William Dickson, the Lumière brothers, and others to create motion pictures.

These new media, which were electronically driven, created for the nation a new means of social encounters, according to Carolyn Marvin.[13] These new inventions, a 1902 publication said, "mark the real epochs of civilization far more truly than do the wars of the nations or the reigns of powerful monarchs."[14] Historian Arthur M. Schlesinger, writing a half century and two world wars later, agreed, referring to radio and motion pictures as "revolutionary inventions," along with automobiles and airplanes in a 1950 *Washington Post* reflection on world-shaking events from the first half of the twentieth century.[15] Within a decade, television would be added to the list of "revolutionary inventions," as well. At the turn of the century, though, all the new means of disseminating information and entertainment were new, newsworthy, and often considered potentially world changing as they moved into the public sphere. Sometimes, the public considered them dangerous, as well, and that made them just as newsworthy because of the fear that these new media were influencing society in dangerous ways.

RECORDED MUSIC

"Talking machines"—called the phonograph, graphophone, or gramophone depending upon the inventor and recording medium—first came to the attention of Americans in the late 1870s. *Harper's Weekly* featured Thomas Edison's phonograph, calling its amazing ability to record the words of people something that would have caused colonial America's witch-hunters to have uncovered a "rich harvest of victims" because of the magical nature of the process and the accuracy of reproduction. The feature then explained how the phonograph used tin foil wrapped around a cylinder as the recording medium to capture a person's words. "Every vibration of his voice is faithfully recorded," the article said.[16]

Talking machines created controversy around the turn of the century, not because of the possibility of recording music, but because they offered the possibility of recording voices. That meant that thousands of jobs could be lost to the recording process. In 1907, about 200,000 stenographers worked in the United States, but according to reports, "The phonograph threatens to do for the stenographer what the automobile has done for the horse."[17] Losing jobs to technology, however, was to become a byproduct of technology in the beginning of the twentieth century. Stenographers may not have lost out as quickly as predicted, but illustrators—powerful for decades in print media—inevitably lost out to photographs. Every theater in the country prior to 1927 had at least one musician who played along with movies, but sound killed that job. Musicians would also suffer because records provided music by a relatively small group of artists compared to the number who performed live around the nation.

Recorded music became the major attraction of the phonograph. The Edison Company, which had been the principal maker of talking machines for years, lost control of the medium to the Victor Talking Machine Company, the business that developed around Emile Berliner's gramophone and flat records—instead of cylinders—that it played early in the twentieth century. Victor popularized the phonograph by making it a piece of furniture—the Victrola[18]—and then by giving away recordings by Enrico Caruso with every one purchased. Music by the opera star was not controversial, but it did increase interest in opera and recorded music. From 1901 through 1915, more than 1.2 million phonographs were sold in the United States.[19] Record sales soared to nearly 107 million by 1919, and more than 6 million American homes had a phonograph by 1922.[20] This fact, coupled with the introduction of radio and its penchant for filling air time with recorded music—approximately 250,000 radios were in American homes within a year of the first official radio broadcast—meant that via the phonograph or through radio broadcasts that recorded music was available to most Americans.[21]

Recorded music also found its way into the lives of millions of Americans at the end of the 1920s with the introduction of "talkies." Edison, Lee De Forest, and others had been working on ways to add sound to film for years, and Edison's assistant, William Dickson, even produced early sound footage in the 1890s, but the idea was scrapped. When Warner Brothers released *The Jazz Singer* in 1927, the sound of Al Jolson singing "Mammy" could be heard in only about one hundred theaters, but from then until 1929, motion picture companies spent millions to upgrade projectors in theaters.[22] The fact that the first sound feature was based around a singer of jazz was no accident. The music by 1927 permeated the nation, but that did not make it acceptable to all Americans. Jazz was, in fact, so controversial that its playing was banned in many places and created perhaps recorded music's greatest controversy. Recorded jazz music definitely affected national life in the 1920s, or as Will Hays, the man in charge of motion picture content oversight, said after the success of talkies and the music they provided, "Now, neither the artist nor his art will ever die."[23]

The problem that jazz and its recordings created stemmed from the origins of the music. Jazz was the product of Africans and the assorted rhythms, sounds, and tunes that they had preserved since the slave trade had forced servitude in Americas from the 1600s into the early 1800s. In the 1890s, commercial record production tapped into the wealth of black performers. Ragtime, with talented musicians like Scott Joplin, had been played since the late 1860s, but it came into wider acceptance in 1893 with the Columbian Exposition in Chicago, an event that marked the four hundredth anniversary of Columbus' New-World discovery. Though blacks were basically relegated to a single day to attend the Exposition, a vibrant African-American community existed in Chicago, and through it, numerous whites who were in Chicago heard ragtime and liked it.[24] As ragtime and then jazz became more prevalent in live performances, record companies began to record a small number of black performers. Music by black performers was initially recorded in different ways, depending upon the target audience black or white. Music was cleaned up for whites and left in a more "pure" state for African American audiences. Records targeted at blacks were called "race records," and at least eight hundred of these recordings were made prior to 1920.[25] In the 1920s, though, the music literally caught fire in America as it became the de facto music of the speakeasies and of Prohibition. It also became extremely popular with white college fraternities, all of which increased the calls for its censorship. The combination, one doctor said, was enough to drive people insane. He said that "the radio, jazz and bad liquor" were directly to blame for most of the 45,000 people currently housed in New York asylums.[26]

Being popular in speakeasies and on exclusive college campuses did not equate to acceptance for jazz. The fact that it had its origins in black culture naturally made it something that millions of Americans felt should be banned because of the divisiveness of race in America. In 1917, the New Orleans *Times-Picayune* called on universal suppression of jazz, saying that all citizens should "make it their civic point of honor to suppress [jazz]. Its music value is nil and its possibilities of harm are great." A year later, the *Times-Picayune* pointed out that the fact that jazz originated with blacks meant it was "the manifestation of a low streak in man's tastes."[27] One magazine even suggested that if jazz were not eradicated that the United States would be overtaken by "inferior" races who would listen to the music and lead "the half-crazed barbarian to the vilest of deeds," which suggested the rape of white women by black men.[28] Even record companies described their race records in derogatory language. "These dances are not ordinary music," Columbia Records said of a record by W. C. Handy. "The 'inside' syncopation, the weird harmonies, unforgettable unique, and the strange use of many of the instruments seem to give us an insight into the real, primal, superstitious, humorous nature of the negro."[29]

The attacks on jazz grew louder as the twenties progressed. An Illinois town banned all jazz records in 1921. A record shop in New York was fined for playing jazz when a funeral passed.[30] In 1922, the *New York Times* ran a story about jazz records that captured nearly all of the anti-jazz sentiments that many Americans had. First, phonographs were blamed for the spread of jazz. Next, the writer attempted to make it sound as if jazz were changing to make it more acceptable to mainstream America, away from its black origins, but that did not help. "Dance music is wrong if it creates nasty steps" regardless of the color of the performers. Then, the piece explained the problems with jazz. It came from "the African jungle" and moved throughout the country producing "the final disintegration of American morals." "And so, the article stated, "the great god Jazz spread over our fair land—until the very electric pianos bowed their allegiance. Every dance hall in Harlem had its whining saxophone, and every telephone operator in South Bend was doing the shimmy."[31] Further attacks on jazz blamed it for medical problems including mental illness and alcoholism. The *Ladies Home Journal* even suggested that jazz would lead to a loss of virginity among girls who would leave the [dance] hall in a state of dangerous disturbance."[32]

One African American leader, Carter G. Woodson, who was responsible for the beginning of Black History Week and the *Journal of Negro History*, believed that jazz was such a detriment to society—black and white—that radical measures had to be taken. In 1933 he wrote in the *Baltimore Afro-American* that he believed that Germany's new leader had the right idea about jazz when he banned it from his country. "Would to God that he had the power," Woodson wrote, "to round up all jazz promoters and performers of both races in Europe and America and execute them as criminals."[33] The problem was that by 1933, jazz was so ingrained into American society that it could not be reversed. Louis Armstrong was selling 100,000 copies a year of his recordings by 1930, and Cab Calloway, the orchestra leader at Harlem's Cotton Club, released his recording "Minnie the Moocher" in 1932. Minnie, a prostitute, represented all the evils of jazz, but the song was so infectious with its melody, scat singing, and overall happy feeling that it became a perfect representation of why so many Americans loved jazz.

With jazz on record, on radio, and in motion pictures, it was now thoroughly a part of American culture. White musicians co-opted the sound, too. As big band musicians such as Paul Whiteman, Benny Goodman, Glenn Miller, and Tommy Dorsey adapted jazz rhythms to typical white dance bands, jazz became mainstream. Efforts to suppress recorded music failed because new media made its dissemination too great to be stopped.

MOTION PICTURES

In 1896, "the Wizard" revealed his latest invention, something he called the vitascope. Thomas Edison, often called "Wizard" (shortened from "the Wizard of Menlo Park") because of his amazing inventions, provided a preview of his newest, which allowed the projection of moving images in large public venues, at his West Orange, New Jersey, laboratories before its New York City premiere. For his demonstration, Edison projected, among several of his "magic pictures,"[34] a piece filled with serpentine dances, "which astonished all who saw the views."[35] Not only did Edison cast the image of the dancer onto a large canvas screen, he tinted many of the frames so that the dancer, Annabelle, moved in a swirl of colors. "The umbrella dancers seemed almost to be creatures of flesh and blood," one reviewer said following the New York showing. "Even blond tresses, stirred by the vigor of the performer's exercise, streamed out as naturally as if a little breeze were toying them."[36] At his initial display of the vitascope, Edison also presented a film by Briton Robert Paul. It was what would be called an "actuality," a film not staged on a set like Edison's soon-to-be famous Black Maria; it was a film of the real world. Titled *Rough Sea at Dover*, the film depicted waves crashing on the beach. "It was," according to one review, "the closest copy of nature any work of man has ever yet achieved."[37] The vitascope was quickly dubbed "a miracle of human ingenuity." Another newspaper asked of Edison, "Can genius go farther?" "The vitascope is a wonder, a marvel, an outstanding example of human ingenuity," the *Los Angeles Times* raved. In their first week in operation in Los Angeles, vitascope showings drew an estimated twenty thousand to the Orpheum with another ten thousand turned away.[38]

The vitascope, which was actually developed by Thomas Armat and Francis Jenkins and licensed by Edison,[39] was not the world's first projection system. French brothers Auguste and Louis Lumière, in 1895, created the cinématographe and a film of workers leaving their Lyons factory. During the next two years, the pair would travel the world creating more than fifteen hundred actualities in cities that most people could only dream of visiting. The cinématographe arrived in the United States a few months after the vitascope's premiere. The Lumière projector and moving images had a number of advantages over the vitascope. "The remarkable feature about it [cinématographe] is that the figures not only go through the action, as in the kinetoscope, but appear and disappear, walk, run, and grow smaller or larger, as seen in perspective or near by," one reviewer noted in comparing the two projection systems and their films.[40] In New York, the cinématographe drew attention away from the vitascope initially with one publication saying that "nothing has ever before taken so strong and lasting a hold" as the cinématographe.[41] "It brings the world with it," another news story said. "It leaves absolutely nothing to the imagination, save sound and color. All the form, no matter how small, and all the action, no matter how swift, are reproduced with the unvarying exactness of the camera."[42]

The cinématographe and vitascope were not the only projection systems developed at the end of the nineteenth century. All battled for the public's coins for admission to see moving pictures, something that Edison's earlier invention, the kinetoscope, had monopolized since 1894. Using the ideas of Muybridge and his zoöpraxiscope projection system and the work of assistant William Dickson, Edison introduced the kinetoscope in 1891 at his laboratory. A box with a peep hole in its top, a continuous loop of film filled with images rotated into view. Edison previewed the invention for a group of women in May 1891. "As they looked through this hole they saw a picture of a man," the news report of the demonstration explained. "It was a most marvelous picture. It bowed and smiled and waved its hands and took off its hat with the most perfect naturalness and grace. Every motion was perfect. There was not a hitch or a jerk."[43] Another paper noted that "Edison promised to do the thing for which men and women have been burned at the stake, for if he succeeds he will practice what was once called the art of sorcery."[44]

By the end of 1894, the kinetoscope was referred to as "perfection." "The idea of looking into a box and seeing reproduced there the exact counterpart of a human being—walking, dancing, and laughing before your very eyes—is a sight almost too wonderful for the mind to comprehend," the *Washington Post* exclaimed.[45]

The possibilities that motion pictures and radio stirred among inventors and the general public foreshadowed the future. In 1898, reports surfaced of a French invention called the telelectroscope. With it, a news story speculated, "one might sit in a room in New York and see instantaneous pictures of what might happen to be going on in Havana harbor. These pictures would be shimmering affairs like the vitascope pictures, except that they would reproduce the objects seen in something like their natural colors."[46] The telelectroscope truly offered possibilities that were conceivable to few at the end of the nineteenth century. Inventors, however, were more interested in harnessing radio waves and celluloid motion images. Moving sound wirelessly, though, was what had captured the world's attention. Broadcasting, a term taken from America's rural heritage of scattering seeds, was to be the next great creation. More than recorded music or motion pictures, the wireless promised to be the most transformative communication medium of the electronic era because, in the first half of the twentieth century, it allowed instantaneous sharing of information between points miles apart as nothing known to humankind.

Motion pictures, however, were of immense interest, also, and used as more than entertainment pieces. Their content, one publication said, "go hand in hand with the daily press in presenting to nightly audiences events which they have seen during the day or read of in the evening papers."[47] Movie goers at the end of the nineteenth century and beginning of the twentieth century were able to view fighting from the Spanish-American War; its hero, Theodore Roosevelt, hunting in Colorado; the funeral of President William McKinley; gripping images of the San Francisco earthquake; and the activities of suffragists and temperance-movement advocates. What resulted when people went to see pictures early in their development, then, was something very similar to what has been called a "visual newspaper."[48] A variety of information appeared on the screen. News and entertainment played back-to-back, just as news and entertainment were scattered through the pages of newspapers, often with a lighthearted story following the report of some tragedy. A perfect example of the visual newspaper showed at a Kentucky carnival in 1900. The moving-picture newspaper began with President McKinley's funeral, "illustrating all the pomp and ceremony attending the mournful event....Five hundred feet of films made when Carrie Nation was on her joint washing crusade in Kansas will also be presented. The robust home defender and her band of wreckers are shown at work with their hatchets and the exhibition undoubtedly will prove a drawing card," the *Paducah (Ky.) Sun* told its readers. Those news items were followed by "A trip to the moon, a laughable Parisian moving picture invention" and the "successful colored animated scenes [of] Cinderella."[49]

At the Kentucky carnival, people were assured that they were seeing exactly what had just played in New York City. The former president's funeral and temperance firebrand Carrie Nation received top billing, just as they would have in a print paper. They were followed by the ground-breaking film of Georges Méliès, *A Trip to the Moon*, which is credited with introducing special effects to motion pictures. Included, too, was an animated feature—a cartoon or comic—something that had become a regular in newspapers courtesy of journalists Joseph Pulitzer and William Randolph Hearst. And just like newspapers, where the major metropolitan dailies often broke the news first, these features of the visual newspaper had already played in the information and cultural center of the nation. They were now in Kentucky, and those about to see them were assured that their viewing was not far behind that of New Yorkers. In addition, everyone who viewed them, whether they lived in New York City or Paducah, Kentucky, could understand what happened on screen better than many could comprehend the information in the newspaper. Motion pictures, presented without a soundtrack, were created so that anyone might understand their content. Illiteracy did not hamper movie attendance, and it was not a factor in understanding motion picture content.

Motion pictures also came into popularity in American society at the same time that the Progressives and Muckrakers controlled an important element of U.S. media. Consequently, American movies also assumed a similar mindset in many instances. Pictures that dealt with social issues and political corruption could be found playing with comedies, just as the film about Carrie Nation and temperance played along with *A Trip to the Moon* in Paducah. A twenty-five-second film from 1900, *How They Rob Men in Chicago*, in one quick scene depicted corrupt police officers. *The Black Hand* in 1906 dealt with extortion by gangsters. D. W. Griffith produced *The Usurer* in 1910 about a hardhearted loan shark who foreclosed on anyone who owed him money only to find that all his money could not save him when he was locked in his vault with it. *The New York Hat* in 1912 provided commentary on life in American towns, while *The Crime of Carelessness* in 1912 dealt with the real-life tragedy of the March 1911 Triangle Shirtwaist Factory fire in New York City that claimed the lives of 146 workers, almost all poor, female immigrants or children. Charlie Chaplin's "the Little Tramp" later in the silent era served as surrogate for the nation's underclass—for powerless immigrants and the poor. Creating motion pictures with a social consciousness naturally made them problematic for some Americans, and protests against movie content grew along with film popularity.

Motion picture problems

By 1905, the nation had entered the nickelodeon boom, the name given to theaters that sprang up around the country in converted storefronts and makeshift facilities. Within three years, nearly ten thousand nickelodeons (the name taken from the Greek for "nickel theater" and so named because of the cost of admission) operated in the United States. That number would grow to around 21,000 theaters by 1916. In order to feed the country's seemingly insatiable appetite for film, twenty to thirty new movies were needed per week by 1910, most produced by one of the twenty companies that now existed to create them. In 1907, *The Saturday Evening Post* reported, daily attendance at nickelodeons topped two million.[50] In New York City alone, 250,000 people attended moving picture theaters during the week with at least a half million catching shows on Sundays.[51] By 1910, approximately 26 million Americans attended at least one movie weekly.[52]

While most Americans may have viewed motion pictures as harmless in their content, some saw film as a corruptor of morals. In 1905, the *New York Herald* decried a motion picture as "an insult to decency." "The whole story," the paper said, "the atmosphere surrounding it, the incidents, the personalities of the characters are wholly immoral and degenerate....It defends immorality. It glorifies debauchery."[53] Even earlier, the *San Francisco Call* promoted an effort to eradicate vitascope and phonograph parlors from the city. They "Debauch the Young and Pander to the Depraved," the headline claimed, a readout adding, "Parlors Teach Boys and Girls a Lasting Lesson in Immorality." The story told of how a young boy was able to view "moving scenes that inflamed his imagination with corrupting thought....Curious people, careful of their nickels, may in each machine receive a sample view, and if satisfied may then purchase the rest of the disgusting exhibition. Thousands upon thousands of dollars have been invested in the business and a tremendous daily revenue is the result. A powerful corporation has been organized to debauch the young."[54]

In March 1907, the *Chicago Tribune* initiated a campaign to control motion pictures. Movies, were "without a redeeming feature to warrant their existence," the paper said, "ministering to the lowest passions of childhood." The *Tribune*'s editorial then called for a law "absolutely forbidding" admission of any child under the age of eighteen to any motion picture theater in the city. Within three months, New York City's police commissioner recommended that all licenses for nickelodeons be cancelled.[55] Chicago, however, put concern into action and created the Police Censor Board late in the year. Every film that played in the city had to receive a screening permit. That meant that a copy of each film had to be supplied to the board, which screened it. If found to be "immoral, obscene or indecent" based upon the perceived morals of the city, regulators denied the film a permit, and it could not be shown in Chicago.[56] New York added its board in 1909. Soon, cities and states across the nation began film boards, and many remained in existence until the

Supreme Court narrowly defined what could warrant censorship in the 1965 case *Freedman v. Maryland*.

Klan rally

Filmmaking in the United States changed forever in 1915 following the release of D. W. Griffith's *The Birth of a Nation*. Until that time, the average American film lasted a little more than twelve minutes. Griffith's epic ran for almost three hours. Griffith based his film on a novel by Thomas Dixon, *The Clansman*. It told the story of the Civil War and Reconstruction with a sympathetic understanding of the South, especially the formation of the Ku Klux Klan. Griffith spent a little more than $100,000 to make the film—when the typical film cost about $20,000 to produce. At its release, about 10 percent of all Americans saw the movie, and it grossed nearly $50 million throughout its run and re-release after World War I.[57]

The Birth of a Nation created controversy immediately because of its content. President Woodrow Wilson after seeing the film reportedly said, "It is like writing history with lightning, and my only regret is that it is all so terribly true."[58] At least thirty-eight Senators and the Chief Justice of the United States, however, "cheered and applauded throughout the three hours required to show the gigantic picture" at a special viewing at the National Press Club in Washington.[59] By the time of the movie's March 3 premiere in New York City, its subject matter had embroiled the film in charges of racism, protests, and calls for censorship, which began after the Los Angeles branch of the National Association for the Advancement of Colored People requested the city's film board ban the movie.[60] Since film boards were comprised almost entirely of whites, few review boards initially banned Griffith's picture, and white audiences for the most raved over it. Still, some white Americans found the film to be perverse. "The portrayal, unjust as it is to the negro, showing him as a cruel, inhuman, almost demented being, cannot help but create prejudice against a race that has a difficult road to travel at best and needs all possible sympathy and understanding from his white neighbor," a letter writer in New York said.[61]

Racial discrimination in 1915 was a part of life in America, and the calls by the NAACP and others to ban the film because of its racial portrayals meant that stopping the showing of *Birth of a Nation* would be much more difficult than if those trying to shut it down could point to scenes of obscenity. Repeatedly, *Birth of a Nation* opened in cities across the country. When it opened in Boston, African Americans and white sympathizers first approached state legislators to ban further showings. The state house arranged a mass meeting at Faneuil Hall with Griffith present to promote free expression for and historical accuracy of his film.[62] When this effort failed, blacks rioted. Boston sent 260 police to the theater after being alerted of the possible protest. Hundreds of African Americans attempted to stop the showing of the movie and failed.[63]

The fact that a motion picture could create such a massive uproar in the nation pointed to the power of visual expression. People stood and cheered and were amazed by the scenes of fighting in *Birth of a Nation*. Others were horrified at the picture's content and portrayal of blacks and its glorification of the Klan. Americans, with *The Birth of a Nation*, truly realized the social and cultural power that film possessed. As Hollywood turned into a film mecca following World War I, more questions about the power and effects of motion pictures began to be asked. In the 1920s, Americans sought answers as they questioned the power of film to effect society.

Youth and movie effects

The 1920s provided Americans with more expendable income than had ever existed. New technology and credit led to a period of extravagance for many during the jazz age with as many as 80 million of the nation's approximately 106 million citizens taking in a movie each week. Even though many in the nation felt that they had been able to legislate morality in a number of ways including Prohibition, the increasing demand for entertainment—especially movies—created the star system in Hollywood and a newfound focus upon popular culture and movie idols. In 1915, the Supreme Court ruled in *Mutual Film Corporation v. Industrial Commission of Ohio* that film did not possess First Amendment rights and that motion pictures were "capable of evil, having power for it, the greater because of the attractiveness and manner of exhibition."[64] The ruling did not stop what many believed to be the excessive, corrupt, and sinful lifestyle of those involved in the movie industry, and they gave California's motion picture town the name Hollywood Babylon.

With more than three-fourths of the population attending a movie every week, a series of events in 1921 only confirmed the increasingly bad name that the movie industry was receiving. Stories of immoral lifestyles and drug deaths were all encapsulated in the sensational story of Roscoe "Fatty" Arbuckle, one of the nation's most popular comedic stars who was arrested on September 10 in San Francisco and charged with the murder of actress Virginia Rappe in Arbuckle's hotel suite. Reports soon told how Arbuckle, who weighed three hundred pounds, suffocated Rappe as he attempted to rape her.[65] The case truly placed Hollywood in the national spotlight. "The publicity has put an unbearable burden of infamy on the motion picture. Every individual within the industry will in some degree feel the stigma of it," one editor said shortly after Arbuckle's arrest.[66] Two hung juries on the murder charge kept the Arbuckle case in front of the nation into 1922 when a third jury exonerated the comedian. From a note, the jury foreman read, "Acquittal is not enough for Roscoe Arbuckle. We feel that a great injustice has been done him. We feel also that it was our only plain duty to give him this exoneration. There was not the slightest proof adduced to connect him in any way with the commission of a crime."[67]

To oversee the industry and to mollify calls for government oversight and censorship of motion pictures, Hollywood decided that it had to police itself and voluntarily created through the Motion Picture Producers and Distributors of America an office to regulate activities. Will Hays, the former postmaster general, was tabbed to run the agency, which almost immediately assumed his name for it. One of Hays' first acts was to ban Arbuckle from motion pictures. The Hays Office, though, was only a bandage applied by Hollywood. Many Americans believed that motion pictures possessed power so great that they could influence the activities of people, especially the young. Researchers, consequently, began compiling information of movie effects on children and adolescents since most young people in the nation went to the movies more than once a week, and many people believed that what was seen on the screen—and heard, too, after the introduction of *The Jazz Singer* in 1927— affected young people with potential long-term detrimental harm to them and the country. Through a decade of concern, the Motion Picture Research Council finally organized researchers to conduct studies to find out just how motion pictures affected young, impressionable Americans. Known as the Payne Fund Studies for the organization that provided the funding, thirteen separate research volumes were compiled from 1928 through 1933 and presented to the nation. For many, the results confirmed their fears. Motion pictures, many of the studies said, have the ability to affect powerfully young people.

Herbert Blumer and Philip Hauser concluded in the study published as *Movies, Delinquency, and Crime* that when young people watched "crime techniques and criminal patterns of behavior" that pictures "may create attitudes and furnish techniques conducive, quite unwittingly, to delinquent or criminal behavior" among them. Another study, *Getting Ideas from the Movies* reported that 80 to 92 percent of young people retained what they saw in motion pictures for at least three months, and 60 percent retained information for at least six months. Ruth Peterson and L. L. Thurstone were even more certain of the lasting implications of film, saying "that motion pictures have definite, lasting effects on the social attitudes of children." When the same attitudes are seen repeatedly, there is a cumulative effect on attitude, they said. Social behavior and personal choice, Blumer would say in another study, were affected for extended periods of time for the young, regular movie attender.[68]

The reaction to the Payne Fund Studies was immediate in the media. "Commercial 'movies' apparently are the modern adolescent's textbook of life and love," one writer said.⁶⁹ The headline of another stated, "Youthful Crimes Laid to the Movies"; while another concluded, "Our Movie Made Children."⁷⁰ In three decades, motion pictures had become a permanent fixture in American life. Even in 1932, in the depths of depression, at least 55 million Americans attended movies at least once a week.⁷¹ Motion pictures changed the pattern of American life in less than three decades and remained news in media as well as a medium believed to be a powerful shaper of culture and mores. Only one other medium during the first third of the twentieth century would possess more power in the nation: radio.

THE WIRELESS

In February 1902, Ray Stannard Baker wrote for *McClure's* that the successful experiments of Guglielmo Marconi to send a wireless message across the Atlantic Ocean meant "there will at last be no escape for the weary from the daily news of the world." But Baker also understood the enormity of the achievement when he asked readers to "Think for a moment of sitting here on the edge of North America and listening to communications sent *through space* across nearly 2,000 miles of ocean from the edge of Europe!"⁷² While some doubted that the man who perfected the wireless telegraph had successfully sent the letter "S" in Morse Code from Cornwall, England, to Newfoundland, the *New York Times* remarked, "If MARCONI says it is true, I believe it."⁷³ The *New-York Tribune* described Marconi's feat in a more superlative fashion. "Nothing more wonderful in the history of the progress of science or in the story of the reduction of the elemental forces of the universe to the uses of mankind has ever been recorded," the paper declared. Marconi's success was the "MOST WONDERFUL ACHIEVEMENT OF THE ELECTRICAL AGE."⁷⁴ Others suggested that the transoceanic and continental telegraph cables "might now be coiled up and sold for junk."⁷⁵

At the beginning of the twentieth century, the race was on to develop a means of communication based on the telegraph but using radio waves bouncing off the atmosphere, not traveling through wires, to send information. People expected "substantial, not to say startling, advances during the first quarter of the twentieth century" related to the wireless.⁷⁶ Marconi, son of an Italian father and Irish mother, was quickly becoming a household name to people on both sides of the Atlantic, just as wireless telegraphy was a term that people were hearing that promised to change communication. Marconi, however, was not the only player in the race to develop a new form of communication, nor the only person to become the subject of nearly endless news stories. His contemporaries—Reginald Fessenden, Nathan Stubblefield, Nikola Tesla, and Lee De Forest, to name several—shared the public spotlight, though Marconi was surely the star of the group. While they all extolled the potential of wireless, they were not the first to propose that invisible waves existed within the atmosphere—or the ether as it was sometimes called—and held great potential for human use.

The belief that electricity could somehow send messages without wires was one of the suppositions made by Joseph Henry and Michael Faraday during the explosion of scientific research and establishment of academic societies of the 1820s. They discovered that when an electric circuit was placed near a magnet they could induce movement or a change in the circuit without any part of the experiment touching another. The agent of change traveled through the air. Later, James Clerk Maxwell and his student, Heinrich Hertz, proved that electric waves moved through space in a spherical form, that is radio waves.[77] This discovery paved the way for Marconi, who in 1896 moved to England and worked to improve all that Maxwell, Hertz, and a growing number of experimenters were doing with radio waves. He was able to send and receive signals over increasingly longer distances. He patented all he created, and in 1899, Marconi sent a message eighteen miles in England and established the Marconi Wireless Telegraph Company in Britain, and the American Marconi Company in the United States.

But what Marconi did that captured the interest of the press and, consequently, people on both continents occurred purely by serendipitous events. In order to prove that he could send signals even farther than eighteen miles, Marconi equipped the lightship *Goodwin* in the English Channel with a wireless telegraph transmitter. He placed the receiver at the South Foreland lighthouse. The ship on April 28, 1899, was struck by a passing vessel in dense fog. The lightship's crew did not realize that the collision had ripped a massive hole in the hull immediately. When they realized they were sinking, the crew quickly knew that making shore would be impossible because of the amount of water being taken on. No one on the ship even remembered "the vertical wire projecting from the 'wireless telegraphy' masthead" that the ship carried until a desperate crew member suggested, "Tell South Foreland that we're sinking." His startled shipmates doubted that the wireless would make a difference, but they sent the emergency message in Morse Code nonetheless: "HELP, WE HAVE BEEN STRUCK; ARE SINKING." Tugboats were immediately dispatched, and the *Goodwin*, despite the fog and the ship filling with water, was found and towed ten miles to safety. "And that's the way wireless telegraphy demonstrated its success and introduced itself to the workaday world last week," a story about the rescue said. "Everything that he [Marconi] had hoped or claimed was satisfactorily realized….The Goodwin lightship incident will give renewed impulse to inventors and experimenters in the new field." It also earned Marconi the title "pioneer."[78] The *Omaha Bee* declared, "All the World Agog Over the Development of Wireless Telegraphy" under the headline "Marconi's Coming Wonder," then stated, "Wireless telegraphy commands the ears of the world at the present time."[79]

The San Francisco Call *of May 7, 1899, created a full-page spread on the rescue of the Goodwin and on other uses of telegraphy.*

Marconi quickly became a celebrity. The press provided biographies of the young, handsome inventor, and people read all about his personal life. One story even referred to him as "A YOUNG DEMI-GOD."[80] The Irish-American press played up the inventor's ancestry, one paper saying that it was Marconi's Irish mother who provided the inspiration for his inventions. "She watched over his studies, encouraged his early bent for electrical invention and did the many things that a woman of insight can do to stir in the boy the capacities that have made Marconi, young as he is, so great a figure in the world's progress."[81] Others realized that the wireless held great potential for the United States, such as creating a way to communicate instantly with the nation's territories in the Pacific Ocean—from San Francisco to Honolulu, to Guam, and finally to Manila.[82]

Wireless competition

The ability to send messages to and from ships pointed to the principal use of wireless as understood at the beginning of the twentieth century. People equated wireless, naturally, with the telegraph. The purpose of the telegraph, whether it was Morse's version or the new wireless version, was to send messages from one point to another. The rescue of the *Goodwin* was but one example. But, moving information from point to point did not have to occur solely to divert tragedy. The wireless could also keep people informed. In late 1899, *The Transatlantic Times*, a wireless newspaper, arrived on board the *St. Paul*, which was sailing in the English Channel, and printed on board ship for passengers.[83] Within a few years, Western Union began sending wireless messages concerning any subject to anyone aboard a steamship at the "sender's risk."[84] Wireless' possibilities appeared limitless.

The new invention was already the focus of dozens of electricians, as they were sometimes called, in America and Europe, but expanded usage of wireless telegraphy intensified the competition surrounding it and expanded the national conversation concerning the invention and its potential. Reginald Fessenden and Lee De Forest quickly became the inventors in the United States that rivaled Marconi the most in terms of press coverage and what they promised to do for and with the wireless technology. Both, in fact, expanded and improved upon Marconi's wireless telegraphy. Fessenden, who had worked as a college professor and an electrical engineer, took a job with the U.S. Weather Bureau, assigned specifically to conduct experiments with wireless telegraphy. Fessenden studied the work of Marconi and believed that a constant signal, not bursts of energy were required to make the wireless more practical. His ideas, in fact, allowed the wireless to be transformed into radio[85] through experiments conducted on the Outer Banks of North Carolina. Fessenden spent more than two years experimenting with sending signals over the sand

and water on North Carolina's coast, and news reports praised his work, saying that Fessenden's wireless system "far exceeds that of the Marconi system and is capable of competing with the land lines."[86] The Canadian-born experimenter contacted ships at least two hundred miles off Cape Hatteras, but what Fessenden did that was truly transforming for the wireless was his "modus operandi" for sending and receiving information: "the Fessenden system will be to wireless telephony," news reports announced—not wireless telegraphy as it had previously been called.[87] Though it would take him several years to perfect, Fessenden sent music to ships on Christmas Eve, 1906.

Fessenden, though, was neither the showman nor the braggart that Iowa-born Lee De Forest turned out to be. De Forest, who grew up in Alabama and earned a doctorate from Yale University, tried unsuccessfully to land a position with the American Marconi Company after graduation. Failing to get the job he wanted, he instead partnered with a Wall-Street investor Abraham White, and the pair formed the De Forest Wireless Company in 1902. De Forest, like Marconi, believed that he should patent everything that he made because he wanted to be rich. In 1902, shortly after beginning to sell stock in his company, De Forest infamously wrote, "Soon, we believe, the suckers will begin to bite."[88] But De Forest just as feverously wanted to be famous and invent something that would change the world. He began referring to himself as "the father of radio" at least as early as 1925.[89] The media soon followed his lead, one 1929 article referring to him as the "inventor who is known as 'The Father of Radio.'"[90] De Forest permanently inscribed the title to himself with his 1950 autobiography, *Father of Radio*.

De Forest and his work only intensified the nation's dialog on wireless as people saw his work and that of Fessenden, Marconi, and others as a competition. "MARCONI IS BEATEN. American Inventor Puts Italian's Device in the Shade," newspapers declared after De Forest performed tests with his wireless apparatus in New York City.[91] De Forest regularly promised that he would make possible the regular sending of information via the wireless "from Maine to the Orient" and across the Atlantic.[92] The race between De Forest and Marconi became one to see who could make the next breakthrough in wireless, and the competition between Marconi American and De Forest Wireless played out on financial pages and in advertisements to purchase shares of the companies. De Forest—with White's pushing—made sure that everything that he did received some sort of media coverage.

Even when everything fell apart for De Forest in 1906, his actions made news. First, Marconi claimed that the De Forest Wireless Telegraph Company infringed upon Marconi patents. New York courts agreed and ordered an injunction against De Forest's company to "forever desist from, directly or indirectly, making or causing to be made, using or causing to be used, or vending to others to be used" any of the elements of wireless on which Marconi owned a patent. The injunction forbad De Forest from even improving upon Marconi's designs.[93] The U.S. Supreme Court upheld the injunction. In Chief Justice Melville Fuller's decision, he ordered the De Forest company to forever desist from advertising or selling any wireless system. Ironically, the U.S. government used the De Forest system for all of its military and commercial enterprises.[94] With the company in ruins, White pulled out, leaving De Forest with no money and a single prototype tube considered of little value.

Next, in what might be considered today as the perfect tabloid complement to De Forest's professional debacle, his much-publicized wedding in 1906—"Wireless Wooing Ends in a Wedding," the *New York World* had announced—came to an even more publicized conclusion. "When Miss Lucille Sherdowne married Dr. Lee De Forest, inventor of the telegraph system that bears his name," the *Washington Times* told its readers in a full-page story complete with images that included a dead Cupid surrounded by vultures, "she became known as 'The Wireless Bride' because her future husband taught her telegraphy, installed a set of instruments in her own particular den and did most of his courting by Wireless—And now comes the sad ending of last winter's greatest romance." The story then asked, "Is there any way of getting a divorce by wireless?" By 1908, De Forest's love life made more headlines when papers revealed his impending nuptials to the granddaughter of woman's rights advocate Elizabeth Cady Stanton.[95]

But 1906 also marked the transforming moment for De Forest and radio because that is the year that he introduced his audion tube. Based off an incandescent light bulb and the work of John Fleming, De Forest added a grid to Fleming's creation, and suddenly he had a tube that enabled wireless signals to be amplified, something he explained in two installments in *Scientific American*.[96] Though it would require much more experimentation before the audion would benefit wireless, it did spark De Forest to think of ways to expand the use of the wireless beyond point-to-point transmission to the idea of broadcasting to an audience rather to individuals. In 1910, De Forest "transmitted direct from the stage of the Metropolitan Opera House New York to over a score of wireless stations some of them miles away," with predictions that now that performances from the Met could be broadcast to luxury liners—"concerts given aboard the great sea-going hotels would soon become a thing of the past." *Modern Electrics* compared the broadcast to the same one as heard through a wired telephone system, noting the wireless version was superior because "we get the received tone in all its beauty" rather than distorted by the medium of the wire.[97]

Wireless expansion

After a number of successful uses of wireless for sea rescues, the government passed the Ship Act of 1910, which required all ships to have a wireless transmitter and receiver. By the teens, wireless needed regulation, though, and the Radio Act of 1912 supplied it. Partly a reaction to the sinking of the *Titanic*, the act looked at wireless still as a point-to-point transmission tool, and it acknowledged that ordinary citizens used the radio telephone, too. But, it underestimated that usage as a letter to the *New York Times* explained. The editor of *Modern Electrics* explained that "close to 400,000 wireless experimenters and amateurs [operate] in the United States alone." The letter also intimated that non-government-controlled wireless transmitting was needed. "The public fully shares this view," the letter said in closing.[98] Just how true the statement was soon played out throughout America.

By World War I, the wireless was practically a household commodity with amateurs, as they were now generally called, providing all types of services. In Kansas, amateurs organized to transmit news. There, "young men interested in wireless telegraphy," a magazine article announced, "will furnish the smaller papers of the state with the news from neighboring towns." Amateurs also brought "Weather reports, market quotations and world news daily by wireless telegraph" daily to farmers throughout the Midwest.[99] As the nation worried about German attacks directly on the United States, wireless operations were viewed as a first line of defense. "The wireless man seems to be omnipresent. He is found in every part of the country," one newspaper explained. "[T]here exists today a formidable defense weapon, which up to now has not been exploited by Uncle Sam. We refer to the thousands of amateur radio stations scattered broadcast through the entire length and breadth of this fair land. There is hardly a hamlet today which does not boast of several amateur wireless stations, as their number is increasing by many hundreds each day."[100] By the time it looked inevitable that the United States would enter the war, amateurs had created a relay system that crossed the continent so that any information of value could be transmitted to national leaders.[101]

This system, however, did not come to fruition. The federal government, instead, deemed the amateurs and their wireless broadcasts as a potential threat to national security. Worried especially about broadcasts by German Americans, the government shortly after declaring war on Germany seized all radio stations in the United States and its possessions worldwide. President Woodrow Wilson ordered the closures "to insure the proper conduct of the war against the imperial German government and the successful termination thereof." Wilson based his decree upon the Radio Act of 1912, which gave government jurisdiction over the medium in times of emergency. The plan included closure of many of the amateur stations and the use of others by the Navy.[102]

Wireless explosion

The end of the Great War marked the return of amateur broadcasters. Before the war, though, a young employee of the Marconi American Company, David Sarnoff, had suggested to his superiors that the wireless could be used as a medium of entertainment that could be carried directly into the homes of Americans on a scale much wider than that of the amateurs in the Midwest. Though the "Music Box Memo" was ignored by Marconi American, the idea of converting the wireless from a point-to-point communicator principally and a broadcast medium for thousands of amateurs was an idea whose time had come. Several major American companies, including American Telephone and Telegraph, General Electric, and Westinghouse, formed the Radio Corporation of America. RCA, with the assistance of the government, bought the American Marconi Company's U.S. operations. Sarnoff assumed a management position. His idea that radio should be a broadcast medium was about to be realized. In the fall of 1920, Westinghouse, working with an RCA engineer named Frank Conrad, established the first licensed broadcast station in Pittsburgh. In November, the station broadcast the presidential election results. Radio, as an official broadcast medium, began.[103] Within three years, Americans had purchased 3 million home receivers. By the end of 1923, nearly 560 broadcast stations covered the nation. The country's largest cities naturally had radio stations, but so, too, did small communities in every corner of the nation. The major corporations that formed RCA were not the only businesses that operated radio stations. Newspapers began them; banks did, too.[104] Department stores were central to the spread of radio. They ran stations in their large stores in Philadelphia, Boston, New York, Chicago, and elsewhere that allowed people to see how a station operated. Conveniently, the stores also sold the wireless, usually on the same floor with the radio broadcast studio.[105]

Radio became, according to many, essential to the operation of the nation. "No vital issue can be decided fairly in this country," *Wireless Age* claimed in its October 1922 edition, "without the use of the radio telephone. Radio can carry into the home nothing more important than the truth about those vital issues to decide which is to determine the course of this greatest of countries."[106] In 1928, Herbert Hoover's campaign managers announced that Hoover's presidential bid would be carried out mainly via radio. "Mr. Hoover will address the voter at home by way of the radio," his campaign announced. "Mr. Hoover is an admirable radio speaker, and can reach thousands of thoughtful people in a way that is most desired to reach them."[107]

By the depths of the Great Depression, radio had made tremendous inroads into the nation. Though print media still saturated the country, radio was the pathway of immediacy and of free entertainment. In 1927, Americans owned 7 million radios. By 1933, that number reached 22 million.[108] With those radios, Americans were able to listen to live broadcasts of the Scopes "Monkey" Trial from Dayton, Tennessee, in 1925 and learned immediately of the 1929 stock market collapse. They would be entertained with the antics of Sam and Henry, later renamed Amos 'n' Andy. Two of three households would have a radio by 1935, and about 70 percent of Americans said they turned to radio for their news as their principal source by the time the nation joined the fighting of World War II. Radio in rural America was more prevalent than telephones, automobiles, or even electricity.[109] Radio, a newsmaker in 1900, was by 1930 a medium shaping the way the nation received information as it also introduced entertainment in ways never experienced in American homes.

TELEVISION

"TELEVISION IS BROUGHT NEARER THE HOME," the headline said that uncharacteristically for the *New York Times* crossed the entire page of eight columns in 1930. Despite the fact that within the text of the story, TV's developers warned that "a few more wrinkles need to be ironed out," the writer envisioned a time—sooner rather than later—when a news reporter would be able fly to some point of interest and the entire "audience will be with him, seeing what he does, and yet the audience will be perfectly safe and comfortable." In 1925, stories about a Russian immigrant, Vladimir Zworykin, and his work with Westinghouse speculated that his photo-electric experiments were leading to "a process for transmitting pictures by radio." "All the processes that are needed for projecting motion pictures are in existence already," Zworykin said. "It will take some years, but we will have the instantaneous or near-instananeous transmission of motion pictures."[110] As early as 1926, however, R. W. De Mott, president of the Radio Magazine Publishers' Association, predicted that television's perfection was imminent. "The engineers know that there are no unsurmountable obstacles in commercial television," he said. "They know, as you and I know, that before long we will be able to sit down at our receiving sets and see as well as hear, the world's series being played on diamonds hundreds of miles from home."[111] David Sarnoff, president of RCA, was a little more cautious, though. "I am confident that within five years you will be able to receive images through space as well as you are able to receive sound through space at the present time," he said in 1930.[112] Television did not broadcast a major-league game until 1939 when the "radio television" made its New York baseball premiere.[113]

Television in the late 1920s and early 1930s worked, though not well, and the technology was not poised to capture the nation. Radio as a broadcast medium was less than a decade old. Television was not yet ready to be mass marketed, and it would not be able to compete with radios in terms of cost, especially in the depression era of the 1930s. Still, Bell Telephone successfully tested television in 1927 by broadcasting the speech of soon-to-be president

Herbert Hoover from Washington to New York. "Herbert Hoover made a speech in Washington yesterday afternoon. An audience in New York heard him and saw him," the lead on the Page 1 story of the *New York Times* announced. "It was as if a photograph had suddenly come to life and begun to talk, smile, and nod its head and look this way and that." The secretary of commerce explained the significance of what he was doing. "Today we have, in a sense, the transmission of sight, for the first time in the world's history," he said and added, "Human genius has now destroyed the impediment of distance in a new respect, and in a manner hitherto unknown."[114] In a year's time, fifteen stations around the United States, were being operated by individuals and corporations like RCA.[115]

By September 1928, a young Mormon man, Philo Farnsworth, was making news with his TV experiments in San Francisco. The *Examiner* of September 3 reported that Farnsworth demonstrated his television with a feature of a man smoking. The smoke from the cigarette was plainly visible, the article said. In it, Farnsworth explained that his television could be easily attached to a home radio set at a cost of less than $100 to manufacture. In the race to perfect television, Zworykin then announced he had created a television that could be viewed by a room full of people. Farnsworth and Zworykin, with David Sarnoff always looming largely in the background, were now locked in a contest to develop television for mass consumption. Stories touted television's readiness, saying it was much further along in process of becoming a medium useful for the home than radio had been in its similar point of development. Farnsworth even broadcast images of the moon.[116]

With RCA promoting NBC television, the Columbia Broadcasting System decided to work on creating TV broadcasting, too, and the two radio networks pushed ahead with experimental stations in 1931 and 1932. Holding most of the patents for television except those belonging to Farnsworth and to whom Sarnoff was forced to pay royalties, NBC became the leader in television, broadcasting President Franklin Roosevelt speaking from the 1939 World's Fair. As the *New York Times* said following the historic broadcast, television, "in quest of its destiny," had now assumed "its role in the art of amusing Americans, and to fit in with the social life of the land."[117] World War II would put TV's role of amusing Americans on hold for nearly a decade. Though few people saw it, NBC and CBS continued to broadcast throughout the war, and the promises of television in the homes of almost all Americans would have to wait.

✳✳✳

The new ways of disseminating information that made the beginning of the twentieth century so exciting captured thousands of inches of copy in the print media that represented the only media prior to the electronic era. As radio threatened the supremacy of newspapers in the late 1920s, the medium often came under attack. The *New York Times* in 1929 derisively commented that "The radio news item is a vibration in the air, without record, without visible responsibility, without that incentive to accuracy that comes with print."[118] The *Times*' comment came months before radio broke the news of Wall Street's crash. In the first three decades of the twentieth century, radio was the most important of the electronic media. It provided people with entertainment and news, all received in the confines of the home. But other media were essential to American life. Motion pictures provided entertainment and news. People saw much of what was happening in the nation. In 1906, for example, footage of the San Francisco earthquake could be viewed in nickelodeons across the nation. Flames, combined with massive destruction, were there for people to watch along with the night's entertainment.

The power of the visual ultimately led to attacks upon cinema and the Payne Fund studies, which concluded that motion pictures possessed the ability to influence greatly, especially the nation's young people. Recorded music, which was an invention of the 1870s and was considered more for entertainment than for information dispersal, still became an issue in the early twentieth century as the phonograph and its other iterations were considered a threat to certain jobs—like that of the stenographer. But recorded music became an issue for society in the 1920s when jazz music suddenly became available. Music, which was rooted in the sounds and traditions of black America, became anathema for mainstream white America. The fact that radio played music only complicated the issue and enhanced the efforts of suppression of the music.

Radio and motion pictures also created radical cultural and social changes in the nation in the 1920s. The immense popularity of recorded music and motion pictures produced cultural icons—stars—unlike anything that had happened in America previously. Yes, performers like Jenny Lind had captured the nation's interest as the Swedish singing sensation did on her 1850 U.S. tour sponsored and promoted by P. T. Barnum. But recorded music and motion pictures because of the way they penetrated society in the 1920s allowed actors and musicians to become a part of everyday life, especially of young people. They repeatedly saw and heard stars, and they emulated them. Magazines especially, such as *McClure's, Life, Photoplay, Motion Picture,* and *Wireless Age,* provided stories, photographs, and examples of the stars' lifestyles that fed a national desire to know about the rich and famous and the opportunity to dress and look like them.

While television only entered the American consciousness in relation to the potential that broadcast media held in terms of the possibilities of sharing information during this era, the idea that such a medium might be feasible had been around almost as long as the phonograph with experiments by Maurice LeBlanc, Paul Nipkow, and John Logie Baird in Europe ultimately creating something that would be called mechanical television, based on Nipkow's spinning disk. The original dated from the 1880s and contained a series of holes around its edge, spiraling to the center. Light and the image from one side passed through the holes. Later the disk was used to produce an electrical signal that could produce an image—though it appeared more as a shadow than anything else.[119] The imaged telelectroscope of the 1890s, with its possibilities of sending live images through the air as the wireless did sound, came closer to describing electronic television that was developed in America in the 1920s and 1930s.

The inventiveness of the era certainly helped shape the nation in terms of thinking about ways to obtain and disseminate information. The ubiquitous nature of the new media—especially motion pictures and radio—guaranteed that they would quickly become essential components of society that people quickly found they did not want to be without. "The fairytales of the past," one writer explained, "are outdone by the realities of the present."[120] Three of every four Americans visited a motion picture theater per week on average in 1930 and two-thirds of all homes and an increasing number of cars had radios by 1935.[121] Add to this all the inventions that accompanied the new media—airplanes, electric washing machines, commercial electric refrigerators, assembly-line automobiles, plastics—even crayons and Kool-Aid—you begin to understand that this 1902 prediction concerning new inventions had not come close to envisioning the technological advancements of the first third of the twentieth century.

"What a century this has been in the annals of communication!" one writer exclaimed in 1927. "In less than three decades greater strides have been made in knitting communities and nations together than in the entire preceding century." Though the article praised all of the inventors of radio broadcasting, from Maxwell to Marconi, Fessenden and De Forest, it looked to television as the medium of the future. "It is as if the eye were carried through space to see what is happening in distant cities."[122] Television offered the promise of media of the future, but it would have to wait for the nation and world to fight a world war, just as radio for the masses had been put on hold to do the same nearly twenty-five years earlier.

NOTES

1. "Telephoning Without Wires," *New-York Daily Tribune*, January 2, 1902, 16; "Wireless Telephony," *The Bee* (Earlington, Ky.), January 9, 1902, 8; "Wireless Telephony Tests," *New York Times*, March 21, 1902, 2.
2. Nathan B. Stubblefield, *St. Louis Post-Dispatch*, January 10, 1902.

3. Carolyn Marvin, *When Old Technologies Were New: Thinking About Electric Communication in the Late Nineteenth Century* (New York and Oxford: Oxford University Press, 1988), 4.
4. *Electrical World*, February 22, 1890, 125, in Marvin, *When Old Technologies Were New*, 11.
5. "A Volcanic Eruption. Fears of a Great Calamity on the Island of Java," *Boston Daily Advertiser*, August 28, 1883, 1.
6. "News by Telegraph," *Daily Evening Bulletin*, (San Francisco), August 28, 1883, 3.
7. See, for example, "Around the World, *The North American* (Philadelphia), August 28, 1883, 1; "The Old World," *Milwaukee Sentinel*, August 28, 1883, 1; "Affairs Abroad," *St. Louis Globe-Democrat*, August 28, 1883, 2.
8. Burton J. Hendrick, "Telephones for the Millions," *McClure's* 44 (November 1914): 45, in Richard R. John, "Recasting the Information Infrastructure for the Industrial Age," *A Nation Transformed by Information*, eds. Alfred D. Chandler Jr. and James W. Cortada (Oxford: Oxford University Press, 2000), 87.
9. John, "Recasting the Information Infrastructure for the Industrial Age," *A Nation Transformed by Information*, 89.
10. Marvin, *When Old Technologies Were New*, 11; John, "Recasting the Information Infrastructure for the Industrial Age," *A Nation Transformed by Information*, 81.
11. For a complete understanding of the development of photography, see Beaumont Newhall, *The History of Photography* (New York: Museum of Modern Art, 1982).
12. Eadweard Muybridge. *Muybridge's Complete Human and Animal Locomotion. All 781 Plates from the 1887 "Animal Locomotion,"* 3 vols. (Mineola, N.Y.: Dover Publications, Inc. 1979), reprinted in "First Motion Pictures," *The Franklin Institute* (2009), http://www.fi.edu/learn/sci-tech/motion-pictures/motion-pictures.php/cts=photography.
13. Marvin, *When Old Technologies Were New*, 4.
14. Trumbull White, ed., *Our Wonderful Progress: The World's Triumphant Knowledge and Works* (1902), 27.
15. Arthur M. Schlesinger, "The 10 World-Shaking Events of the Half Century," *Washington Post*, January 1, 1950, B1.
16. "The Phonograph," *Harper's Weekly*, March 30, 1878, 249–50.
17. Frederic J. Haskin, "Possibilities of the Phonograph," *Washington Herald*, March 15, 1907, 7.
18. Dane Yorke, "The Rise and Fall of the Phonograph," *American Mercury*, 27 (September 1932): 6.
19. Leonard DeGraaf, "Confronting the Mass Market: Thomas Edison and the Entertainment Phonograph," *Business and Economic History* 24, 1 (Fall 1995): 91.
20. Jana Hyde, "The Recording Industry," in Wm. David Sloan, ed., *The Age of Mass Communication* (Northport, Ala.: Vision Press, 1998), 309.
21. David Morton, *Sound Recording: The Life Story of a Technology* (Baltimore: Johns Hopkins University Press, 2006), 64.
22. Morton, *Sound Recording*, 76.
23. Will H. Hays, *See and Hear* (New York: Motion Picture Producers and Distributors of America, 1929), 48.

24. Edward A. Berlin, *King of Ragtime: Scott Joplin and His Era* (New York and Oxford: Oxford University Press, 1995), 11–12.
25. Tim Brooks and Richard Keith Spottswood, *Lost Sounds: Blacks and the Birth of the Recording Industry, 1890–1919)* Urbana: University of Illinois Press, 2004), 6–9.
26. "Says Radio and Rum Drive Men Crazy," *New York Times*, March 6, 1925, 6.
27. *Times-Picayune* (New Orleans), June 17, 1917; *Times-Picayune* (New Orleans), June 20, 1918.
28. *Musical Leader* 45 (March 1923): 259.
29. Quoted in Brooks and Spottswood, *Lost Sounds*, 420.
30. "Voliva Bans Jazz Records," *New York Times*, January 11, 1921, 9; "Jazz Upsets Funeral," *New York Times*, May 19, 1922, 8.
31. Helen Bullitt Lowry, "Putting the Music Into the Jazz," *New York*, February 19, 1922, 8.
32. *Ladies Home Journal*, February 1927, 38.
33. Carter G. Woodson, Has Jazz Been a Help or a Hindrance to Racial Progress?" *Baltimore Afro-American*, October 14, 1933, 18, quoted in Leonard Ray Teel, "The Jazz Rage: Carter G. Woodson's Culture War in the African-American Press," *American Journalism* 11, 4 (Fall 1994): 348–49.
34. *Los Angeles Times*, July 14, 1896.
35. "Edison's Latest," *Omaha (Neb.) Daily Bee*, April 19, 1896, 16.
36. "Edison's Latest Invention," *New York Times*, April 26, 1896, 10.
37. *New York Herald*, April 24, 1896, 11, in Charles Musser, *Before the Nickelodeon: Edwin S. Porter and the Edison Manufacturing Company* (Berkeley: University of California Press, 1991), 63.
38. *Los Angeles Times*, July 5, 1896, 8; *Los Angeles Herald*, July 12, 1896, 12; *Los Angeles Times*, July 7, 1896, 6; *Los Angeles Times*, July 1, 1896, 1; quoted in Musser, *Before the Nickelodeon*, 82–84.
39. "Mr. Armat's Great Invention," *Richmond Dispatch*, May 6, 1900, 14.
40. Esther Singleton, "Life in Picture Films," *Washington Post*, July 5, 1896, 12.
41. *Clipper*, July 11, 1896, 12, quoted in Musser, *Before the Nickelodeon*, 87.
42. "Cinematographe at Willard Hall," *Washington Post*, January 1, 1897, 12.
43. "The Kinetograph," *Salt Lake Herald* (Salt Lake City, Utah), May 29, 1891, 1.
44. "Editorial Notes," *San Francisco Call*, May 22, 1891, 4.
45. "About People You Know," *Washington Post*, December 30, 1894, 5.
46. "The Telelectroscope," *Washington Post*, April 25, 1898, 8.
47. *Leslie's Weekly*, July 6, 1899, quoted in Musser, *Before the Nickelodeon*, 163.
48. Robert C. Allen, "Contra the Chaser Theory," *Wide Angle* 3, 1 (1979): 4–11, in Wanda Strauven, ed., *The Cinema of Attractions Reloaded* (Amsterdam, Netherlands: University of Amsterdam Press, 2006), 405.
49. *Paducah (Ky.) Sun*, April 28, 1902, 8.
50. *Encyclopædia Britannica*, 2009 online ed., s. v. "History of the Motion Picture," http://www.britannica.com/EBchecked/topic/394161/history-of-the-motion-picture; Tim Dirks, "Timeline of Influential Milestones and Important Turning Points in Film History," *AMC Filmsite* (2009), http://www.filmsite.org/milestones1900s_2.html.
51. John Collier, "Censuring Motion Picture Shows," *New-York Daily Tribune*, May 9, 1909, 6.
52. Dirks, "Timeline of Influential Milestones."

53. "Dramatic Indecency," *New York Herald*, October 31, 1905.
54. "Establishments Licensed to Debauch the Young and Pander to the Depraved," *San Francisco Call*, March 31, 1899, 12.
55. Kenneth S. Lynn, *Charlie Chaplin and His Times* (New York: Simon and Schuster, 1997), 108.
56. *Proceedings of the City Council of the City of Chicago*, November 4, 1907, 3052, in Lee Grieveson and Peter Krämer, eds., *The Silent Cinema Reader* (New York: Routledge, 2004), 137.
57. Jack C. Ellis, *A History of Film* (Englewood Cliffs, N. J.: Prentice-Hall, 1990), 28.
58. Melvyn Stokes, *D. W. Griffith's The Birth of a Nation: A History of the Most Controversial Motion Picture of All Time* (Oxford: Oxford University Press, 2007), 111.
59. "Movies at Press Club," *Washington Post*, February 20, 1915, 5.
60. Stokes, *D. W. Griffith's The Birth of a Nation*, 129.
61. Annette Wallach Erdmann, "A Woman's Protest," *New York Times*, March 21, 1915, C2.
62. "Citizens Plead at State House For Film Ban," *Christian Science Monitor* (Boston), April 19, 1915, 4.
63. "Race Riot at Theater," *Washington Post*, April 18, 1915, 2.
64. *Mutual Film Corporation v. Industrial Commission of Ohio*, 236 U.S. 242 (1915).
65. "Arbuckle Is jailed on Murder Charge in Woman's Death," *New York Times*, September 12, 1921, 1, 3; "Arbuckle Dragged Rappe Girl to Room, Woman Testifies," *New York Times*, September 13, 1921, 1.
66. William A. Johnston, *The New World*, quoted in Stuart Oderman, *Roscoe "Fatty" Arbuckle: A Biography Of The Silent Film Comedian, 1887–1933* (Jefferson, N.C.: McFarland, 2005), 172.
67. Oderman, *Roscoe "Fatty" Arbuckle*, 192; Jennifer Rosenber, "The 'Fatty' Arbuckle Scandal," http://history1900s.about.com/od/famouscrimesscandals/a/fattyarbuckle_2.htm.
68. *Motion Pictures and Youth: The Payne Fund Studies* (New York: Macmillan, 1933). Quotes taken from David Copeland, "Media Effects: Powerful or Minimal?" *Contemporary Media Issues*, eds. Wm. David Sloan and Emily Erickson Hoff (Northport, Ala.: Vision Press, 1998), 9–10.
69. Eunice Barnard, "In Classroom and On Campus," *New York Times*, November 19, 1933, 24.
70. "Youthful Crimes Laid to the Movies," *New York Times*, May 12, 1934, 13; E.C. Sherburne, "Our Movie Made Children," *Christian Science Monitor* (Boston), June 22, 1933, 14.
71. Tino Balio, *Grand Design: Hollywood as a Modern Business Enterprise, 1930–1939* (Berkeley: University of California Press, 1995), 14.
72. Ray Stannard Baker, "Marconi's Achievement. Telegraphing Across the Ocean Without Wires," *McClure's Magazine* (February 1902): 299.
73. "Marconi," *New York Times*, January 15, 1902, 8.
74. "Signor Marconi's Triumph," *New-York Tribune*, December 15, 1901, 1.
75. Carl Snyder, "Wireless Telegraphy and Signor Marconi's Triumph," *Review of Reviews*, 25 (February 1902): 173, quoted in Susan J. Douglas, *Inventing American Broadcasting, 1899–1922* (Baltimore and London: The Johns Hopkins University Press, 1987), 63.

76. "Wireless Telegraphy," *New-York Tribune*, October 1901, 8.
77. Douglas, *Inventing American Broadcasting, 1899–1922*, 10–13.
78. "Wireless Telegraphy Saves Life. Practically Used for the First Time in History," *San Francisco Call*, May 7, 1899, 19.
79. *Omaha (Neb.) Daily Bee*, May 12, 1899, 7.
80. "Marconi and his 'Wondership,'" *Richmond Dispatch*, November 2, 1902, 16.
81. "Guiding Spirit," *Kentucky Irish American* (Louisville), March 15, 1902.
82. "Wireless Telegraphy Is Now Claimed to Be Letter Perfect," *Weekly Rocky Mountain News* (Denver), June 22, 1899, 7.
83. Herbert Wallace, "Revolution in Marine Travel," *Salt Lake Herald* (Salt Lake City, Utah), March 25, 1900, 1.
84. "Marconi Wireless Telegraph to Incoming and Outgoing Steamships," *Journal of the Telegraph*, February 20, 1904, 1.
85. Douglas, *Inventing American Broadcasting*, 45.
86. "Professor Reginald A. Fessenden and Our Government's Wireless Telegraph," *Deseret Evening News* (Salt Lake City, Utah), May 31, 1902.
87. "Messages Over Wireless Waste," *The Times* (Richmond, Va.), April 27, 1902, 1–2; "Fessenden's 'Wave Detector,'" *New-York Daily Tribune*, May 4, 1902, 8.
88. "Diary of Lee De Forest, quoted in Douglas, *Inventing American Broadcasting*, 93.
89. *Christian Science Monitor* (Boston), December 3, 1925, 2.
90. "Two are Convicted in Tipster Drive," *New York Times*, September 14, 1929, 9. See, also, "Lee DeForest, 'Father of Radio,' Still Works 8-Hour Shift Daily, *Washington Post*, April 4, 1937, 70.
91. "Marconi is Beaten," *Hopkinsville (Ky.) Kentuckian*, October 3, 1902, 7.
92. *Washington Times*, June 29, 1904, 4; "Wireless to Span the Ocean," *New-York Daily Tribune*, November 28, 1909, 8.
93. *Los Angeles Herald*, April 1, 1906, 12. The Marconi Company ran advertisements like this one in papers across the country to stop the sale of De Forest stock and to promote the decisions of the New York court and the U.S. Supreme Court.
94. "De Forest Hit by Supreme Court, *Salt Lake Herald* (Salt Lake City, Utah), March 26, 1906, 1.
95. "Wireless Wooing Ends in a Wedding," *New York World*, February 17, 1906, 5; "Will They Get Divorce by Wireless?" *Washington Times*, October 14, 1906, magazine, 9; "De Forest Will Wed," *Washington Post*, February 12, 1908, 3.
96. Lee De Forest, "The Audion.–I: A New Receiver for Wireless Telegraphy, *Scientific American* 1665 (November 30, 1907): supplement 348–350; and Lee De Forest, "The Audion.–II: A New Receiver for Wireless Telegraphy, *Scientific American* 1666 (December 7, 1907): 354–356.
97. "Opera by Wireless," *Ogden (Utah) Morning Examiner*, March 20, 1910, 14; "Radio Telephone Experiments," *Modern Electrics*, May 1910, 63.
98. Hugo Gernsback, "400,000 Wireless Amateurs," *New York Times*, March 29, 1912, 12.
99. "News Out of the Air," *Electrical Experimenter*, May 1914, 18; "How Radio Brought the News to the Farm," *Electrical Experimenter*, July, 1917, 189.
100. "20,000 American "Watchdogs,'" *San Francisco Chronicle*, January 30, 1916.
101. "Amateur Wireless Crosses Continent," *New York Times*, March 8, 1917, 8.
102. "Wilson Orders All Radios Seized; Amateur Stations Will Be Closed," *Washington Post*, April 7, 1917, 4.

103. Anthony R. Fellow, *American Media History* (Belmont, Calif.: Thomson, 2005), 255.
104. Tom Lewis, "'A Godlike Presence': The Impact of Radio on the 1920s and 1930s," *OAH Magazine of History* 6 (Spring 1992), http://www.oah.org/pubs/magazine/communication/lewis.html.
105. See Noah Arceneaux, "Department Stores and the Origins of American Broadcasting, 1910–1931" (Ph.D. diss., University of Georgia, 2007).
106. Quoted in Douglas Craig, "Political Waves: Radio and Politics, 1920–1940," in Michael C. Keith, ed. *Radio Cultures: The Sound Medium in American Life* (New York: Peter Lang, 2008), 237.
107. "Hoover Planning Quiet Campaign Via Radio and Movies if Nominee," *Christian Science Monitor* (Boston), May 8, 1928, 1.
108. Christopher Sterling, *Electronic Media: A Guide to Trends in Broadcasting and Newer Technologies 1920–1983* (New York: Praeger, 1984), 216.
109. Ronald R. Kline, *Consumers in the Country: Technology and Social Change in Rural America* (Baltimore: Johns Hopkins University Press, 2000), 5.
110. "A Shadow Howls by a New Device," *New York Times*, October 21, 1925, 25.
111. "See Radio Drawing Nations Together," *New York Times*, May 11, 1926, 18.
112. Orrin E. Dunlap Jr., "Television Is Brought Nearer the Home," *New York Times*, May 25, 1930, 137.
113. "Games Are Televised," *New York Times*, August 27, 1939.
114. "Far–Off Speakers Seen as Well as Heard Here in a Test of Television," *New York Times*, April 8, 1927, 1, 20.
115. David E. Fisher and Marshall Jon Fisher, *Tube: The Invention of Television* (San Diego: A Harvest Book, 1997), 37–46.
116. "Television for Home Use," *Wall Street Journal* (New York), November 19, 1929, 20; "Television Declared Ready to Broadcast, Starting Sectionally, With Relays Later," *New York Times*, August 11, 1933, 18; "Moon Makes Television Début In Pose for Radio Snapshot," *Christian Science Monitor* (Boston), August 24, 1934, 3.
117. Orrin E. Dunlap, Jr., "Today's Eye-Opener," *New York Times*, April 30, 1939, 12.
118. "'News' By Radio," *New York Times*, April 19, 1929, 18.
119. James Von Schilling, *The Magic Window: American Television, 1939–1953* (Binghamton, N.Y.: Haworth Press, 2002), 2–3.
120. White, ed., *Our Wonderful Progress*, 9.
121. Christopher H. Sterling and John M. Kittross, "The Radio Receiver," *Stay Tuned: A History of American Broadcasting*, 3rd. ed. (Mahwah, N.J.: Lawrence Erlbaum, 2002), 201.
122. Waldemar Kaempffert, "Seeing by Wire and Radio Opens New Highways in the Field of Communication," *New York*, April 17, 1927, 25.

8. My Notebook Still Carries Bloodstains: The World Wars and the Cold War

In the winter of 1945, Allied troops pushed toward Berlin. The landing on the beaches of Normandy, though it cost thousands of lives, led to the liberation of Paris and—ultimately—the end for Germany. The Allies successfully recovered after a final German offensive in Belgium known as the Battle of the Bulge failed and then focused on the Nazi homeland. Richard Tregaskis, a reporter for the International News Service who nearly died from a wound suffered in Italy that required placing a metal plate in his head, traveled with the troops into Germany. He called the final war efforts in Europe a "block-by-block and house-by-house struggle." When Tregaskis wrote of moving through the streets of the German city of Geilenkirchen with American troops fighting and dying every step of the way, he made a simple but powerful observation that encapsulated everything about the war and a journalist covering it up close and personal: "And my notebook still carries bloodstains as a reminder of the occasion."[1]

War creates a special relationship between the press and the military, an uneasy truce, according to Michael Sweeney. "News provides American citizens with information," he says. "For the soldier, however, news is primarily a tool or weapon."[2] World War I, World War II, the Korean War, and the concurrent and continuing Cold War that followed the second world war did not always support the premise of the uneasy truce between the government, military, and media, but the information that media provided always served as a tool to shape thought. As such, what media presented to the American public very much shaped public ideology and the course of the nation. It provided the impetus for the nation to join the war efforts in 1917 and 1941—though the sinking of the *Lusitania* and the bombing of Pearl Harbor were both catalytic events that would make declarations of war logical in one case and mandatory in the other.

Because information can be used as a tool or weapon, controlling it became important to U.S. leaders. President Woodrow Wilson did not move to have the government assume control of all amateur radio broadcasters in 1917 because he felt that they could be valuable in assisting the country's war effort. He seized them because he was afraid that ham operators who were German immigrants might use their wireless operations to provide revealing information to the enemy. The amateurs had already established a network that stretched across the nation and had offered the government use of that network to keep the nation in constant contact from Atlantic to Pacific. That offer, however, was not accepted. When you consider, however, that approximately 13 million recently arrived first- and second-generation immigrants came from the countries fighting in Europe and retained ties to their mother countries, the government may well have had some validation for wanting to keep German-Americans under scrutiny. Approximately 62 percent of the 13 million (8 million) had German ties, and another 4 million were Irish-American. Both groups tended to side with the Central Powers in the conflict because of longtime animosity toward Britain.[3]

The United States in the years before World War I was decidedly isolationist, and most American media outlets in 1914 favored neutrality for the country as war erupted in Europe. "Let all sane men go in and shut their doors," editor William Allen White's *Emporia Gazette* said. "In a world war mad, we have the peace that passeth understanding. By God's grace we should keep it."[4] But, "Honest, unbiased news simply disappeared out of the American papers along about the middle of August, 1914," historian C. Hartley Grattan observed.[5] This happened for a number of reasons, but a major one was the direct result of Britain's effort to control information that the United States received from Europe. German telegraph cables ran from Europe to the United States, France, Spain, and South America. In August, the British severed these cables, leaving only ones from Britain spanning the Atlantic. That meant that the information that Americans received about the fighting was tinted with the British point of view. One month into the Great War, nearly all American newspapers printed in English, one historian noted, editorially parroted England's line of propaganda.[6] "It all turns on which side gets the news in first; for the first impression sticks," Count Johann Heinrich von Bernstorff, the German ambassador to the United States at the time, observed.[7]

Twenty-five years later, the nation was still reticent to join the fighting in Europe after Germany sent its "lightning war" machine into Poland in 1939 and then rapidly against other helpless European nations. Just as they had in 1914, Americans noted that what was happening in Europe was not an assault upon the United States. Americans even argued over whether they should provide the outmanned Allied forces with supplies. Just as in World War I, though, media information soon tilted in opposition of the Axis forces, especially as American media broadcast and wired stories to the States about the

German bombings of England and of its submarine attacks on merchant ships. And, just as had happened in World War I, the media became a tool for war.

By the time World War II began, though, many journalists expected their work to be censored and to be used as propaganda. "Censorship and war go together just as censorship and dictatorships do," George B. Parker of Scripps-Howard Newspapers said.[8] Censorship would become a principal feature of media during the world wars, and the use of media as a propaganda machine would as well, though such terminology would never be used openly to describe media content. The day after the bombing of Pearl Harbor, for example, Walt Disney offered his studio, his employees, and his stable of stars—Mickey Mouse, Donald Duck, and Goofy—to the government to produce motion pictures or anything to help the nation now at war. Journalists during World War II became in some ways extensions of the military and the war effort. They wore uniforms with officer rank. They moved with the troops. They flew on bombing missions. Their stories reflected pride in the American fighting man and the country's efforts in fighting.

In both world wars, controlling public opinion assumed a central role in government thinking. In World War I, the government employed Edward Bernays, George Creel, and other "advertising men" and journalists to help shape opinion. Many were principal proponents of the newly burgeoning field of public relations, but all believed that the government needed to control information available to the public in some way. Bernays, a decade after the war's conclusion, explained, "The American government and numerous patriotic agencies developed a technique which, to most persons accustomed to bidding for public acceptance, was new" and "appealed to the individual by means of every approach—visual, graphic and auditory."[9] Early in America's involvement in World War II, Archibald MacLeish—lawyer, editor, author, Pulitzer Prize winner, and assistant director of the Office of War Information in the Roosevelt administration—was even more direct about the role that media would play as the United States rapidly escalated its war efforts globally. "The principal battle ground of this war is not the South Pacific. It is not the Middle East. It is not England or Norway, or the Russian steppes. It is American opinion."[10]

Korea, however, posed a different type of war in a changed global balance of power. By 1950, the world that most Americans believe existed consisted of us and them, the free world versus the world of communism. The nuclear age ushered in with the bombings of Hiroshima and Nagasaki was quickly followed in 1949 with the Soviet Union's successful tests of nuclear weapons. The threat of a giant mushroom cloud that could annihilate the world became reality. Korea represented a chance to stop the spread of communism and more nuclear opportunities for the Soviet Union. Media in many cases supported the nation's effort in Korea and the mindset of stopping communism wherever it threatened to overtake a free country. But Southeast Asia was a very different place from Europe, something that the Asian theater in World War II proved. Understanding the "Asian mindset," way of life, in short, almost everything about these new cultures was difficult for Americans. Because there was so much fear in the United States concerning communism, the government had already put into place restrictions on information, and President Harry Truman was wary of excessive media information concerning the nation's nuclear abilities. This fear would culminate in the Atomic Energy Act of 1954, which made it illegal to publish any information concerning "the design, manufacture, or utilization of atomic weapons; the production of special nuclear material; or the use of special nuclear material in the production of energy."[11]

America's full complement of print and broadcast media existed to cover the Korean conflict, though television was just becoming a medium that people knew. Whether it was trusted cannot be certain. Television news for the most part was still read by ex-radio or ex-newspaper men. Its visuals from any location were delayed because the film had to be developed just as that from still cameras did. Generally, it required four to five days from shooting to presentation to the public with footage from the Korean peninsula. Americans also continued to see the news via newsreels in theaters. The same had been true for World War I and especially World War II. Millions of Americans visited theaters each week, and a newsreel was still a part of the package from 1950-1953, the years of the Korean conflict. As the nation moved into the 1950s, though, television would assume an increasingly important role in providing information and shaping thought, especially as the Cold War intensified.

THE GREAT WAR

President Woodrow Wilson called it "a war to end all wars." The *New York Times* called the Great War "the first press agents' war."[12] World War I was the bloodiest conflict the world had ever known, and, according to Ross Collins, it was directly the cause of the even more bloody World War II and most of the violence worldwide that occurred for the remainder of the twentieth century.[13] By war's end, nearly 5.5 million Allied soldiers were dead with another 3.4 million casualties among the Central Powers. The death toll among civilians reached about 6.5 million.[14] More than 116,500 Americans died either fighting or from some other war-related injury.[15]

When war erupted in Europe in 1914, the United States media had some reporters in Europe, but the ability to relay information back to the United States and access to the front lines was curtailed greatly by the Allies. In addition to the cutting of German telegraph cables by the British, England and

France limited access for media to the fighting. Initially, journalists were arrested if found in the war zone. To provide information, the British appointed a military officer who regularly provided stories to the press. Later, British military authorities granted press credentials to five British journalists and one American, Frederick Palmer, to cover all the European fighting. The French allowed the press to travel with troops and write stories using only information authorized by French military officials. Then, the communiqués had to be written in French only, which eliminated front-line access for all of the press except the French.[16]

Because the United States was dependent upon sources other than firsthand accounts for so much of the information about the war, especially prior to America's entry in 1917, propaganda from the Allied countries—especially England—tinted the lens of understanding for Americans. Many members of the media bought into the atrocity accounts that regularly arrived from Europe. Even though American journalist Irvin Cobb, who traveled with the German army when it invaded Belgium in 1914, reported never having seen any of the horrid acts that were supposedly inflicted upon civilians by German solders—the spearing of infants, the chopping off of hands and other mutilations, unprovoked murder of civilians—these kinds of stories became commonplace in U.S. publications.[17] The British gave atrocity stories credence when the Bryce Committee, named for its chairman and noted scholar James Bryce, authenticated them using depositions taken from more than twelve hundred witnesses, most of whom were Belgians who had escaped to England before the invasion. The report said that all the things that Cobb reported never having seen happened—and more—did occur as German soldiers "became filled with lust for blood and spoils in the wake of battle."[18] When a German U-boat sank the *Lusitania* in 1915 with at least one thousand civilians onboard, more than one hundred of them Americans, many writers said the atrocity stories concerning the Huns had been validated. We are in "a war between civilization and barbarism, between humanity and savagery; between the light of modern times and darkness of the years that followed the collapse of Rome," Frank Simonds concluded in the *New York Tribune*.[19] "Germany believes that war has no morals," a Kansas paper said. "The rape of Belgium, the murder of women and children on the Lusitania ... these are the barbarities of German's warfare that have turned the world from her cause in horror."[20]

In reality, Cobbs' story for the *Saturday Evening Post* about not having seen brutal activities by German soldiers was probably closer to the truth than anything in the Bryce Committee report or the propaganda coming out of Britain or France. Most historians admit that some atrocities happened as part of World War I, but the Allies successfully controlled the information reaching the United States about the war prior to the nation's entry in 1917. That did not mean, though, that Americans on the whole did not believe that they were getting the true story from Europe. It means that by controlling media information, the Allies were able to shape U.S. public opinion. "The American likes to form his own opinions and so only requires facts," German ambassador Bernstorff would later say. "It is quite incredible what the American public will swallow in the way of lies if they are only repeated often enough and properly served up."[21]

Cutting telegraph cables in 1914, however, did not allow for complete control of information coming from Europe. Wireless made it possible to send information across the Atlantic, something that Guglielmo Marconi and Lee De Forest accomplished in the first decade of the twentieth century. Late in 1914, the government declared that radio waves were being used by the Germans as weapons of war. "The spread of official news by wireless broadcast over the globe for anyone who can pick it up is a new method of political propaganda which has been introduced through the European War," the Associated Press reported. The cutting of the German telegraph cables required innovation, the story said, and "the German high-power wireless began sending out news which was destined primarily for the United States and for German colonies." As a means to negate German wireless propaganda, the British and French decided "to combat this German wireless news campaign by spreading reports of their own" via wireless.[22] Many Americans directly blamed wireless for the sinking of the *Lusitania*. Immediately after its sinking, speculation arose that coded messages from German sympathizers in the United States were transmitted by wireless to Germany and then to its submarine force operating in the Atlantic. Speculation quickly became verified information. "The German Government was informed that the *Lusitania* had absolute contraband on board when she sailed from New York," Carl L. Schurz, a lawyer in the city reported in an effort to justify the sinking. A cable from London that ran in papers the same day said that "there is no doubt that the German Admiralty received a wireless message from New York, giving the date of the *Lusitania*'s sailing, and that a number of submarines were told to torpedo her at any cost."[23] This use of wireless alone was enough for the government to begin closely monitoring broadcasts and to justify takeover of all wireless transmitters when the country entered the war two years later.

The obvious use of media to control and sway public opinion in the United States by warring European factions would have no doubt been enough to cause the nation to seek a means of controlling to some degree its own media once war was declared on Germany on April 6, 1917. But, the United States was still in the era of muckrakers, though their influence was waning. Experience had taught Americans that the press could effect great changes among the public and even in government. The amount of legislation that had been the direct result of the investigative journalism of muckrakers was fresh on the minds of politicians, journalists, and probably most of the public as well. That is why the government turned to a series of press controls and to "advertising men" as the reality that the United States would enter the war unfolded. Before World War I ended, the United States would prosecute more than nineteen hundred people for speaking and writing against the war with at least one hundred publications banned from using the mails for distribution, most notably through the Alien and Sedition Acts and the Espionage Act, which was aimed directly at German Americans.[24]

But, voluntarism, not coercion was viewed by the government as the better way to achieve control of media content, so the Wilson administration created the Committee on Public Information on the recommendation of Josephus Daniels, a newspaperman who was serving as secretary of the navy, one week after the war declaration. Daniels believed that the country needed a way to diffuse war-time information so that "every citizen is given the feeling of partnership" with the government's war efforts.[25] The CPI became the government's promotional arm for the war. To achieve its purpose, the government hired men who were at the forefront of using the media to mould and shape thought on a variety of subjects. George Creel, who had been a journalist in Denver, was selected to head the CPI. According to Newton Baker, secretary of war, "The whole business of mobilizing the mind of the world as far as American participation in the war was concerned was in a sense the work of the Committee on Public Information."[26] Creel believed the best way to control information was to "make every paper in the land its own censor, putting it up to the patriotism and common sense of the individual editor to protect purely military information of tangible value to the enemy."[27] Creel and the CPI were extremely successful. Nearly everyone who created information that was shared with the public, from journalists and film makers to teachers and priests, aided, knowingly or unknowingly, the propaganda effort of World War I.[28]

The CPI created a comprehensive program of information control. This included a government newspaper, the *Official Bulletin* and a company, Official Film News, that created newsreels for theaters because Creel believed that "the motion picture had to be placed on the same plane of importance as the written or spoken word."[29] The CPI produced thousands of news releases for the press including a column called "The Daily German Lie," which was one of the government's propaganda efforts aimed at tainting anything that might be said in a positive light about Germany. The CPI created what one historian called "perhaps the most gigantic propaganda campaign in all American history,"[30] or as Creel said, America's war preparations and plans were created "behind a wall of concealment built upon the honor of the press and the faith of the individual editor."[31] The CPI's efforts at information dissemination did not stop with America's borders. Using the wireless, the CPI established offices throughout the world, including Tokyo and Beijing in Asia and Bern, Lisbon, Madrid, Paris, and Rome in Europe. With these communiqués, the CPI attacked Germany, but, more importantly, emphasized the United States' efforts to end the war, American self-sacrifice, and general goodwill toward the Allies.[32]

Just as historian David Ramsey credited the American Revolution to the pamphlets and words printed in the years before fighting began, the public relations people who worked for the CPI gave their efforts a central place in the successful fighting of World War I. Bernays, writing forty years after he served as bureau chief for the CPI, noted that at the time he never realized just what an impact the office had on the war effort and upon Americans' view of it. "I didn't think of the phrase 'the words that won the war,' but I reflected on the idea these words signified," Bernays said. "I didn't realize then what I saw and heard was in reality a consequence of communications."[33]

On the Western front, the CPI closely monitored what journalists prepared for dissemination in the United States. In order to travel with the American Expeditionary Forces, as America's fighting unit was called, journalists were required to appear before Secretary Daniels, swear not to create any information that might assist the enemy, keep an autobiographical diary about their activities on the front, wear an arm band marked with a "C" to connote they were media correspondents, wear military uniforms, and pay $1,000 for food and shelter and $2,000 as a security bond.[34] Voluntarism was the principal element here even though reporters agreed to all of the above because the CPI promoted patriotism. If that did not work, Frederick Palmer, the only American reporting in Europe officially early in the war, became the CPI's official press officer in Europe. He kept tabs on all reporters, censoring them heavily if he felt it necessary, and monitoring all movement, making sure reporters never wandered off on their own to collect information for stories. The results were stories that praised the AEF and the Allies, demonized the enemy, and omitted any information that might, in Palmer's estimation, be detrimental to the war effort.

Reporters became a part of the story, too. When the ship on which *Chicago Tribune* writer Floyd Gibbons sank after being torpedoed as it sped for Liverpool from New York, for example, the reporter got just what he'd hoped—the chance to write a first-person account of an attack on a passenger vessel by a German U-boat. In February 1917, just prior to the U.S. declaration of war, Gibbons boarded the *Laconia*. Gibbons had always intended this story to be a first-person one, even injecting his own quip, "What do you say are our chances of being torpedoed?" early in the copy. After that, Gibbons described the explosion from the torpedo, how the passengers and crew abandoned ship, and how they rowed in lifeboats for shore, always keeping himself in the forefront of the story. But, he began the story by asking whether what he had experienced was real or just a dream. "But," he said, "dream or fact, here it is."[35] Stories such as Gibbons' on the *Laconia* were powerful ones because they gave a sense of realism to the atrocities of war and to the brutality and disregard for innocent lives by the Germans. Later, Gibbons lost an eye to sniper fire as he tried to save an American officer. His heroics and his eye patch became part of the stories of bravery of Americans and of what reporters would do for the war effort.[36]

World War I allowed the government to utilize the media in ways never imagined. As researcher Harold Lasswell wrote a decade after the United States' entry into the war, "There is no doubt about the superb qualifications of newspapermen for propaganda work....They have a feeling for words and moods, and they know that the public is not convinced by logic, but seduced by stories."[37] The Allies and the American government knew that Lasswell's observations were true because of past examples of media setting the agenda and direction of their respective countries and the United States. Because the print media were still the ones that reached into every American home, they became the weapon of choice to sway public opinion and get the nation solidly behind the Allies and in utter disgust of the Central Powers. The wire services, especially, became the optimal conduit for effecting widespread change in the United States. On average, if the Associated Press created a dispatch and all newspapers within the AP network used it, about 30 million Americans would be able to read it and pass along its ideas to others. "Here," one magazine writer said, "is the most tremendous engine for Power that ever existed in this world."[38] Through the CPI, the United States was able to develop a brand of nationalism never known before in the country because the CPI's goal was to reach every American and draw each one closer to the government and the war effort. Civil liberties were really considered secondary during the war and immediately afterward as Supreme Court cases such as *Schenck v. United States* proved.[39] While propaganda and information control would be a part of the next world war, the nation learned a great deal about using media in the two decades between the wars. In World War II, the nation had more laws in place to control media and information but used them less because voluntary censorship was the rule.

World War II

President Franklin Roosevelt said December 7, 1941, the day that the Japanese attacked Pearl Harbor was "a date which will live in infamy." Four years later, more than 405,000 Americans had died in a war that saw American military men fighting around the globe.[40] For the media, World War II was a chance to exhibit that country always came first. Ten days after the Pearl Harbor attack, the president said, "All Americans abhor censorship, just as they abhor war. But the experience of this and of all other nations has demonstrated that some degree of censorship is essential in war time, and we are at war."[41] Media agreed. In almost every situation, the nation's media voluntarily censored its information, always keeping the nation's welfare central to its reporting. And, Roosevelt repeatedly emphasized the importance of media and "The constant free flow of information among us" to ensure an informed populace.[42] The nation's media had never been more prepared to cover conflict, either. Before the war ended, the nation's press outlets had sent more than seventeen hundred reporters, photographers, videographers, and sound technicians to supply the nation with information.[43]

The president's rhetoric of a free press with the necessity of self control by the media became the basis of information dispersal during World War II. Let the people know what's going on, but be mindful that what you say—if not weighed against national security—could prove disastrous. Nothing better illuminates this than some of the slogans created by the administration's Office of War Information: "Loose Lips Might Sink Ships" and "Free Speech Doesn't Mean Careless Talk." Media responded. It is "amazing ... to see the press and radio asking for rather than standing solidly against such a thing as censorship," Stephen Early, a former United Press and the Associated Press reporter and FDR's press secretary, said just two weeks after Pearl Harbor.[44] The media simply followed the maxim created by Byron Price, the former head of the AP and director of the Office of War Information. Price said that editors should "ask themselves with respect to any given detail, 'Is this information I would like to have if I were the enemy?' and then act accordingly."[45]

The media's solid support of voluntary censorship and its global coverage of World War II were easily visible and audible. Patriotism became a key component of media. Hollywood, for example, produced a host of pro-American, flag-waving motions. Movies about World War I heroes glorified the American fighting man, and Jimmy Cagney's *Yankee Doodle Dandy* gave the country a new song to sing in support of the troops. To make sure that movies supported the war effort, the OWI provided Hollywood studios with a guideline for picture content. "Will this picture help win the war?" the "Government Information Manual for the Motion Picture Industry" asked. Will it harm the war effort by creating a false picture of America, her allies, or the world we live in?"[46]

While patriotism achieved through voluntary media action and recommended guidelines was what the Roosevelt administration hoped would occur during World War II, the government made sure it had in place sufficient laws and agencies to ensure press control in case information became too forthcoming from media. Shortly after Roosevelt assumed office in 1933, he and his administration began instituting policies that would make it possible to control the media if needed during times of national crises or to keep the government apprised of the information being shared throughout the nation by media. With the Division of Press Intelligence, workers daily monitored the press, synthesized its coverage, and supplied that synthesis to the administration. A year later, the Federal Communications Act gave government the ability to assume control of radio if a national emergency warranted it. As the specter of U.S. entry into the war in Europe loomed over the nation, Congress passed the Alien Registration Act, better known, though for its sponsor, Representative Howard Smith of Virginia. The Smith Act made it illegal to "knowingly or willfully advocate, abet, advise or teach the duty, necessity, desirability or propriety of overthrowing the Government of the United States or of any State by force or violence, or for anyone to organize any association which teaches, advises or encourages such an overthrow, or for anyone to become a member of or to affiliate with any such association." The law also required all alien residents in the United States over 14 years of age to file a comprehensive statement of their personal and occupational status and a record of their political beliefs.[47] Within four months a total of 4,741,971 aliens had been registered.

After the bombing of Pearl Harbor, the FDR administration put into place offices to control information. The first was the Office of Censorship, which requested that media outlets voluntarily submit stories for approval before presenting them to the public. When reporters—both print and broadcast—traveled with the military, their information had to pass through military censors. In June 1942, the administration created the Office of War Information. It was the most active of government offices in terms of media. The OWI filled its offices with media professionals who wrote press releases and even took control of media outlets in some cases. With these controls, the administration could, if needed, clamp down upon the media if needed, but, as noted, FDR preferred to promote free speech in the context of voluntary control of content. Media more than complied in most cases, though there were a few instances where the government had to take action against the press.

The fighting man's storyteller

Dozens of journalists' names became household words during World War II, but none commanded the attention and respect of Indiana-born Ernie Pyle. Pyle made the fighting man a hero. He focused upon the small events that made a world war real. Pyle referred to his understanding of the war as one taken from "a worm's-eye view."[48] Pyle was already a household name in the United States prior to the war. As a reporter for the Scripps-Howard newspaper chain, Pyle traveled the country writing stories about everyday life in the United States. That experience gave him an "in" with soldiers no matter where they lived because Pyle knew something about their hometowns, their states, their lives before they became soldiers. Before the war, but especially during it, Ernie Pyle was read by millions.

When the war began in Europe, Pyle left for England and reported on the bombings there. Later, he would travel with American troops in France, North Africa, and into Italy. On the beaches of Normandy, he wrote that "it seems to me a pure miracle that we ever took the beach at all," adding "Ashore, facing us, were more enemy troops than we had in our assault waves."[49] Pyle's most famous column, though, was written as the Allied troops took control of Italy. His story on the death of Captain Henry T. Waskow of Belton, Texas, was the kind of story that the OWI and every American loved because it elevated the actions of infantrymen—the "mud-rain-frost-and-wind boys"[50] as Pyle called them—to the mythological level of Washington and the other Founding Fathers who fought for American independence in the Revolution. The story was so powerful that the *Washington Daily News* increased its font size so that the story filled the entire front page.[51]

"In this war I have known a lot of officers who were loved and respected by the soldiers under them. But never have I crossed the trail of any man as beloved as Captain Henry T. Waskow, of Belton, Texas," Pyle said in his lead. After introducing the body of the dead captain to the story, Pyle told of how, one by one, soldiers came to pay their respects. He minced no words and let readers experience exactly what the soldiers said and did. "One soldier came and looked down, and he said out loud, 'God damn it.' That's all he said, and then he walked away.... Then a soldier came and stood beside the officer, and bent over, and he too spoke to his dead captain, not in a whisper but awfully tenderly, and he said, 'I sure am sorry, sir.' Then the first man squatted down, and he reached down and took the dead hand, and he sat there for a full five minutes holding the dead hand in his own and looking intently in the dead face. And he never uttered a sound all the time he sat there. Finally he put the hand down. He reached over and gently straightened the points of the captain's shirt collar, and then he sort of rearranged the tattered edges of his uniform around the wound, and then he got up and walked away down the road in the moonlight, all alone."[52]

When Germany surrendered in May 1945, Pyle had already left for the Pacific theater so that he could experience first-hand the war there. Many believed that Pyle would return to the States now, but he decided that he needed to cover the war in the Pacific. Pyle joined troops as they landed on Okinawa on Easter Sunday, 1945. He wrote of picnicking on the beach since the Japanese did not oppose the troops' landing. As the troops moved deeper onto the island, fighting intensified, but on April 18, the day Pyle died, a lone sniper, one that no one even imagined was there, put a bullet through Pyle's head. He died, the only one to die during this movement of troops. Pyle died only ninety days before the war would end with the atomic bombing of Hiroshima and Nagasaki. "No man in this war had so well told the story of the American fighting man as American fighting men wanted it told," President Harry Truman said in a statement printed in Pyle's obituary. "More than any other man, he became the spokesman of the ordinary American in arms doing so many extraordinary things."[53] When someone looked inside Pyle's pocket, they found an unedited and unpublished column he had written about the imminent victory expected in Europe. "The catastrophe on one side of the world has run its course. The day that it had so long seemed would never come has come at last," he said. "This is written on a little ship lying off the coast of the Island of Okinawa, just south of Japan, on the other side of the world from Ardennes. But my heart is still in Europe, and that's why I am writing this column. It is to the boys who were my friends for so long. My one regret of the war is that I was not with them when it ended."[54] As Evans Wylie wrote for *Yank*, "The island [Okinawa] probably will be remembered only as the place where America's most famous war correspondent met the death he had been expecting for so long."[55]

Ernie Pyle made Americans feel better about the role the nation played in World War II. A general believed that Pyle "helped our soldiers to victory."[56] He certainly set the tone for what the nation expected in strong, positive reporting about the war by making the nation proud of its fighting forces.

This is London

If Ernie Pyle was the most popular journalist of the war, CBS' Edward R. Murrow was close behind, though he would become even more popular in the 1950s when he switched from radio to television. Murrow's platform for reaching the American public was larger than that of any print journalist. Radio's audience numbered in the tens of millions daily with at least 83 percent of U.S. households owning at least one radio early in America's involvement in the war.[57] As war approached and became reality in 1941, Americans purchased approximately 13 million radios, and nearly two in three Americans said they depended on radio as their principal news medium.[58] Murrow traveled to Europe in 1938. There, he began his eyewitness reports of the growing tensions on the continent by covering the annexation of Austria by Germany. But he captured the nation's attention with his live broadcasts from London in 1940 as the German air blitz attempted to bomb the island country into submission just as Nazi battalions had done to Poland, the Netherlands, Belgium, and France. With his trademark "This … is London," Murrow provided the country with the sounds of nightly bombings as his "ear witness" radio reports described the city, British resilience, and their determination to survive and conquer.

Just as the nation had leaned toward isolationism when the Great War began, Americans maintained isolationist tendencies when World War II began in 1939. Murrow's reports from London, however, helped change the minds of Americans. In fact, more than 50 percent of Americans favored increasing aid to the Allies following his reports compared to just 16 percent prior to Murrow's "This … is London" broadcasts.[59] This is exactly what Murrow had hoped would happen. He wanted Americans to be incensed at German aggression and ruthlessness. He believed that he could use radio to shape public opinion and to push policymakers to support the Allies. He used his journalism, as one journalist-turned-historian has said, "to push the United States away from its self-deluding notion that it could go its own way, safe amidst the world's storms" because Murrow "did not see a need to divorce journalism from patriotism," which is exactly the way England's Winston Churchill and President Roosevelt thought.[60] In 1941, Murrow returned to the States where he said that America would not be able to avoid entering the war. He called it "a new kind of war, a war that is twisting and tearing the social, political, and economic fabric of the world." And, he said that though it was the job of journalists "to prevent our own prejudices and loyalties from coming between you and the information which it was our duty to impart," Murrow could not hide his contempt for what Adolph Hitler and his military had done and the fact that he believed that Britain must survive if democracy were to survive.[61]

After the United States entered the war, Murrow returned to Europe. In 1943, he gave the nation a look at bombing from the Allied perspective as he flew with bombers in raids on Berlin and other targets in Germany. "Berlin was a kind of orchestrated hell, a terrible symphony of light and flame," Murrow told his listeners.[62] Murrow, just as Pyle and other journalists did, drew admiration from a nation eager for war information because people realized that Murrow was putting his life on the line to provide information. Radio broadcasts brought this home even more than newspaper and magazine accounts because the bombs, the sirens, and the machine gun fire were audible via the broadcast medium. When Murrow toured the Buchenwald prison camp—the first of the concentration camps to be liberated—in 1945, he could scarcely believe what he had to describe for the public. "Permit me to tell you what you would have seen, and heard, had you been with me on Thursday. It will not be pleasant listening," Murrow said as he began his report. He then described briefly the scene outside the camp in Weimar, Germany, but then decided that a first-person account was the best way to describe all the horrors of Buchenwald, that and allowing those in the camp who were able to speak to do. "I pray you to believe what I have said about Buchenwald," Murrow pleaded. "For most of it I have no words. Dead men are plentiful in war, but the living dead, more than twenty thousand of them in one camp….

If I've offended you by this rather mild account of Buchenwald, I'm not in the least bit sorry."[63] Murrow and other broadcasters gave the nation the sounds of the war in a way that had never before been available. The immediacy aspect of radio—Murrow in particular—greatly shaped the national reaction to the war and those fighting it.

Disney, Donald, and the silver screen

Since early in the 1920s, the motion picture industry had been under scrutiny. People did not like the lifestyles of some Hollywood actors, and they believed that the content of many films was detrimental to the moral compass of the nation. The Hays Office kept an eye on the industry and monitored it closely. Many Americans believed that motion pictures greatly affected the thoughts, emotions, and actions of young people. The Payne Fund studies that were published in the early 1930s only affirmed this idea. Even in the Depression, more than 55 million people attended at least one motion picture every week during the days of deepest economic problems, and young people often attended multiple times. World War II gave Hollywood the opportunity to change its image and to use its power for shaping national conversation and action.

Hollywood studios, before America's entry into the war, began working on films that projected patriotism, the American soldier's fighting abilities, and support for the countries that would make up the alliance opposed to the Axis powers of Germany, Italy, and Japan. After Pearl Harbor, Hollywood went into total war mode. About one-third of all Hollywood employees—from directors and producers to stars and cameramen—entered the military. There, most had their unique skills used in the war effort. Others, like Jimmy Stewart who enlisted in March 1941, used both their Hollywood abilities and other skills in the war effort. Stewart became a pilot in the Army Air Corps, made promotional features for the government, and finally flew numerous bombing raids on Germany beginning in 1943. He enlisted as a private but ended his active military career as a colonel and a chief of staff with the Eighth Air Force.[64]

In Hollywood, studios turned out two different kinds of motion pictures during the war, those made for servicemen—often training films—and those made for viewing on the homefront and by soldiers for relaxation. The latter might have been traditional recreational films, but they generally included elements that were educational or inspirational for the audiences. Almost always, they were uplifting of American ideals and the nation's war effort.[65] Motion pictures became the perfect venue for promoting the war and keeping homefront morale high. By the time the United States entered World War II, some estimates say that the average American went to the cinema three times a week.[66] By the end of the war, more than 90 million Americans attended a show at least once a week.[67] Newsreels, regular features that always appeared before the main feature, devoted almost all of their content to the war—nearly 75 percent[68]—and this content presented the nation's war effort—as well as Hollywood's part in fighting the forces of totalitarianism and evil—in a way that promoted America's sacrifice, war effort, and determination just as the OWI wanted.

All studios worked to support the war, but none, perhaps, dedicated more effort to the cause than Walt Disney Studio. By the time President Roosevelt had completed his "day of infamy" speech to Congress on December 8, 1941, Walt Disney was in contact with Washington offering his studio to the government's service. By June 1942, 80 percent of the output from Disney studios consisted of government projects that Disney produced on a cost-only basis. Disney charged the government $12 per foot for the films, while an identical style film produced for commercial release cost the studio $65 per foot.[69] A year later, about 94 percent of what Disney produced was war related.[70] At first, the Disney productions were mainly training films to be used for producing war-related products or for military use, but Disney was also producing shorts for public consumption in theaters. Early in 1942, Disney released *The New Spirit*, which featured Donald Duck and focused on filing income tax. "Donald will in his inimitable dialect give a little lecture on patriotism, too, to try to get across the thought that tax-paying in a democracy is a privilege rather than a burden," one news story explained.[71]

By 1943, Disney was producing elaborate shorts for theaters and at least one full-length film, *Victory Through Air Power*, which promoted the idea—obvious from its title—that air superiority was the only way to ensure victory. *Victory Through Air Power* combined all of Disney's trademark qualities for animation to promote Major Alexander P. de Seversky's book on the use of airplanes and super-bombers to win wars. "If 'Victory Through Airpower' is propaganda, it is at least the most encouraging and inspiring propaganda that the screen has afforded us in a long time," one reviewer said.[72] But it was Donald, the mischievous, non-pants-wearing duck that became Hollywood's voice for America. "Who would have thought," Theodore Strauss wrote, "that one day Donald Duck would be giving the retort perfect to Herr Doktor Goebbels? ... But there it is. Donald, who used to be just another noisy neighbor, has by some odd token of fate become a sort of ambassador-at-large, a salesman of the American Way." And, the article explained, "He has become one of this country's No. 1 propagandists."[73]

Disney was cognizant of the power of his medium. He knew that motion pictures had the ability to reach far beyond the printed word, and he knew that his animated characters were loved and accepted by almost all Americans. "The war has taught us that people who won't read a book will look at a film," Disney said. "It's shown that you can take knowledge out of a dusty tome somewhere, and wrap up the effort of many teachers in one can of film. You can show that film to any audience and twenty minutes later it has learned something—a new idea, or an item of important information—and it at least has stimulated further interest in study." Disney knew, as did the OWI, that film was the medium of entertainment for the nation. He realized that he could take his message and that of the government, package it in entertainment, and affect the thinking of a nation. Though later studies would disprove the ideas of the Payne Fund studies about long-term changes that motion pictures could create among their viewers, Disney and the OWI depended on those studies' validity. "Mass education is coming," Disney said as he looked ahead in 1943 and predicted what Hollywood was doing. "It's coming because it's a necessity. Democracy's ability to survive depends on the ability of its individuals to appreciate their duties as citizens and to comprehend the complex problems of the changing world we live in."[74] That was Disney's goal with thirty shorts his studios produced plus the feature *Victory Through Air Power*. The nation had, by 1944, "become highly film conscious," Jack Warner, head of Warner Brothers Studios, said.[75]

Motion pictures, Thomas Doherty said in his study of film and World War II, produced the most powerful impact and lasting consequences on Americans' understanding of World War II—more than the stories of Ernie Pyle or the radio broadcasts of Edward R. Murrow. At the cinema, all the elements joined together. The newsreels, he believes, were not timely—like the radio broadcast—nor did they provide the in-depth urgency of print journalism. What cinema did was verify what was read and heard and provide coherency through moving images. "The newsreels made sense of the thing," he said. Pearl Harbor, no doubt, was nearly incomprehensible to the American public. But in February 1942, Movietone released its newsreel of the attacks, taken by Al Brick, who was ironically shooting footage for a movie for Twentieth-Century Fox. After seeing the rain of machine-gun fire on the U.S.S. *Arizona*, burning buildings, and airplanes in pieces did everything that had been heard on December 7 and then read about in the following days truly become comprehensible.[76] After these newsreels, a feature that kept people focused on the "good fight" at home or on some story with a positive message or smiling at an animated short with a propagandistic twist meant Americans had spent an evening absorbing a message that told them to support the nation's war efforts as they viewed and heard, in twenty-four frames per second, exactly why.

Fighting racism

Americans may have appeared united in the fight against the Axis Powers, but the United States of that period was divided completely along racial lines. In World War I, the country refused to draft African Americans. In World War II, blood for transfusions was separated. A black man's blood was never given to a white man. Even the most compelling of stories—if they discussed blacks—generally demeaned them. Floyd Gibbons' story of the sinking of the *Laconia* included this sentence: "The jibbering bullet headed Negro was pulling directly behind me and I turned to quiet him as his frantic reaches with his oar were hitting me in the back."[77] It was a typical sentence to describe anyone of color—an incoherent, odd-looking individual without the ability to act in a rational way or as one with any intelligence. One needs to remember that this was the time when *The Birth of a Nation* was earning praise from whites and setting records for box-office receipts. Supreme Court justices and members of Congress stood and cheered when they saw the film for the first time. Even President Wilson received a special screening of the movie at the White House, calling it "writing history with lightning."[78]

African Americans, of course, deplored the racism that controlled their everyday lives, but most black leaders felt that the danger to the country from without was greater than the dangers that racism posed from within and asked that African Americans support the war effort. About 370,000 black soldiers served in the military during World War I.[79] "We of the colored race have no ordinary interest in the outcome. That which the German power represents today spells death to the aspirations of Negroes and all darker races for equality, freedom and democracy," W.E.B. DuBois said in the NAACP publication *The Crisis*. "Let us not hesitate. Let us, while this war lasts, forget our special grievances and close our ranks shoulder to shoulder with our own white fellow citizens and the allied nations that are fighting for democracy."[80]

All black leaders, however, did not believe that war in Europe trumped bigotry and mistreatment at home. Robert S. Abbott, the publisher of the *Chicago Defender*, was one. Abbott acknowledged the importance of fighting the enemy from without, but he was deeply concerned about the enemy within, racism. Abbott, who noted wryly in his paper that "Woodrow Wilson (white) declared war on Germany today,"[81] believed that the plight of blacks was too important to take a back seat even to world war. "The main issues are the protection of our citizens, and by citizens we do not mean exclusively WHITE citizens," Abbott said, adding that "black citizens need protection right here at home, the upholding of our honor, the preparation for our safety and the securing of our prosperity."[82]

Abbott's *Defender* grew in circulation during and after World War I—some estimates say to nearly 300,000 copies per issue by 1920—but the growth of the black press between the wars gave African-American newspapers even more power when World War II began. The attack on Pearl Harbor was an affront to Americans, regardless of their skin color. For African Americans, though, the war posed an opportunity to make real changes in the way the nation thought, and it all started with a cafeteria worker from Wichita, Kansas, named James Thompson who saw in the war effort an opportunity to fight another enemy simultaneously—racism. In a letter to one of the nation's prominent African-American newspapers, the *Pittsburgh Courier,* Thompson suggested a "Double V" campaign to be supported by African Americans and all believers in democracy. The first V was, of course, for victory over Germany, Japan and Italy—"enemies without." The second V stood for victory against prejudice in America—"enemies within." The *Courier* promoted the Double V campaign throughout the war, and it quickly caught on with other newspapers and with most blacks in the United States. Even some whites endorsed the Double V campaign, which can be considered the opening action in the fight for civil rights that would grow into a national movement at the end of the 1950s.

Thompson posed excellent questions for a segregated country: "Should I sacrifice my life to live half American? Will things be better for the next generation in the peace to follow? Would it be demanding too much to demand full citizenship rights in exchange for the sacrificing of my life.   Is the kind of America I know worth defending? Will America be a true and pure democracy after this war? Will colored Americans suffer still the indignities that have been heaped upon them in the past?" Thompson closed his letter by saying, "In conclusion let me say that though these questions often permeate my mind, I love America and am willing to die for the America I know will someday become a reality."[83] The *Courier* followed up with a "full speed ahead" charge to all black Americans. "We, as colored Americans, are determined to protect our country, our form of government and the freedoms which we cherish for ourselves and the rest of the world, therefore we have adopted the Double 'V' war cry—victory over our enemies on the battlefields abroad. Thus in our fight for freedom we wage a two-pronged attack against our enslavers at home and those abroad who would enslave us. WE HAVE A STAKE IN THIS FIGHT....WE ARE AMERICANS TOO!"[84]

In many ways, the war was a catalyst for improved race relations. Black newspapers were under scrutiny by the government because some—especially J. Edgar Hoover, head of the Federal Bureau of Investigation—believed that elements of the Double V campaign were seditious and therefore violated the Espionage and Smith acts. Francis Biddle, the attorney general, thought that some of what the black press promoted was harmful to the war effort, and he and Hoover seriously considered shutting them down. But Biddle, who was a supporter of racial equality, instead met with John Sengstacke, who now ran the *Chicago Defender* following the death of his uncle, Robert S. Abbott, in 1940. The two reached an agreement in 1942: The black papers would scale back their attacks on U.S. racism in lieu of more positive articles about the war effort. The government, in turn, would provide a more positive image of the African-American war effort. Jobs became more plentiful for blacks, too, on the homefront. The Double V campaign continued, but Sengstacke, speaking for the black press said, "The Negro press throughout the country, although they very properly protest, and passionately, against the wrongs done to members of their race, are loyal to their government and are all out for the war."[85]

Newsreels, however, ignored African Americans in the war effort. Through 1943, one agency determined, black troops had been shown in only three newsreels. When you consider that five different companies produced about six hundred newsreels per year, the African American fighting man was nonexistent in newsreels. Black publications, however, pointed the omission out incessantly. *The Crisis* believed the exclusion racism was meant "to keep from white America the news that the Negro minority is doing its par in the war."[86] In 1944, Paramount News began showing black soldiers and content in its newsreels. Another company, the All-American Newsreel, created newsreels with black content in 1942. These played in theaters with African-American clientele only and were estimated to been seen by about 4 million people every week.[87] Many in Hollywood realized the racism that had continually been displayed both in motion pictures and in the newsreels. That led to the release of *Americans All* in 1944, which advocated integration and equality for African Americans and religious toleration wherever needed. In the film, the narrator reminded those watching that "the United States has called for and has received in full measure the help of the Negro," while pointing out "the injustice of denying to the Negro the rights of American citizenship while expecting him to shoulder its ultimate responsibility—that of defending his country with his life."[88]

The effort by the black press truly made a difference. President Truman established the Committee on Civil Rights in 1947, and the U.S. military became the first officially desegregated element of American society immediately after the war when the president signed Executive Order 9981, which established the President's Committee on Equality of Treatment and Opportunity in the Armed Services.

THE COLD WAR AND KOREAN CONFLICT

Though the fighting in Korea was never a declared war and viewed as a "police action" by President Truman, more than 5,000 Americans either died fighting or from war-related causes.[89] "If news is what affects the lives of people as well as what interests them," Douglas Daniel observed, "then the Korean War was the biggest news story of the new decade [1950s]."[90] The Korean War was part of what would become known as the Cold War, the confrontation of ideologies and sometimes of soldiers between the Communist bloc led by the Soviet Union and the democracies of the West, led by the United States. The Cold War was truly an extension of the world problems that were never resolved by World War I, and it continued until the fall of the Berlin Wall in 1989. In the late 1940s, 1950s and into the 1960s, though, no one was sure if the Cold War would mean World War III. Many believed that it did, and the Cuban Missile Crisis of 1962 brought the country and world as close to nuclear war as perhaps it ever has. As the United States and the Soviet Union positioned missiles in Turkey and Cuba respectively, the two superpowers postured, neither appearing willing to back down. President John Kennedy and Premier Nikita Khrushchev, however, ultimately avoided a showdown and returned to the status quo of a balanced us versus them in terms of the world situation.

The tenuous nature of the world situation in 1950, however, made support of Korean intervention by the United States almost mandatory. "At this critical hour in the history of the world, our country has been called upon to give of its leadership, its efforts, and its resources to maintain peace and justice among nations" is the way President Truman stated the issue in a broadcast in September 1950.[91] The Soviet Union's control over eastern European countries and the division of Germany and of Berlin all pointed to an effort in the minds of most Americans of a totalitarian machine intent upon gobbling up as much of the world as possible. Communism had to be stopped, if that meant fighting North Korea, China, and the Soviet Union over Korea's fate, then that is what had to happen. Media supported this theory, too. "The lesson of Korea is grimly simple. Despite innumerable warnings, we did not do enough to deter Soviet aggression or to contain Soviet imperialism," the *Saturday Evening Post* in 1950 stated. "The real Kremlin goal is to make the living death of the slave society the universal condition of mankind, from the shores of the Atlantic to the islands of Japan, from the icy cliffs of Spitsbergen to the bright sands of Cape Comorin." Writers Joseph and Stewart Alsop called upon World War II to make another point: that "the Kremlin plays the power-and-terror game upon a larger stage than Hitler ever dreamed of." The Alsops mused that "we, as leaders of the free world, have ignored the inherent menace of Soviet war preparation," adding "Everything that can conceivably contribute to the strength of the free world must now be done without hesitation, without question, without regard to politics or cost."[92] Historian John Fousek says that magazines like *Saturday Evening Post* as well as other media organizations bought into the government's presentation of the world situation and served as high-powered public relations arms for the State Department. The media, consequently, portrayed the Korean War as the first shots in what would be the next world war, one that might last for generations.[93]

Media were not as unified in the nation's effort in Korea as they were with World War II, a sign of what was to come in the mid-1960s with Vietnam. Even though the military was officially desegregated, the country remained very much black and white. Some elements of the media called America's involvement in Korea "'imperialism' against colored people across the Pacific." But this statement by the *Pittsburgh Courier* could also be considered an extension of the Double V campaign that fought fascism abroad and racial inequality in America.[94] For many Americans, though, racism was secondary to another problem in the nation, the spread of Communism.

Fighting Communism abroad was one thing, but the real danger of the Cold War was Communism within. Many believed that media were the chief means of spreading Communism in the United States. With a means into every home via magazines, newspapers, radio, motion pictures, and now television, those who wrote stories and scripts, provided programming, starred in productions, or somehow had the ability to address so many in the nation at one time were looked upon with suspicion by an increasingly paranoid nation. Television and motion pictures, considered the most powerful of media because they combined audio and video as a means to address the public, were the most suspect media. Also, Hollywood had long been perceived as a moral influencer of the nation and had prompted studies and oversight since the 1920s. In 1950, a book titled *Red Channels* was published. A private work by three former FBI agents, the book listed organizations and individuals suspected of having links to the Communist party. Even though the book was a private publication and often based in rumor and innuendo, those whose names appeared in the listings were blacklisted in the industry, forced out of their jobs, and refused work because hiring them made a show, company, or media outlet appear to be "Red" friendly. *Red Channels* never worried about whether those in its listings had any real ties to the Communist party; punishing even innocent people was deemed OK because it would teach others a lesson.

It was in this atmosphere that media decided to break the chains of fear and suspicion that seemed to grip the nation, which was being spurred on by Wisconsin senator Joseph McCarthy. Though many journalists and media outlets were a part of an effort to disprove McCarthy's claims, Edward R. Murrow and CBS Television combined to end McCarthy's reign of accusations that Communists had infiltrated all elements of American life, even Congress. But, the junior Senator from Wisconsin used media to achieve his purposes as much as Murrow and others used media to combat him.

McCarthy fired the first salvo in his assault on Communism in Wheeling, West Virginia, in 1950. Speaking to a gathering of Republican women but knowing there would be press coverage, he gave media a prepared copy of his address. McCarthy told those gathered "that the fate of the world rests with the clash between the atheism of Moscow and the Christian spirit throughout other parts of the world." As he continued, McCarthy charged that Communist spies had already infiltrated the State Department. "I have here in my hand a list of 205 that were known to the Secretary of State as being members of the Communist Party and who, nevertheless, are still working and shaping the policy in the State Department....The modern champions of Communism have selected this as the time, and ladies and gentlemen, the chops are down—they are truly down."[95] After this speech, McCarthy became increasingly bold as he attacked people, accusing them of Communist affiliation. Print media, radio, and television served as McCarthy's tools to spread his charges of how Communism, through what appeared to be your typical Americans by looks but with a Communist ideology pushing their actions, was undermining the American way of life, democracy, and, ultimately, the entire free world. McCarthy took his inquisition on the road, speaking wherever he could. The more charges he made, the more media attention he gained. The more media attention McCarthy garnered, the wilder and more unverified his charges became.[96]

The Red Scare of the 1950s purported by McCarthy appeared likely for a number of reasons. In 1948, Alger Hiss, a State Department official who had been a key player at the 1945 Yalta Conference where the main Allied powers devised their final strategy to bring down Germany and who helped in the organization of the United Nations, was accused of having Communist ties. In January 1950, he was convicted of perjury and sentenced five years in connection to his comments about Communist ties to the House Committee on un-American Activities.[97] If such a key figure could hide his Communist ties, what about other, less well-known government workers, McCarthy and others figured? The Korean conflict obviously exacerbated the situation. Communism was on the move in other parts of the world. Why should Americans expect infiltration not to take place in the United States as with Hiss? Many of the nation's most popular publications bought into the idea wholly. "The Russians do not grab merely real estate. They also grab people. And *this is where you come in*," *Reader's Digest*, one of the nation's most widely read magazines, announced. "The communist masterplot … is focused on everyone, everywhere. It proceeds step by step and region by region. By its all encompassing timetable *sooner or later it has to reach you.*"[98]

Congress, then, to a certain extent bought into the possibility, too. In September 1950, it passed the McCarran Internal Security Act—named for senior Nevada Senator Pat McCarran—that required any group with Communist ties to be registered.[99] With the government and the country fearful and media following the senator and publicizing all that he charged, McCarthy appeared to be the most powerful man in America. He used the media to heighten the fear surrounding his charges by timing his press releases that contained "headline-catching statements" to coincide with the news cycles on the East Coast. That way, he could catch the leads in the afternoon dailies and ensure updates for later among print and broadcast outlets.[100] After his initial press releases, McCarthy would then elaborate. The original, one reporter said, dropped a hint of accusation. "Then he gives out a name. Third, he gives his version of what the name said or did. And the press carries all three." He was, another contemporary said, "master of the press."[101]

By the time Murrow took on McCarthy full force, the Wisconsin senator had been on the attack for at least four years and attacked by a number of media outlets. The *Washington Post*, for example, referred to McCarthy's efforts as "Sewer Politics." "Rarely has a man in public life crawled and squirmed so abjectly," the editorial said.[102] When McCarthy attacked another government worker from the Senate floor, the *Christian Science Monitor* said "an American citizen has been cruelly maligned and the mutual trust of the American people further undermined."[103] The *Wall Street Journal* said that in his re-election bid, "McCarthy is running on a platform to abolish the bill of rights so that he will not be hampered in his pursuit of some deluded people—more to be pitied than blamed—who embraced Communism."[104]

But, television in 1954 had become a medium of necessity in the United States. Newspapers still had a saturation among households of more than 100 percent, and radio could be found in 94 percent of all homes, but the new medium was rapidly becoming a household commodity. Where only 172,000 TV sets were in use in the United States in 1948, by 1954, Americans owned 35 million sets, with 413 stations broadcasting from 273 different cities. Only the most remote parts of the country lacked access to a TV station.[105] McCarthy surely used television, but it was television that was able to expose his inconsistencies, falsehoods, and end his campaign of accusations and red-baiting with Murrow at the helm of his "document for television," *See It Now*. Begun in November 1951, *See It Now* provided the nation with a new type of program. In the fifth edition, *See It Now* used edited footage of McCarthy as a helpless victim immediately contrasted with McCarthy as a bulldog on the attack. Murrow then commented on the two cuts with an obvious bias against McCarthy's tactics. It demonstrated that *See It Now* with Murrow was "a program with a point of view"—Murrow's.[106]

As Murrow prepared for his major attack on McCarthy, *See It Now* ran shorter pieces about the senator and his Communist exposé campaign leading up to the March 9, 1954, show devoted totally to McCarthy, his methods, and his accusations. "Tonight," Murrow said to begin the broadcast, "*See It Now* devotes its entire half hour to a report on Senator Joseph R. McCarthy, told mainly in his own words and pictures." Murrow almost immediately offered the senator a chance for rebuttal in a subsequent production, and he acknowledged that what viewers were about to see would be controversial and asked permission from viewers that he be allowed to read from a prepared script in order to avoid appearing to add bias to the report.[107] Murrow generally read what he reported on the program anyway, but he did not want there to be any chance for someone to find impropriety with the report, even though Murrow and CBS executive Fred Friendly detested McCarthy's methods and agenda. Then, in video and audio footage with some still images, *See It Now* picked apart McCarthy and his allegations.

Murrow, of course, was anything but unbiased in his commentary on the evidence he presented. Just as he had in his broadcasts from London during World War II, Murrow believed that through his journalism that he needed to effect change. "This is no time for men who oppose Senator McCarthy's methods to keep silent, or for those who approve," Murrow said in his closing commentary. "The actions of the junior senator from Wisconsin have caused alarm and dismay amongst our allies abroad and given considerable comfort to our enemies, and whose fault is that? Not really his. He didn't create the situation of fear; he merely exploited it, and rather successfully. Cassius was right: 'The fault, dear Brutus, is not in our stars but in ourselves.'"[108]

McCarthy, who had a 50-percent approval rating among Americans before the broadcast, saw it drop immediately and continue in a downward spiral. Where he had once bullied colleagues on the Senate floor, McCarthy faced censure in December. He died a broken man three years later.[109] Immediately after the broadcast, CBS was flooded with approving telegrams and telephone calls, about fifteen to one in supporting the *See It Now* presentation.[110] Television demonstrated how powerful it could be as a medium that could affect the public. "It was crusading journalism of high responsibility and genuine courage" was the way that one reporter described the medium after the *See It Now* broadcast. Another declared, "Right there television came of age."[111] Murrow believed this, too. McCarthy, Murrow said, was "the creature of the mass media. They made him. They gave nationwide circulation to his mouthings."[112] Because McCarthy had the eyes and ears of the nation through media, Murrow was saying, he was able to convince millions that his claims of Communist infiltration into multiple elements of society was true and presented a real danger to the nation.

Murrow won numerous press awards for the broadcast. It was deemed "a new type of TV programming" that provided a new kind of reality for viewers.[113] It pushed television into the sphere of print and radio in terms of its power to reach and affect people. However, *See It Now* and Murrow to a certain extent became collateral damage of McCarthyism. Even though the program signaled the Wisconsin senator's downfall and the production received accolades, Murrow became the focus of an FBI investigation, ordered by McCarthy's close friend J. Edgar Hoover, and the Aluminum Company of North America, the program's sponsor, was also pressured to drop the show. It did so two months after the McCarthy program aired. CBS canceled *See It Now* in 1958.[114]

***

The cataclysmic wars that rocked the world in the first half of the twentieth century would have changed the course of nations regardless of media, but media became powerful tools during the wars. World War I demonstrated exactly how media could be used for propaganda. World War II did the same, but the U.S. government realized that voluntary censorship that embraced, for the most part, truth was a more powerful weapon than media that created and manipulated information to effect public opinion. The fact that in World War I the government created the Committee on Public Information and in World War II the Office of War Information so closely after the United States began war efforts attests to the power media had obtained by the twentieth century. Presidents realized this, too. Herbert Hoover declared that he would focus on radio campaigning in 1928, and FDR immediately turned to radio for his "Fireside Chats" as a means to quell worry over the Depression. Public campaigning was still important, but the new broadcast media meant that millions of Americans could be reached without spending thousands to tour the country.

In both world wars, journalists played a critical role in creating a positive image for the American war effort and of the Allies. Some reporters became the news. Floyd Gibbons, who sailed on the *Laconia* in hopes of the ship being torpedoed so that he could write a story, got his wish—and his story. Edward R. Murrow had no reservations about creating radio pieces from London that evoked sympathy for England. He said that "radio used as an instrument of war is one of the most powerful weapons a nation possesses."[115] The same had been said of newspapers in the Revolution, magazines in the progressive reforms of the 1910s, and would be said of television, especially in the turbulent times that America was experiencing at the time of Murrow's death in 1965.

NOTES

1. Richard Tregaskis, "House to House and Room to Room," *Saturday Evening Post*, February 24, 1945, 18, 101.
2. Michael S. Sweeney, *The Military and the Press: An Uneasy Truce* (Evanston, Ill.: Northwestern University Press, 2006, 5.
3. George Brown Tindall and David Emory Shi, *America: A Narrative History*, 5th ed., 2 vols. (New York: W. W. Norton, 1999), 1,124.
4. "Lucitania Incident," *Emporia (Kan.) Gazette*, May 10, 1915.
5. Quoted in Michael Kunczik, *Images of Nations and International Public Relations* (Mahwah, N.J.: Lawrence Erlbaum Associates, 1997), 44.
6. Stewart Halsey Ross, *Propaganda for War: How the United States Was Conditioned to Fight the Great War of 1914–1918* (Jefferson, N.C.: McFarland, 1996), 20.

7. Count Bernstorff, *My Three Years in America* (New York: Charles Scribner's Sons, 1920), 54.
8. George B. Parker, "Censorship, You–and Us," in Laird B. Anderson and Wm. David Sloan, eds., *Pulitzer Prize Editorials: America's Best Writing, 1917–2003*, 3rd ed. (Ames: Iowa State Press, 2003), 60.
9. Edward Bernays, *Propaganda* (New York: Horace Liveright, 1928), 27.
10. Quoted in Gerd Horten, *Radio Goes to War: The Cultural Politics of Propaganda During World War II* (Berkeley: University of California Press, 2003), 52.
11. Atomic Energy Act of 1954, 42 U.S.C. § 2011.
12. "The Press Agents' War," *New York Times*, September 9, 1914, 8.
13. Ross F. Collins, *World War I* (Westport, Conn.: Greenwood Press, 2003), 2.
14. *Twentieth Century Atlas–Death Tolls*, 2005, users.erols.com/mwhite28/warstat1.htm.
15. "Fact Sheet: America's Wars," *United States Department of Veterans Affairs*, November 2008, http://www1.va.gov/opa/fact/amwars.asp.
16. Sweeney, *The Military and the Press*, 37–38.
17. Sweeney, *The Military and the Press*, 42.
18. "Bryce Committee's Report on Deliberate Slaughter of Belgian Non-Combatants," *New York Times*, May 13, 1915, 6.
19. Frank H. Simonds, "The Lusitania Anniversary," *New York Tribune*, May 7, 1916.
20. "Lusitania Tragedy," *Emporia (Kan.) Gazette*, May 10, 1915.
21. Bernstorff, *My Three Years in America*, 54.
22. "War News Bureaus in Wireless Rivalry," *New York Times*, November 30, 1914, 2.
23. "Says Germany Got Tip on Contraband," *New York Times*, May, 11, 1915, 7; "Wireless From Here to German Admiralty," *New York Times*, May 11, 1915, 8.
24. Margaret A. Blanchard, *Revolutionary Sparks: Freedom of Expression in Modern America* (New York and Oxford: Oxford University Press, 1992), 75–76.
25. Quoted in Wm. David Sloan, ed., *The Age of Mass Communication* (Northport, Ala.: Vision Press, 1998), 403.
26. Newton Baker, "Forward," in George Creel, *How We Advertised America* (New York and London: Harper & Brothers, 1920), xiii.
27. Creel, *How We Advertised America*, 18.
28. Stéphane Audoin-Rouzeau and Annette Becker, *14–18: Understanding the Great War* (New York: Hill and Wang, 2002), 109–110.
29. Quoted in Thomas Doherty, *Hollywood, American Culture, and World War II* (New York: Columbia University Press, 1993), 88.
30. Selig Adler, *The Isolationist Impulse: Its Twentieth Century Reaction* (London: Abelard-Schuman, 1957), 37.
31. Creel, *How We Advertised America*, 24.
32. Petrina Dewitt, "Committee on Public Instruction," *Germany and the Americas: Culture, Politics, and History*, ed. Thomas Adam (Santa Barbara, Calif.: ABC Clio, 2005), 252.
33. Edward L. Bernays, *Biography of an Idea: Memoirs of Public Relations Counsel Edward L. Bernays* (New York: Simon & Schuster, 1965), 170.
34. James D. Startt, "The Media and National Crises, 1917–1945," *The Media in America: A History*, ed. Wm. David Sloan, 5th ed. (Northport, Ala.: Vision Press, 2002), 322.
35. Floyd Gibbons, "The Sinking of the Laconia," *Chicago Tribune*, February 26, 1917.

36. See Edward Gibbons, *Floyd Gibbons–Your Headline Hunter* (New York: Exposition Press, 1953).
37. Harold D. Lasswell, *Propaganda Technique in the World War* (New York: Alfred A. Knopf, 1927), 31–32.
38. Charles Edward Russell, *Pearson's Magazine*, April 1914, 30.
39. *Schenck v. United States*, 249 U.S. 47 (1919). Charles Schenck was arrested for his anti-war activities and for distributing pamphlets that urged resistance to the draft. A unanimous court upheld his conviction. Several statements from Justice Oliver Wendell Holmes' decision have become part of the American terminology when speaking of free speech and its limitations. "The question in every case is whether the words used are used in such circumstances and are of such nature as to create a clear and present danger that they will bring about the substantive evils that Congress has a right to prevent" (249 U.S. 52).
40. "Fact Sheet: America's Wars," *United States Department of Veterans Affairs*, November 2008, http://www1.va.gov/opa/fact/amwars.asp.
41. "Statement by the President, December 16, 1941, Record Group 80, File 89-1-8, National Archives.
42. Betty Houchin Winfield, *FDR and the News Media* (Urbana: University of Illinois Press, 1990), 231.
43. John Hohenberg, *Foreign Correspondence: The Great Reporters and Their Times* (New York: Columbia University Press, 1964), 383.
44. Quoted in Patrick S. Washburn, *World War II, The European Theater, Greenwood Library of American War Reporting*, ed. David A. Copeland, 8 vols. (Westport, Conn.: Greenwood Press, 2005), 5:251.
45. Office of Censorship, Press Release, January 14, 1942, in Washburn, *World War II*, 251.
46. "Government Information Manual for the Motion Picture Industry," in Clayton R. Koppes and Gregory D. Black, *Hollywood Goes to War: How Politics, Profits, and Propaganda Shaped World War II Movies* (New York: Free Press, 1987), 66.
47. Alien Registration Act of 1940 18 U.S.C. § 2385.
48. Ernie Pyle, *Here Is Your War* (New York: Henry Holt, 1943), 304.
49. Ernie Pyle, *Cleveland Press*, June 12, 1944.
50. Pyle, *Here Is Your War*, 247.
51. Frederick S. Voss, *Reporting the War: The Journalistic Coverage of World War II* (Washington, D.C.: Smithsonian Institution, 1994), 102.
52. Ernie Pyle, "The Death of Captain Henry Waskow," Scripps Howard Newspaper Alliance, January 10, 1944.
53. Quoted, in Bradley Hamm and Donald L. Shaw, *World War II, The Asian Theater, Greenwood Library of American War Reporting*, ed. David A. Copeland, 8 vols. (Westport, Conn.: Greenwood Press, 2005), 6:180.
54. Ernie Pyle, "V-E Day, Scripps Howard Newspaper Alliance, April 1945.
55. Evans Wylie, "Death of Ernie Pyle," *Yank*, May 18, 1945.
56. James Tobin, *Ernie Pyle's War: America's Eyewitness to World War II* (New York: Free Press, 1997), 2.
57. Voss, *Reporting the War*, 102.
58. *Golden Age of Radio 1935–50*, March 2007, http://history.sandiego.edu/gen/recording/radio2.html.

59. Michael S. Sweeney, *From the Front: The Story of War Featuring Correspondents' Chronicles* (Washington, D.C.: National Geographic Press, 2002), 181.
60. Philip Seib, *Broadcasts from the Blitz: How Edward R. Murrow Helped Lead America Into War* (Washington, D.C.: Potomac Books, 2006), x.
61. Edward R. Murrow, "Report to America," *CBS News*, December 2, 1941.
62. Edward R. Murrow, "The Cookies—the Four-Thousand-Pound High Explosives—Were Bursting Below Like Great Sunflowers Gone Mad," *CBS News*, December 3, 1943.
63. Edward R. Murrow, "Where Are They Now? The Liberation of Buchenwald," *CBS News*, April 15, 1945.
64. See Starr Smith, *Jimmy Stewart: Bomber Pilot* (St. Paul, Minn.: Zenith Press, 2005).
65. Doherty, *Hollywood, American Culture, and World War II*, 61.
66. George H. Roeder Jr., *The Censored War: American Visual Experience During World War II* (New Haven, Conn.: Yale University Press, 1993), 4.
67. Thomas W. Bohn and Richard L. Stromgren, *Light and Shadows: A History of Motion Pictures* (Mountain View, Calif.: Mayfield Publishing, 1987), 236.
68. Sweeney, *The Military and the Press*, 97.
69. Thomas F. Brady, "Donald Doesn't Duck the Issue," *New York Times*, June 21, 1942, X3.
70. "Disney Net for Year $431,536," *Motion Picture Herald*, November 6, 1943, 16.
71. "Quack, Quack, See Mr. D. Duck, Expert on Income Tax Returns," *Christian Science Monitor* (Boston), January 23, 1942, 5.
72. T.M.P., "The Screen," *New York Times*, July 19, 1943, 13.
73. Theodore Strauss, "Donald Duck's Disney," *New York Times*, February 7, 1943, X3.
74. Strauss, "Donald Duck's Disney."
75. "Films Must Have Part in Peace Talks–Warner," *Motion Picture Herald*, September 23, 1944, 30.
76. Doherty, *Projections of War*, 229–32.
77. Gibbons, "The Sinking of the Laconia."
78. "Movies at Press Club," *Washington Post*, February 20, 1915, 5; "Movies at the White House," *Washington Post*, February 10, 1915, 4; "Writing history with lightning" quoted in Wyn Craig Wade, *The Fiery Cross: The Ku Klux Klan in America* (New York and Oxford: Oxford University Press, 1998), 126.
79. Collins, *World War I*, 196.
80. W. E. B. DuBois, "Close Ranks," *The Crisis*, July 1918.
81. In *The Black Press: Soldiers Without Swords*, Produced and directed by Stanley Nelson, 86 min. (California Newsreel, 1999), videocassette.
82. Robert S. Abbott, Strict Accountability," *Chicago Defender*, March 31, 1917.
83. James G. Thompson, "Dear Editor," *Pittsburgh Courier*, January 31, 1942.
84. "The Courier's Double 'V' For a Double Victory Campaign Gets Country-Wide Support," *Pittsburgh Courier*, February 14, 1942.
85. *Amsterdam Star-News*, February 20, 1943, quoted in Blanchard, *Revolutionary Sparks*, 203. For a complete understanding of the confrontation between the black press and the government, see Patrick S. Washburn, *A Question of Sedition: The Federal Government's Investigation of the Black Press During World War II* (New York: Oxford University Press, 1986).
86. "Omissions from Newsreels," *The Crisis*, February 1944, 39.
87. Doherty, *Projections of War*, 220–21.

88. *March of Time*'s *Americans All* (July 1944), in Doherty, *Projections of War*, 224–25.
89. "Fact Sheet: America's Wars," *United States Department of Veterans Affairs*, November 2008, http://www1.va.gov/opa/fact/amwars.asp.
90. Douglas K. Daniel, *The Korean War*, in *The Greenwood Library of American War Reporting*, ed. David A. Copeland, 8 vols. (Westport, Conn.: Greenwood Press, 2005), 6: 250.
91. Quoted in Bruce Cumings, *The Origins of the Korean War*, 2 vols. (Princeton, N.J.: Princeton University Press, 1990), 2:542.
92. Joseph Alsop and Stewart Alsop, "The Lessons of Korea," *Saturday Evening Post*, September 1950, quoted in John Fousek, *To Lead the Free World: American Nationalism and the Cultural Roots of the Cold War* (Chapel Hill: University of North Carolina Press, 2000), 174–75.
93. Fousek, *To Lead the Free World*, 175.
94. Fousek, *To Lead the Free World*, 180.
95. Frank Desmond, "M'Carthy Charges Reds Hold U.S. Jobs," *Wheeling (West Virginia) Intelligencer*, February 10, 1950.
96. Blanchard, *Revolutionary Sparks*, 246.
97. See Allen Weinstein and Alexander Vassiliev, *The Haunted Wood: Soviet Espionage in America—The Stalin Era* (New York: Modern Library, 1999) and Fred J. Cook, *The Unfinished Story of Alger Hiss* (New York: William Morrow, 1958).
98. L. Stowe, "Conquest by Terror," *Reader's Digest*, June 1952, 137, quoted in Joanne P. Sharpe, *Condensing the Cold War:* Reader's Digest *and American Identity* (Minneapolis and London: University of Minnesota Press, 2000), ix.
99. Robert Griffith, *The Politics of Fear: Joseph R. McCarthy and the Senate* (Lexington: University of Kentucky Press, 1970), 117–18.
100. Edwin R. Bailey, *Joe McCarthy and the Press* (Madison: University of Wisconsin Press, 1981), 70.
101. Quoted in Bailey, *Joe McCarthy and the Press*, 84.
102. "Sewer Politics," *Washington Post*, February 14, 1950, 10.
103. "A Crying Need," *Christian Science Monitor* (Boston), April 25, 1952, 24.
104. "Refuge for Traitors," *Wall Street Journal* (New York), September 2, 1952, 6.
105. Bailey, *Joe McCarthy and the Press*, 176.
106. A. M. Sperber, *Murrow: His Life and Times* (New York: Freundlich Books, 1986), 370. *See It Now*, December 16, 1951.
107. *See It Now*, March 9, 1954.
108. *See It Now*, March 9, 1954. A transcript of Murrow's commentary and the *See It Now* episode on McCarthy are contained in Fred W. Friendly, *Due to Circumstances Beyond Our Control* (New York: Random House, 1967), 41.
109. Blanchard, *Revolutionary Sparks*, 261.
110. Val Adams, "Praise Pours in on Murrow Show," *New York Times*, March 11, 1954, 19.
111. Jack Gould, "Television in Review: Murrow vs. McCarthy," *New York Times*, March 11, 1954, 38; John Crosby, "Murrow-McCarthy Case Places TV in a New Role," *Washington Post*, March 12, 1954, 41.
112. Edward R. Murrow, "Television and Politics," Guildhall, London, October 19, 1959, quoted in Sperber, *Murrow*, 471.
113. *Variety*, March 17, 1954, quoted in Sperber, *Murrow*, 465.

114. Bob Edwards, *Edward R. Murrow and the Birth of Broadcast Journalism* (Hoboken, N.J.: John Wiley, 2004), 121, 128; Bailey, *Joe McCarthy and the Press*, 199.
115. Edward R. Murrow, *This Is London* (New York: Simon and Schuster, 1941), 76.

9. The Whole World Is Watching: The Journalism of Change, 1950s-1970s

In 1968, the United States was in the midst of massive upheaval. The president, Lyndon Johnson, announced he would not seek a second term. War in Vietnam, which most Americans believed was going well for the country, took a deadly turn with the Tet offensive by the North Vietnamese. Two national figures, Martin Luther King, Jr. and Robert F. Kennedy, were gunned down. Rioting in the nation's cities erupted again, just as it had for four previous summers. Increasingly, young Americans, disenchanted with what was happening in the nation, took to the streets in protest. On August 28, 1968, outside the Hilton Hotel in Chicago, thousands of them gathered in Grant Park to protest at the Democratic National Convention. Millions of Americans turned on their televisions expecting to see Hubert Humphrey win the Democratic presidential nomination. But, increasingly, their screens were filled with protesters and police clashing on Michigan Avenue. As a soundtrack to the images, viewers could hear the protesters chanting repeatedly: "The whole world is watching. The whole world is watching."

In the decades following World War II, Americans watched as media presented what was happening in the country. Incredible changes politically, culturally, and socially created unrest in the United States not known since the Civil War. Media advancements fueled those changes in ways never experienced by Americans. Television's mercurial penetration into the nation's psyche was unprecedented. From just a few thousand TV sets in existence in the United States in 1946 to tens of millions by the middle of the 1950s, television became an integral part of American life faster than any medium yet to be introduced. Marshall McLuhan in his book *Understanding Media* concluded in 1964 that "the medium is the message," meaning that television, because of its ability to provide sound *and* pictures to people anywhere in the United States became as important as the information that the medium provided. By the time of the Democratic National Convention in 1968, in terms of the

United States, the protesters were correct; the whole world was watching since about 95 percent of all U.S. homes had at least one TV set.[1]

When the 1950s began, about 10 million TV sets were in use in the United States. Despite their cost, Americans eagerly bought them. A tabletop seventeen-inch, black-and-white set cost, on average, about $250. Prices for a console model with a twelve-inch screen averaged about $450, which means Americans were paying the equivalent of between $2,200 to about $4,000 in 2010 dollars for the latest media technology at the mid-point in the twentieth century, roughly equal to what people paid for cutting-edge technology, large screen, high definition televisions in the first decade of the twenty-first century. By 1960, the cost of a tabletop set remained the same, while a console sold for about $100 more, but had a twenty-three-inch screen versus the twelve-inch one of a decade earlier. In 1960, color tabletops could be bought for $495, and roughly 87 percent of all American homes had at least one television, according to the U.S. Census Bureau. That percent would continue to rise.[2]

Television news in the 1960s and 1970s—essentially—gave all Americans the same story. With three national networks only, the script did not vary greatly from CBS to NBC to ABC in terms of what appeared on the nightly news. With 50 million viewers or more each night, and television listed as the principal source of news from 1963 onward,[3] similar story lines, images, and commentary made television central to Americans' understanding of the nation's situation, something that Maxwell McCombs and Donald Shaw would prove concerning media in general with their pivotal agenda-setting studies, published in 1972. Media, they concluded, set the agenda of the nation by providing the topics about which people thought, talked, and sometimes acted or reacted.[4] Though McCombs and Shaw did not claim that media told people what to think, in many cases, that is exactly what occurred, though.

Television, of course, was only the latest addition to media that Americans used. Other media continued to play a vital role in the nation and played critical roles in establishing agenda and guiding national thought. Newspapers dominated the advertising dollar and saturated the nation with daily subscriptions per households equaling or exceeding 100 percent throughout the 1960s. Nearly nine thousand different magazines were published yearly in 1970, and about seven thousand commercial radio stations were licensed in 1969, a 40-percent increase over the number of stations serving the nation in 1960, according to the Federal Communications Commission *Annual Report* of 1970.[5] The nation had never before had so many options for information from so many media sources, and never before had Americans wholeheartedly opted to own or to use so many different sources of information, entertainment, and sources of consumer information, that is, advertising. But, because of the momentousness of the events of the era, almost all media tended—when it came to news—to focus, naturally, on those critical occurrences. Because the audience was much less fragmented than it would become in the last two decades of the twentieth century and into the twenty-first, any unified message from media strengthened its impact on people and the country.

Another medium greatly affected the nation, too, recorded music. If "the whole world is watching" was the soundtrack for 1968, it was only one cut on a double album that played to millions of young Americans from the end of the 1950s through the 1970s. Recorded music became a powerful influencer because it offered the ability of providing a political, social, or cultural message with a catchy rhythm, tune, and beat. That message was spread via record, radio, and television, and it often was unintelligible to older Americans who found the sound of rock 'n' roll to be nothing but noise. As musicians increasingly became socially conscious, they used their music to speak out against what they perceived as the ills of the nation. Younger Americans listened, agreed, and used recorded music as a way to comprehend the news they saw on television and read about in newspapers and magazines.

The momentous events of the 1950s, 1960s, and 1970s have become some of the best known among Americans in terms of understanding media's impact on society because almost all media tended—when it came to news—to focus, naturally, on those critical occurrences. The civil rights movement and the war in Vietnam, for example, were greatly influenced by media coverage, particularly by television. As David Vann, mayor of Birmingham during the early 1960s' civil rights protests, said, "It was a masterpiece of the use of media to explain a cause to the general public of the nation. Because in those days you had fifteen minutes of national news and fifteen minutes of local news, and in only marching one block they could get enough news footage to fill all the newscasts of all of the television stations in the United States."[6] Journalist Peter Braestrup said that in 1968, following the Tet offensive, the media message "most simply put, was: DISASTER IN VIETNAM!" This, he says, shaped the political reaction in Washington, whether what was happening in Vietnam was really a disaster or not. What Americans would ultimately receive from television concerning Vietnam was "worst case" speculation.[7]

In the 1960s and 1970s, then, the United States witnessed what one report called the development of a true "national media." Because the message remained constant, the report concluded, people became increasingly dependent upon this national news that was based in television, and that message, the report said, tended to create a lack of confidence in the government and skepticism concerning established institutions in the country.[8] NBC News anchor David Brinkley said in 1969 that television simply showed the nation a "mirror" of itself and that people did not like what they saw.[9] While Brinkley's observations were true to an extent, a powerful media still chose what images to show, what sounds to play, and what subjects on which to comment. And, even though national media have rightly garnered the most attention in terms of shaping the nation during the '60s and '70s, smaller media outlets still functioned in powerful ways. Sometimes, as Horace Carter's *Mount Tabor (N.C.) Tribune* did, they were able to shake the foundations of a community forever.

TAKING ON THE KLAN

When Joseph Pulitzer and William Randolph Hearst began their newspaper competition late in the 1890s, their papers reached circulation sizes of more than a million copies per day. Dozens of other newspapers in the United States had circulation figures well in excess of 100,000 daily. Though papers such as the *World* and *Journal* were printed for a New York clientele, the power of the papers stretched across the nation. The nation's newspapers subscribed to wire services, like the Associated Press and United Press, which provided another means for sharing the same stories with Americans everywhere. Then, men like E. W. Scripps and Frank Munsey began consolidating newspapers, creating chains that served larger and larger areas with the same paper and information. With the introduction of electronic media, the commonalty of information increased even more. People in Paducah, Kentucky, watched the same motion pictures that people in New York City and Los Angeles did. In the second half of the 1920s, radio networks begun by NBC and CBS provided news and entertainment programming that all Americans shared. With television, the nation now followed mass media that gave them the same stories, images, and ideas because of their massive size and ability to share that message across time and space.

But, during this massive growth of media, small media outlets continued to flourish and serve communities in ways the large media outlets were unable to do. These outlets, just as they had since the beginning of the press in the American colonies, took on issues of significance to the people they served. That is exactly what a young newspaperman in a small town in the southeastern part of North Carolina did in 1950 through 1952. Horace Carter, after serving in the Navy during World War II, moved to Tabor City, North Carolina, and began a newspaper for the community. In July 1950, the Ku Klux Klan, which was comprised of members from North and South Carolina, began a recruiting drive in Mount Tabor. About thirty cars filled with robed klansmen drove through the town, targeting sections where African Americans lived. KKK members effectively controlled many aspects of life in the region surrounding Mount Tabor. The Klan was especially effective at building upon the fears of Communism that Senator Joseph McCarthy would arouse to enlist membership, and using this fear, to continue efforts at denying African

Americans any semblance of equality. Carter, along with a few other newspaper owners, decided to end the Klan's denomination.10

The *Mount Tabor Tribune* was a small newspaper, run by Carter and a few other linotype and office workers, but Carter believed that it was his and his newspaper's responsibility to eradicate Klan presence from the community. "My duty as the only newspaperman in Tabor City stared me squarely in the face," he said. "I must fight the Klansmen with all the power that my tiny press could muster. That meant that I too would be the victim of their wrath."11 Following the Klan's first visit to Tabor City, Carter wrote, "Any organization that has to work outside the law is unfit for recognition in a country of free men....It takes a united front to combat lawlessness. It takes all the law abiding people as a unit to discourage and combat a Ku Klux Klan that is totally without law. The Klan, despite its Americanism plea, is the personification of Fascism and Nazism. It is just such outside-the-law operations that lead to dictatorships through fear and insecurity."12

Immediately, irate Klan members attacked Carter. They threatened to burn down his home and newspaper office. They promised that they would end local business advertising in his paper and start boycotts of companies that did advertise in the *Tribune*. The threats worked in some cases as businesses pulled their ads from the *Tribune*, while some people cancelled their subscriptions. Carter continued his campaign against the Klan, though. The situation exploded in late August in nearby Myrtle Beach, South Carolina, when Klan members attacked a black nightclub. The two sides exchanged gun fire, and a Klan member was killed. When he was taken to the hospital and his KKK robe removed, hospital workers discovered that the dead man was a sheriff's deputy and newly elected magistrate for the county.13 The gun battle led to the arrest of the Grand Dragon of the Klan and to renewed efforts by Carter to demonstrate that the Klan in Southeast North Carolina and Northeast South Carolina was composed of sordid characters who were worse than those they attacked both physically and verbally. "Klansmen on the whole are not fine, Christian, God-fearing Americans," who, Carter declared, keep their membership a secret, just as the Communist Party does. That is why, he concluded, the Klan "carry on its activities in the dead of night."14

Both the Klan and Carter continued their activities. The Klan initiated a campaign of flogging and whipping blacks and those who sympathized with them. Death threats aimed at Carter continued as did his editorials.15 Finally, with the Federal Bureau of Investigation infiltrating the Klan, the activities ended. When that happened, Carter had achieved his goal of breaking the Klan's hold on his section of North and South Carolina. "I doubt seriously that it [the Klan's control of the region] would ever have been broken," Carter would later say about the newspaper effort against the KKK. "I think it's one of the purposes that perhaps we served, to get enough attention focused on it that it did get the FBI interested. And if we hadn't I don't believe the local law enforcement would have ever broken it."[16] Through editorials, stories, and personal threats, the paper, along with the nearby *Whiteville (N.C.) News Reporter*, ended what the *Washington Post* called a "terror outbreak" and through reports ensured the arrest and conviction of more than one hundred Klansmen for assorted crimes.[17]

EQUALITY, ONE SEAT AT A TIME

A decade after the *Mount Tabor Tribune* took on the Klan and about two hundred miles northwest in North Carolina, four college students decided to test a segregation policy that was never questioned, merely accepted by people throughout the South, and, in fact, most of the nation. The four had no intention of using media to spread their cause, but that is exactly what happened; local media rushed to cover the story, and national outlets immediately grabbed it. Within days, what happened in Greensboro, North Carolina, on February 1, 1960, was news throughout the United States. On that Monday, four students from North Carolina Agricultural and Technological College confronted segregation head-on over an afternoon cup of coffee. The four, Franklin McCain, Ezell Blair Jr., Joseph McNeil, and David Richmond, walked into the F. W. Woolworth's department store in the heart of Greensboro's business district and sat down at the lunch counter and asked for a beverage. Lunch counters served blacks, but they had to come to the end of the counter, order their food or drink, and then leave. The stools at the lunch counters were reserved for whites only. When the four arrived, they sat down, expecting to be arrested or beaten up. They were ignored, instead, by the white customers and by the African-American waitress who worked there. They sat there for an hour, until the counter closed, and then returned to A&T.[18]

The next day, twenty black students returned to take seats on the stools. After the first protest, the *Greensboro Record* ran a story about it on the front page of the inside section of the paper. The afternoon daily told readers that workers at Woolworth's "did not serve the group and they sat from 10:30 a.m. until after noon. White customers continued to sit and get service."[19] Stories about the sit-ins in the Greensboro media immediately drew state and national attention. The initial sit-in "started spontaneously," the *New York Times* reported, but three days later, students from other predominantly black colleges had joined the protest. Whites joined with blacks, too, with about eighty-five people in total coming to the Woodworth's and to other businesses that served lunch to those who worked in the business district. By day four, white reaction led to confrontations, and Woolworth's closed the lunch counter at the end of the week rather than serve blacks or be the scene of violence, triggered by a bomb threat on the store.[20] The sit-ins and store reaction to them spread in Greensboro. "Lunch counters at two downtown variety stores, hit last week by a mass sit-down demonstration by Negro students protesting segregation of the eating places, will be closed when the stores open today," the *Record* reported.[21] The closings were soon extended indefinitely.

National media coverage initiated sit-ins throughout the South, obviously, much faster than word-of-mouth transmission could have. Within two weeks, similar sit-ins were taking place in about a dozen North Carolina cities and in Virginia and Tennessee. Before the month ended, sit-ins had taken place in eight states with about seventy thousand people taking part. The Southern Christian Leadership Conference headed by Martin Luther King Jr. joined in, though King admitted that neither the SCLC or the NAACP had anything to do with the original sit-ins. It was, in fact, reports about the sit-ins, spread by media, that led to the widespread protests. Because video footage was picked up nationally from Greensboro station WFMY, the sit-in movement spread more rapidly than it would have had it depended solely upon word-of-mouth transmission. One Atlanta student, for example, said that "she ran home to see it on television" after word of the Greensboro protests reached Spelman College in Atlanta.[22] In Nashville, Tennessee, where sit-ins began with more than 120 protesters on February 13, the *Nashville Tennessean* ran seventy stories in fourteen weeks on sit-ins and protests in the Tennessee capital.[23]

Once protesters realized the power of media coverage for their cause, they actively sought it, using media as a way to portray themselves positively while the whites who attacked them were often described negatively. According to Gene Roberts, leaders of the sit-ins actively sought to befriend white reporters, who got to know the protesters. This, in turn, Roberts says, led to a more positive depiction of protesters in the media than might have happened.[24] But the passive, nonviolent methods of protesters may have been all that was needed to evoke the positive image in the media. Early in the coverage of North Carolina sit-ins, the *New York Times* told of eggs tossed at demonstrators and of "a sullen crowd of white youths" who shoved the managers of stores, fought with media. The *Reporter* in Nashville described how white "hoodlums" spit on, punched, and stuffed cigarette butts down the collars of black protesters. It told of how whites "pulled three Negro boys from the counter and started beating them. The three Negroes did not fight back."[25] The same media images appeared throughout the nation, which is why David Vann would say three years later that the civil rights movement was "a masterpiece of the use of media."

The sit-in images produced amazingly fast results considering how long segregation of all elements of U.S. society had existed, especially in the South. By the summer of 1960, Greensboro desegregated all public facilities, and blacks were served at the Woolworth's lunch counter on July 25, the first being the black waitress—Geneva Tisdale—who refused to serve the four original protesters.26 Major department store chains—like Woolworth's—had allowed individual stores to set policies on service before the sit-ins began, but in October, four major department stores—including Woolworth's—ordered that luncheon facilities would be integrated, affecting about 150 stores in 112 cities.27 While not all media outlets supported African-American efforts, many did. The *Washington Post*, for example, pointed out the fallacy of the segregated lunch counters by quoting Florida Governor LeRoy Collins, saying, "if a man invites the public generally to trade at his store, 'I think then it is unfair and morally wrong for him to single out one department and say he does not want or will not allow Negroes to patronize that one department.' This is a question of basic moral justice, and ultimately throughout history moral considerations have usually triumphed."28 The violence that met sit-in protesters actually created a financial boon for store owners. "No store in the South which has opened its lunch counters to Negroes has reported a loss of business. Managers have reported business as usual or noted an increase," a report from Atlanta stated. It also reported that where segregation continued business had been lost "as much as 65 per cent in one store in Charlotte, N.C."29 One Northern media outlet, however, correctly described what lay ahead, that "the lunch counter movement is only part of a long series of nonviolent protests that may be expected from educated and dedicated young Negroes seeking equal treatment in all the public phases of community life."30

Media were now poised to bring this movement to the nation and so were civil rights leaders, though the gains would not come as quickly nor with as little violence as accompanied the sit-ins. Civil rights leaders urged Martin Luther King, Jr. to construct his messages so that they would fit into neat, short, TV sound bites. The goal was to make sure those messages appeared on the *CBS Evening News* or *The Huntley-Brinkley Report* on NBC. Media, in turn, flooded the battleground of the civil rights efforts—Alabama and Mississippi, especially, but wherever the cause turned—with reporters, photographers, and cameramen. Even civil rights' harshest opponents—like Alabama Governor George Wallace—prepared their messages to provide good quotes for print media and titillating sound bites for television. Being "camera ready" became essential throughout the nonviolent portion of the civil rights struggle.31 "Alabama has tried to nurture and defend evil, but evil is choking to death in Alabama's dusty streets. We are moving toward the land of freedom," King said after leading 25,000 marchers from Selma, Alabama, to the state's capital, Montgomery.32 Wallace irreligiously told Alabamans, reporters, and TV cameras at his 1963 inaugural address, "I draw the line in the dust and toss the gauntlet before the feet of tyranny, and I say segregation now, segregation tomorrow, segregation forever."33 Later, the images of what Wallace promised played out in media as the nation saw blacks being attacked by police dogs and being blasted by fire hoses that sprayed with enough force to remove bark from trees. The confrontations of the civil rights movement filled the networks' evening news, front pages of newspapers, and spreads in magazines.

As the movement continued, the nation was able to view all the confrontations and nastiness that accompanied the push to obtain equality among the races in the United States. And that, according to writer Jon Wiener, is amazing in itself because before the mid-1950s and the *Brown v. Board of Education* Supreme Court case and Rosa Parks' decision not to give up her bus seat for a white person, mainstream U.S. media rarely paid any attention to anything that dealt with African Americans.[34] That all changed, however. From the brutal deaths of James Cheney, Andrew Goodman, and Michael Schwirmer in Mississippi to the burning of Watts, Detroit, and dozens of other cities in race riots to the assassination of Martin Luther King, Jr., media provided coverage. Using media to present a message became central to the effort. "If it hadn't been for media," John Lewis, a leader of the Nashville sit-ins, the Student Nonviolent Coordinating Committee, and later a Georgia representative to Congress, said, "the civil rights movement would have been like a bird without wings, a choir without a song."[35]

DEBATES, ASSASSINATION, AND JFK

The 1960 presidential election provided the nation with something new: TV debates. Television made its presidential election premiere in 1948 with limited coverage of the Republican National Convention, but few watched. By 1960, with a nation nearing 90 percent saturation of households with televisions, things were different. Plus, the race provided great interest to the public. Vice President Richard Nixon wanted to remain in Washington and move into the White House following the popular Dwight Eisenhower. Nixon had proved eight years earlier just how powerful a weapon television could be for a politician. Accused of accepting illegal campaign funds and gifts, Nixon turned to television to explain the events and to argue that "the decision, my friends, is not mine" to stay on the Republican ticket. Instead, the 38-year-old first-term California senator left it to those viewing. "I am going to ask you to help them decide. Wire and write the Republican National Committee whether you think I should stay on or whether I should get off. And whatever their decision, I will abide by it," Nixon announced at the end of what would come to be called the "Checker's Speech" because of references to a gift given to his children, a dog they named Checkers.[36] Overwhelmingly, the public responded in support of Nixon. His speech, which cost the RNC about $75,000, was the first direct appeal to the voting public by a politician. Nixon saw how well it worked. He should have realized how powerful a presidential debate could be, but he didn't.

Nixon was facing another junior U.S. senator, John F. Kennedy of Massachusetts. Kennedy, a relative unknown nationally before an inspiring speech at the 1956 Democratic National Convention, was a World War II war hero, Harvard graduate, and member of one of the nation's more well-known families. He was also Roman Catholic, and many Americans—especially in the South—worried what that might mean for a president who might consider his faith and the pope more important than duty to the nation. The two candidates squared off in front of television cameras for the first time on September 26, 1960, with an estimated audience of 70 million Americans watching on television and millions more tuning in on radio. What the radio audience could not see was a tanned and rested Kennedy and a tired-looking and haggard Nixon with his typical "5 o'clock" beard shadow and dark circles under his eyes. They could not tell that Kennedy looked calm and assured, wearing make-up made to withstand the lights of television while Nixon refused to wear any, opting for a powder that was supposed to hide the beard shadow that instead made him look worse. They could hear the answers that the men gave, and afterwards, those who listened to the first debate gave the edge to Nixon. Those who watched them on television believed Kennedy held the upper hand.

Heading toward the November election, Kennedy chose to use clips from the first debate for advertising; Nixon selected radio clips for his ads, both using elements from the medium through which they appeared the strongest. But Kennedy's campaign pieced together even stronger visual images as an editor clipped shots of Nixon reacting to what Kennedy said. In those shots, Nixon was seen scowling, sweating, smiling, and sometimes nodding approval of what Kennedy said.[37] Kennedy would later say, "We wouldn't have had a prayer without that gadget," referring to television.[38] Though Nixon was much better prepared for the remaining three debates, the first one left an indelible mark on Americans. After the election, more than half of all voters in a Roper survey—about 57 percent—said the debates had influenced their opinion on the election, and about 6 percent reported that the debates solely were the determining factor for them.[39] When you realize that Kennedy defeated Nixon by only 118,550 votes—a margin of less than 1 percent of the total vote—you also realize the significance of the debates in giving Kennedy the election. That 1 percent translated into 303 electoral votes for Kennedy to 219 for Nixon, with 280 needed to capture the election.

With the first debate, television executives realized the power that the medium offered for politics at about the same time that politicians realized it, too. "When I left the control room that night I said to myself, 'My God, we just elected a president of the United States, and it isn't even election day,'" Don Hewitt, director of the debates, executive producer of the *CBS Evening*

News with Walter Cronkite, and later the creator of *60 Minutes*, said. "In the middle of this thing, the politicians are looking at the television executives and thinking, 'Those guys have a reach into everybody's living room.' The television guys are looking at the politicians and saying, 'They're a source of unlimited advertising dollars. That's a bottomless pit.' That night changed the face of American politics. That was the night that television and politics eyed each other, flirted with each other, got engaged, and eventually got married. And because of that you cannot hold office in the United States of America—or even think about running for office—unless you've got the money to buy television time. Politics in America is now a money game, and it all weaned off the night of the first television debate."[40] Politicians would continue to stump the nation for votes, but now they had a means of reaching every voter in their living rooms, and that fact was not lost on politicians, political parties, TV networks, or the American public.

Death of a president

When John Kennedy was murdered in Dallas, Texas, on November 22, 1963, the event changed television forever. The assassination created massive overhaul in media news coverage, and those changes, in turn, affected the nation because of the role television would take in providing breaking news, updates, and continuous coverage of events. Until September 1963, the networks presented national news in a fifteen-minute segment each night. On September 2, CBS expanded its news to thirty minutes with an interview of President Kennedy by Walter Cronkite at the president's summer home in Hyannis Port, Massachusetts. Eleven weeks later, the networks increased news coverage to twenty-four hours in an unprecedented effort to provide the nation with the tragic news that began with the president's murder and ended with his funeral. In between, television would provide the nation with footage of the bizarre arrest of Lee Harvey Oswald and his subsequent murder as he was being transported away from a Dallas police station. For four days, from Friday's assassination through Monday's funeral, television gave the nation all the information it could garner—commercial free. Television, researchers would later discover, "served as a catalyst to speed up adjustment to the finality of the President's death and serve to renew faith in the future."[41]

At points, an estimated nine of every ten Americans watched TV coverage of the tragedy with about 93 percent of all households with televisions watching the funeral procession to the National Cemetery in Arlington, Virginia. A.C. Nielson reported that the average American home during that extended weekend watched for more than thirty-one and a half hours. About 166 million Americans watched the coverage for more than eleven hours a day. Coverage was sent to twenty-three other countries.[42] Television during the

Kennedy assassination was able to do something media had never before accomplished: It carried viewers to the scene of the news—live.[43] Television on location before the assassination required film to be shot and developed just as thirty-five millimeter still camera film was. That meant that thirty minutes to an hour had to pass before images from the scene could be shown to viewers. Television could broadcast live from stations and send it live via microwave relays, something it had done since 1951. And, the president's assassination required live coverage. That meant that the networks had to figure out a way to give the nation coverage from remote locations that wasn't recorded.

When Oswald was to be transferred that Sunday morning, the networks were prepared, having moved bulky cameras into place and connected them to local affiliates via telephone lines. As Oswald appeared and Jack Ruby burst from the crowd and fired his pistol into Oswald's abdomen, the nation witnessed it in real time. Millions of Americans became eyewitnesses to the shooting. "The event, now witnessed by camera as well as by the human eye, emphasized the media's presence," Barbie Zelizer observed. "Reporters would replay the murder across media with the assistance of tapes, recordings, and photographs. Through technology, their reactions would become embedded in the story's retelling."[44] When the ambulance left the station to carry Oswald to the hospital, live coverage was forced away from Dallas and back to the Capitol. Television utilized all of its resources to ensure that the nation could watch Kennedy's funeral live on Monday by bringing in cameras from affiliates along the East coast and lining them along the route to the cemetery. Now connected to the studios as the cameras had been in Dallas, television gave the country a way to be there without leaving home. Doing so changed the country forever.

VIETNAM

President Richard Nixon, never overly fond of media, announced in 1971, "Our worst enemy seems to be the press!"[45] His predecessor and the president most blamed for America's involvement in Vietnam, Lyndon Johnson, came to view media similarly. "[T]here was a great deal of emotional and exaggerated reporting," he said, "in our press and on television. The media seemed to be in competition as to who could provide the most lurid and depressing accounts." Ultimately, Johnson concluded that in terms of the American effort in Vietnam, the media from 1968 on made it appear as if "we must have suffered defeat."[46] America's involvement in Vietnam stretched from the Eisenhower administration into that of Gerald Ford, and for much of that time, media supported the nation's involvement. That changed gradually, however, after the Tet offensive. From that point until the United States pulled out of Vietnam in 1975, media often questioned the war effort. Until Tet, William

Hammond, senior historian with the U.S. Army's Center on Military History, says television reports pitched the war as "our side" against "their side." After Tet, he said, that analogy disappeared.[47]

At the end of January 1968, events in Vietnam changed drastically. Vietnam was supposedly in a truce to celebrate the holiday for the lunar new year. That's when, however, the North Vietnamese coordinated a massive attack on all areas of the South. The initiative caught the U.S. military off guard. On the night of January 31, NBC reported, "Two hundred and thirty-two GIs killed and 900 wounded makes one of the heaviest weeks of the Vietnam war, and it is not a week; it is just over two days, the past two days, some of the worst that we have known in Vietnam."[48] News reports out of Saigon, South Vietnam's capital, showed fighting in the streets. Viewers of the nightly news reports watched as Americans worked to retake the U.S. Embassy, which had fallen to the enemy. They saw bodies; they saw wounded soldiers; they saw images that contradicted the Johnson administration's "optimism that seemed to flow without stopping from Saigon."[49] As one reporter observed, "naturally the tone of the coverage changed."[50]

The Tet offensive sent shockwaves throughout the United States because the visual images from it, whether they were photographs or video footage, created a lasting impression upon Americans. No better example exists than the execution of a captured Vietcong by General Nguyen Ngoc Loan, chief of the South Vietnamese National Police. Associated Press photographer Eddie Adams caught the event at the exact moment the shot killed Bay Lop. At the same time an NBC cameraman filmed the execution, and it played that night on the *Huntley-Brinkley Report*. Howard Tuckner introduced the footage saying, "Government troops had captured the commander of the Vietcong commando unit. He was roughed up badly but refused to talk. A South Vietnamese officer held the pistol taken from the enemy officer. The Chief of South Vietnam's National Police Force, Brigadier General Nguyen Ngoc Loan, was waiting for him."[51] After that, the video played, and it showed Lop crumpling to the street, blood pumping out to the hole in his head. Then, it stopped. John Chancellor of NBC news called it "Rough justice on a Saigon Street."[52]

Adams would later say in connection with the photograph, "Still photographs are the most powerful weapon in the world."[53] He, and numerous media outlets at the time of the execution, pointed out that what General Loan did was not as horrific as it appeared. The *New York Times* tempered the shock that the execution photograph created by running a smaller photograph under it of a South Vietnamese soldier carrying the body of one of his children who was killed by Vietcong. The caption explained that the Vietcong— referred to as "terrorists"—"beheaded an officer and killed women and children." The story that accompanied the photographs jumped to Page 12.

There, readers saw the execution photograph again along with two others, including Lop's body lying on the street with Loan's statement, "They killed many Americans and many of our people," as a rationale for the execution.[54] Despite efforts to rationalize what happened, the images of the execution, according to Susan Moeller, "were seared into people's brains."[55] The photograph became one of the most reproduced media elements of the war, something that antiwar advocates used to point out the horrors of the war and that war supporters used to demonstrate how an enemy without principles could be dealt with swiftly.[56]

Harry McPherson, counsel to President Johnson, would later explain that the contradictions of information—a positive spin from officials in Saigon and from inside the administration that often said that there was a "light at the end of the tunnel" in terms of the war versus what could be seen in television footage—created confusion in the United States. "Even though we were reading many of those cables and going down there for what reassurances as we could get [from administration updates], we were also watching the American Television," McPherson said. "And American television was showing a different sight. That sense of the awfulness, the endlessness of the war. The unethical quality that did not recognize that when a man was taken prisoner he was not to be shot at point-blank range. The terrible sight of General Loan raising his revolver to the head of a captured Vietcong and killing him. They were awful contradictions, the cables on the one side, the television on the other. It was very disturbing."[57]

U.S. reaction to what was being reported from Vietnam led to some quick changes. First, television increased its video footage of U.S. casualties, civilian casualties, and of massive destruction. Images of American casualties jumped more than 183 percent per week in nightly news reports from the beginning of Tet until the end of March 1968. Civilian casualties and destruction of cities increased more than four times the previous average of such images.[58] From 1968 through 1973, about 25 percent of all air-time stories were about Vietnam.[59] Polls revealed that 63 percent of Americans disliked LBJ's handling of the war, a number that stood at 43 percent before Tet. Most importantly, about 74 percent of the American public said it lacked confidence in U.S. military policies, which had stood at 54 percent before Tet. Media commentary moved to roughly fifty-fifty in terms of support and opposition to U.S. involvement in Vietnam during this time, as well.[60]

While some observers such as William Hammond and Peter Braestrup concluded that media coverage of Tet made little difference in American attitudes toward the war,[61] the constant images of death—especially of U.S. troops and of Bay Lop's execution—and the violence of the North Vietnamese attacks, surely played a pivotal role in the changing numbers related to the war. The three networks gave viewers images from rice patties, grasslands, and jungles and then compared that footage to Johnson administration policy. Seeing how they fit was difficult for the typical American.[62] Media coverage of Tet—if it did nothing else—increased the anguish, shock, and uncertainty that Americans felt for the war.[63] And then, CBS anchor Walter Cronkite headed for Vietnam.

Cronkite, who earned credentials as a war reporter in World War II, had become the nation's most reassuring and trusted journalist. He had long supported the war effort, but beginning in 1965 with Morley Safer's report on the burning of Vietnamese villages and the seemingly unexplainable way older people and children were treated and displaced, Cronkite began to question whether the United States was acting in the best interest of Vietnam. In February 1968, Cronkite decided to travel to South Vietnam and observe the situation. What he saw left Cronkite a changed man in terms of the Vietnam war. When he returned, he prepared commentary to be read on February 27 at the end of a half-hour special on Tet. "To say that we are closer to victory today is to believe, in the face of the evidence, the optimists who have been wrong in the past," Cronkite said. "To say that we are mired in stalemate seems the only realistic, if unsatisfactory, conclusion. But it is increasingly clear to this reporter that the only rational way out then will be to negotiate not as victors, but as an honorable people who lived up to their pledge to defend democracy and did the best they could."[64] The reaction to Cronkite's announcement rocked the nation all the way to the White House. Tom Johnson, a presidential aide to Lyndon Johnson, remembered the president's reaction: "LBJ said to many of us. If I've lost Walter Cronkite, I've lost the war."[65]

Cronkite's pronouncement and the negative tenor to much of the coverage may have affected U.S. politicians and opinion leaders more than it did the typical American, and that is important. "We are in the wrong place, and we are fighting the wrong kind of war," one senator observed.[66] The Johnson administration decided that it needed to mount a public relations campaign to affect media coverage, something the president considered "a critically important dimension of fighting the war," according to Walt W. Rostow, special assistant for national security affairs.[67] Johnson's PR campaign produced limited results despite the crucial nature of media coverage to the war. Having a strong backing from the American people, Johnson felt, was critical. The enemy realized this, too. "The war was fought on two fronts," Vietcong leader Vo Nguyen Giap observed. "The most important was American public opinion." General William Westmoreland, commander of America's forces in Vietnam agreed, saying news coverage "attained an importance almost equal to the actual combat operations."[68]

Media coverage from Tet forward may not have sent most Americans into the war protest ranks, but it did seem to galvanize those who were making decisions on financial and military support of the war. Johnson, in fact, did believe that media were turning the nation against the war effort by showing horrific events, providing a steady dose of casualty numbers, and by giving protesters excessive coverage. Johnson complained that "our newspapers and our TV programs and our radio commentators informed us fully about the protesters and the 'peaceniks' who invaded the Pentagon.... Unfortunately, a student carrying a sign or a protester wearing a beard, or an attention-seeker burning a draft card in front of a camera can get more attention—and more billing—than all ten thousand of these volunteers" who enlisted in the military.[69] And the coverage, especially on television, shaped the public's opinions about it because, in addition to what came out of Vietnam, the nation's political leaders and other experts on the war were constantly seen there discussing it and letting media coverage affect decisions.

Television became the key to the messages of the Vietnam era. Coverage of any activity was imperative. "Have you ever seen a boring demonstration on TV?" antiwar leader Jerry Rubin said at the end of the 1960s. "Television creates myths bigger than reality. Demonstrations last hours, and most of that time nothing happens. After the demonstration we rush home for the six o'clock news. The drama review. TV packs all the action into two minutes—a commercial for the revolution."[70] All elements involved in the Vietnam war—from the president to protesters on the streets—assumed that the whole world was watching, and that changed everything.

THINKING SMALL, SELLING LOTS

In 1960, the advertising agency of Doyle Dane Bernbach introduced a new advertising campaign in the United States for German auto manufacturer Volkswagen. It was simple in its concept but massive in its implications. "Think small," the advertisement said, accompanied by a photo of the tiny-by-1960-standards VW Beetle on an empty page. The advertisement evoked multiple thoughts, but, ultimately, it promoted feeling comfortable with your individual decisions that might go against the mentality of the majority that was buying up gargantuan-sized cars. It was a way to get into the psyche of the American public by intimating that thinking small in terms of purchasing the mini VW Beetle meant that you were smarter than everyone else who was buying huge cars that were hard to park, terrible on fuel economy, and expensive.[71] Before Volkswagen stopped selling Beetles in the United States in 1979, more than 30 million were sold, and in 1968, more Beetles were sold than any single automobile had ever sold in the United States,[72] which surely does not equate to individualism, but it does equate to what advertisers hoped to accomplish in the 1960s: use the angst, emotions, and individualism of the age as a means of selling products. Advertising saturated the nation because that's what print and broadcast media did. Advertising fueled the Cold

War economy, and advertisers used whatever they found that worked to achieve sales for clients.

Advertisers began ramping up the notion of consumerism even before World War II ended. While Americans had been urged—commanded—to scrimp, save, and make do early in the war, as the contest appeared to be tilting in the Allies' favor, advertisers started gearing product promotion in some media toward acquiring goods and services that would make life easier in postwar America. Items to make it easier for women returning to the job of homemaker started to appear in magazines like *Saturday Evening Post* and *Ladies Home Journal*. Americans were told that the technology that was helping to win the war would be redirected toward making life easier once the war ended. In the 1950s, these ideas played out. Consumers were driven by the desire to buy so that they would have the newest conveniences and to "keep up with the Joneses" as Americans migrated toward the suburbs. It worked. In 1950, Americans spent about $5.7 billion on products. By 1960, that figure was nearly $12 billion, an increase of more than 110 percent.[73] This led a contemporary observer to conclude that by 1952, advertising had assumed a dominant role in media that had begun to shape cultural standards. Advertising, through media, had become one of the few societal institutions that was able to exert control on people throughout the nation.[74] Homes, automobiles, televisions, and fast food became items that most interested people.

Advertising in the 1960s, according to Thomas Frank, played upon the turmoil of society and of rebellion, of being different. The consumerism of the 1950s demanded that you fit in with the crowd. Advertisers in the 1960s took a different approach. You consumed products in order to meet individual wants and needs. You did it to stand out and to be an individual. A new Oldsmobile Toronado, for example, was not for the average man as its advertising proclaimed. That meant it was for the unique individual.[75] Of course, in the end, Americans consumed, which is exactly what advertisers wanted.

In the 1960s, TV advertising changed. In the 1950s, many programs aired with a single sponsor. By the mid 1960s, though, this kind of advertising had been replaced almost completely by multiple sponsorships for programs with advertising time being sold to anyone who would pay the price of the spot. Advertising on network television had become such an expensive endeavor that most companies could not afford the cost of sponsoring the whole program. Advertising also changed so that spots were more often thirty seconds in length, sometimes twenty or even ten.[76] This meant that viewers were exposed to many more products than ever before. Americans had grown accustomed to broadcast advertising being limited to few products per program, now they were bombarded with multiple advertising per commercial break and by dozens of products on any typical night of TV viewing. The advertising blitz changed American buying patterns. Total advertising revenues by the end of the 1960s, *Advertising Age* estimated, topped $19.5 billion, an increase of more than 58 percent during a decade rife with social, political, and cultural unrest.

Advertisers also began using new tactics to reach audiences in addition to promoting individualism. The use of psychographics, as described by Vance Packard in his book *The Hidden Persuaders*, meant that media sent messages in advertising to consumers that appealed to them by using the discoveries in the growing field of psychology. Advertisers also began to use other "tricks," such as subliminal messaging, which used quick messages or imaging to send a subconscious message to consumers in order to persuade them to buy.[77]

But tricks weren't necessary in some cases, especially in targeting women. Before 1950, women comprised less than 30 percent of the U.S. work force. By 1970, that percentage had risen to more than 43 percent, which equaled about 31.5 million women.[78] Advertisers took aim at women and fashioned advertisements to reach them on any number of levels. Despite the fact that women were increasingly out of the home working, advertising still focused on selling to women as homemakers, especially in the 1950s and into the 1960s. By the 1970s, though, advertisers were targeting females as "superwomen."[79] A line from a perfume commercial became the catchphrase: "I can bring home the bacon, fry it up in a pan, and never let you forget you're a man."[80] Advertisers used the feminist movement, too, to target women, advertisements for Virginia Slims cigarettes being but one example. With the slogan, "You've come a long way, baby," the advertisements played upon equality between the sexes, and the ads represented women as successful in all endeavors—especially in business. But, advertisers knew that the women in the ads represented more, using models with flawless features to provide women with an ideal to which they should aspire.[81] Advertisers would continue to build upon these perceptions of women in media product promotions and also begin to target children more specifically than ever with products. Advertising targeted at young people, according to the advocacy group Action for Children's Television, became so blatant in its promotions and false claims that the group lobbied—unsuccessfully—to have all advertising removed from TV programming targeted at children. By the mid-1970s, children aged 2 to 14 directly influenced about $20 billion of items purchased in the United States.[82]

The thirty years that ushered in the age of television witnessed unprecedented use of media to sell goods to society. By the end of the 1970s, spending on advertising topped $54.7 billion annually, according to *Advertising Age*. That, not taking inflation into account, means that spending on advertising increased by almost 870 percent from 1950 to the end of the 1970s.

YOUR SONS AND DAUGHERS ARE BEYOND YOUR COMMAND

Recorded music had been problematic for many Americans in the 1920s because of the influence of jazz, which was the product of African Americans. Though jazz eventually became mainstream because it was co-opted by white performers, African Americans continued to make music, though much of it went unnoticed by white America. After World War II, it resurfaced as an element of national culture. As Alan Freed and other disc jockeys began playing what Freed called "rock 'n' roll" on radio and featuring a new generation of black artists, recorded music again became controversial. The new sound produced by black artists such as Chuck Berry, Fats Domino, Little Richard, and others played as the civil rights movement in the 1950s was greeted with negative reactions by outspoken white community leaders, just as the same people did during the 1920s with jazz. Bridgeport, Connecticut, for example, outlawed rock 'n' roll dance parties in 1955. Authorities blamed the public disruptions on the musicians who then influenced the young people in attendance.[83] Hollywood used rock music as the soundtrack to *Blackboard Jungle*, a 1955 film about juvenile delinquency, which equated by association rock 'n' roll and problematic behavior by young people. But, just as jazz was embraced by many in the 1920s, especially young people, rock 'n' roll found the same acceptance in white society in the 1950s. It accounted for more than 50 percent of total record sales in the decade.[84] Radio and television helped spread the music so that it reached into even the most isolated and least-populated parts of the country. Black performers were still a problem for mainstream white America, which led to Sam Phillips' famous quest to find "a white boy who can sing colored." For many, Elvis Presley was not much better, but by the early 1960s, rock 'n' roll was a staple on commercial radio and increasingly on television with programs such as *American Bandstand*, hosted by Dick Clark.

Efforts to suppress African American performers and the "wiggle-your-hips" type of music of Elvis Presley ushered popular music into the 1960s, but that all changed with the British invasion led by the Beatles, Rolling Stones, and a number of other bands. That music, however, did not initially deal with the growing turmoil that affected the United States. Song as protest, though, had been used by artists such as Woody Guthrie and Pete Seeger, especially to discuss conditions for America's working class. Music as a social compass, however, had never been an integral part of popular music until Bob Dylan burst onto the scene in 1964. Coming from the folk movement, which had a large fan base but not one that could be heard regularly on radio or its artists seen to any extent on television, Dylan provided young people with songs and a message that gave social relevance to lyrics. "I want to hold your hand" may have been the first step in what was on the minds of most teenagers, but it did not address the growing issues that the nation faced. With the cold war, the assassination of a president, civil rights, and the escalating war in Vietnam as the issues steadily consuming the nation, Dylan's music as well as that by an increasing number of other artists, discussed the nation's problems and used clever lyrics, a catchy beat, and a song that most adults simply considered noise as a way to spread messages, comment on the ills of society, and influence the younger part of U.S. society in ways that perhaps no other medium had been able to accomplish.

In January 1964, Dylan released *The Times They Are a-Changin'*. Its lyrics addressed the growing generational divide that America faced and the impending problems the country would face. Dylan later told music critic Cameron Crowe, "This was a song with a purpose."[85] Speaking to politicians, Dylan told them not to be impediments to change. "There's a battle outside And it's ragin'. It'll soon shake your windows And rattle your halls." He admonished parents: "And don't criticize What you can't understand. Your sons and your daughters Are beyond your command. Your old road is Rapidly agin'." He closed by predicting, "The order is Rapidly fadin'. And the first one now Will later be last For the times they are a-changin'."[86] The song and the album, one writer observed, "stirred up intense yearning, provoked introspection, and capitalized on the tidal currents of social unrest."[87] The *New York Times*, after the album's release, called Dylan a "gifted, driven power, in the face of the inequities and injustices he sees around him." What worried the *Times*, however was "the unsettling imagery, and the philosophy expressed. Dylan is an important figure in popular American culture. Doesn't it reflect something about the current crop of teen-agers and collegians that they have found an idol in a moralist with poetic vision such as Bob Dylan?"[88] Dylan's protest music, the *Times* later stated, became the "credo for the discontented."[89]

Soon, other musicians followed Dylan's use of popular music to discuss what was happening. Simon and Garfunkel, in the Dylan folk tradition, was one example. Simon's song, *A Church Is Burning* from *The Paul Simon Song Book* album, discussed the bombing of the 16th Street Baptist Church in Birmingham on September 15, 1963, which took the lives of four young girls. The church served as the center of protests in the city. As the decade progressed, recorded music focused more on the war in Vietnam. Songs such as Buffalo Springfield's *For What It's Worth*, *Fortunate Son* by Creedence Clearwater Revival and *I-Feel-Like-I'm-Fixin'-To-Die Rag* by Country Joe and the Fish were aimed directly at the Vietnam war, protests against it, and how it represented the social and cultural divisions in the nation. Barry McQuire's *Eve of Destruction* provided some of the most pointed lyrics:

> The Eastern world it is exploding, violence flaring and bullets loading,
> You're old enough to kill, but not for voting,
> You don't believe in war, but what's that gun you're toting?
> And even the Jordan River has bodies floating
> And you tell me, over and over and over again my friend,
> Ah, you don't believe we're on the Eve of Destruction.[90]

Volunteers by Jefferson Airplane served more as a call for widespread rebellion: "Look what's happening out in the streets, Got a revolution, got to revolution.... Hey, now it's time for you and me. Got a revolution, got to revolution.... We are volunteers of America, volunteers of America."[91] Most African-American artists avoided recording songs that dealt with the country's troubles, but Marvin Gaye's *What's Going On*, the Temptations' *Ball of Confusion*, and Edwin Starr's *War* provided powerful messages about the state of the country and issues that affected all Americans but African Americans mainly in the songs by Gaye and the Temptations. Songs that described and commented on the nation's problems were mixed in on the radio with typical songs about a host of subjects, but the music of protest infiltrated the airwaves. It constantly provided the sinew that held together protests and the young people of the nation. Its effects on society could be seen everywhere, from TV shows to advertising, from hair styles to clothing, from cultural norms to political consciousness.

In 1960, Americans bought 51.8 million albums,[92] but an amazing 1.5 million 45 rpm records were sold from September 12 to October 10 of the same year.[93] In 1960, record sales totaled $6 million in the United States and rose to $1.6 billion by the end of the decade, this in an era when an album cost $4.98 on average and a 45 rpm single, 98 cents.[94] It was also an era where the majority of sales were of singles in its pivotal middle years, which means that tens of millions of copies of records were sold each year. The 1960s also marked the first time that record sales surpassed the revenues of all forms of entertainment.[95]

Even with these massive sales numbers, most music was not bought, but listened to on radio. Rock music, whether played on the radio, on records, on eight-track tapes, or live by professional musicians, garage bands, or individuals with guitars served as the connecting element for the Baby-Boom generation born following World War II. As *Rolling Stone* magazine said in 1967, "You know rock and roll is more than just music; it is the energy center of the new culture and youth revolution."[96]

Yeah, Yeah, Yeah

Music with a social conscious was critical in defining the nation in the 1960s, but so, too, were the Beatles. From the day after their first appearance on *The*

Ed Sullivan Show on February 9, 1964, the Beatles changed the United States. A Sunday night fixture for America, *The Ed Sullivan Show* drew a weekly audience of around 50 million, but on the night of the Beatles' appearance, more than 74 million tuned in to see and hear them, which translated into 58 percent of all households with television, according to A.C. Nielson, and the highest-rated show since the rating service was established.[97] Though recorded music was the Beatles' initial medium, television provided the nation with the total package—the songs, the hair, and the charm. Within twenty-four hours of the performance, male hairstyles in the nation changed from greased Brylcreem cuts to bangs. Soon, there would be Beatle wigs, boots, lunchboxes, record players, posters—anything that could be marketed with Beatles' images on them or allusions to the Fab Four.

But it was the music that served as the root of influence. From 1964 through 1969, the Beatles topped one-third of the weekly charts of popular music. Their music spent twenty weeks at Number 1, forty-four weeks in the Top 5 and sixty-seventy weeks in the Top 10.[98] They produced nineteen albums in seven years. They affected the cultural mores for the decade and the outlook of an entire generation.

The Beatles sold about 30 percent of all recordings sold by Capitol, and from 1964 through 1969, they sold no fewer than 3 million albums per year, selling more than 9 million in 1964 and 7 million in 1969, the years that book-ended their period of recording.[99] These numbers do not include sales of 45 rpm records or the number of times Beatles' songs were played on the radio. The Beatles became cultural icons and influencers without touring except for 1964 through 1966. Instead, they used all available media to create an image that was both accessible and changing and constantly in the consciousness of American youth.[100]

"The Beatles embraced rock out of their need for self-expression. That drive became the basis of the most ecstatic (and political) explosion in the history of youth culture," one 1968 writer said. "They were a threat, and they never pretended to be otherwise....Long before they began to articulate their ideas on drugs, religion, and social justice, they were writing anthems for liberated youth in the guise of dance-hall music. The implications of their songs were not apparent because rock often masks it depth with the mundane." The meaning, the writer said, "lies between the lines."[101] Though much of popular recorded music of the 1950s through 1970s might have been mundane, a significant portion of it provided a voice that shaped the nation and continues to do so.

ONE SMALL STEP

The Cold War produced more than fear of nuclear annihilation in the United States. It produced competition with the Soviet Union, and nowhere was this more obvious than in the space race. When the Soviets sent Sputnik into orbit on October 4, 1957, Americans realized that the Soviet Union was beginning—technologically, at least—to surpass the United States. The ability to send a satellite into space meant that the Soviets might also be able to deploy nuclear weapons in the same way, and that frightened people in the United States and Europe. Later, the U.S.S.R. would send a payload into space carrying a dog. All of this pushed the United States into overdrive to catch up. Doing so, President Kennedy told Congress on May 25, 1961, was an "urgent national need."[102] The United States launched its first satellite, *Explorer*, in January 1958.[103] During the next dozen years, the two countries engaged in a race that ended with astronauts Neil Armstrong and Buzz Aldrin walking on the Moon.

In 1962, President Kennedy announced, "We choose to go to the moon. We choose to go to the moon in this decade and do the other things, not because they are easy, but because they are hard, because that goal will serve to organize and measure the best of our energies and skills, because that challenge is one that we are willing to accept, one we are unwilling to postpone, and one which we intend to win, and the others, too. It is for these reasons that I regard the decision last year to shift our efforts in space from low to high gear as among the most important decisions that will be made during my incumbency in the office of the Presidency."[104] With that speech, Kennedy focused the nation's attention on the newly established space program, and the media, television especially, accepted the president's challenge.

Life magazine, however, worked out exclusive rights to tell the astronauts' personal stories, something the magazine told throughout the 1960s.[105] As NASA planned for its first manned space mission, media put together "one of the largest coordinated news teams ever assembled, to provide combined live, video-taped and audio coverage" at Cape Canaveral in Florida.[106] Television provided the greatest media impact. President Kennedy, who learned before taking office the power of television through his debates with Richard Nixon and then, after election, through news conferences, knew that public support of the nation's space effort was critical to its success. He pinned much of that hope on television, which he considered "the most powerful and effective means of communications ever designed."[107] The networks, for the most, bought into Kennedy's assessment and supported NASA. Don Hewitt of CBS believed that as long as TV coverage helped NASA—and NASA provided its amazing images to the networks—that Congress would keep the space program well funded.[108] One writer explained that "the TV medium [was] merely a fantastically efficient mode of global distribution" for the space race and the landing on the Moon in 1969. "It was nothing short of an electronic miracle that the pictures and—for the most part—the sound between the surface of the moon and mission control in Houston enjoyed a clarity that boggled the mind."[109]

Perhaps no figure associated with television did more to promote and support the nation's space quest than CBS anchor Walter Cronkite. Early in NASA's space plans, Cronkite endorsed Kennedy's goal of putting Americans on the Moon. He became absorbed in providing coverage and in understanding elements of how the rocketry, capsules, and space flight worked. Through the CBS news programs *Eyewitness* and *The 20th Century*, Cronkite early on gave NASA and space extended coverage. He even asked—given his age and health—what was the possibility of his traveling to the Moon.[110] His enthusiasm, no doubt, fueled CBS's commitment to coverage. CBS probably spent more on coverage than the other networks, and the culmination of coverage occurred in July 1969 when *Apollo 11* headed for the Moon and a lunar landing. "Mr. Cronkite for many years," the *New York Times* would say, "has been something of a one-man phenomenon in space coverage."[111]

In a medium fueled by competition, Cronkite almost singlehandedly led CBS to the front of TV news ratings, and that was never more apparent than in coverage of the lunar mission. With the *Apollo 8* mission to orbit the Moon, approximately 40 million Americans watched the splashdown, but CBS coverage, unlike the other two networks, was provided commercial free.[112] When *Apollo 11* left for the Moon, Cronkite initiated ironman coverage. He stayed on air for twenty-four and one-half hours, more than seventeen of them straight. As a result, CBS grabbed 45 percent of the national viewing audience, compared to 34 percent for NBC and 16 percent for ABC. As Cronkite watched the scenes from the Moon, he could not contain his excitement. He exclaimed, "Oh boy! Hot diggity dog!" Americans obviously liked the way Cronkite covered space and the Moon mission especially.[113]

Though Americans' interest in the Mercury, Gemini, and Apollo space programs waxed and waned, the competition among TV networks tended to maintain the nation's interest. The networks throughout the decade worked to produce more elaborate sets and worked to ensure they had the best commentary possible. While Cronkite and television were the driving force in this interest, all media focused on the race to the Moon. Almost 3,500 press passes were issued to cover *Apollo 11*'s liftoff.[114] In the end, however, television received the greatest credit in promoting NASA's endeavors and in keeping the nation's interests focused on Kennedy's vision of putting a man on the Moon. "Eight years of covering space missions have provided the American television industry with many of its finest hours," the *Washington Post* observed following *Apollo 11*'s Moon landing. "This time television provided the greatest show off the earth."[115] Newsman Reuven Frank had observed that TV's finest function was "the transmission of experience." "Using this standard," the *Washington Post* said, "the television industry has never had a finer eight days in its brief history." In the forefront of that coverage, the *Post* believed, was "the solid and stolid commentary of a Walter Cronkite."[116]

THE RESIGNATION OF A PRESIDENT

On August 9, 1974, the *Washington Post* proclaimed in a towering two-word, Page 1 headline, "Nixon Resigns." The resignation of a president was unique to American politics, and it followed a landslide victory by Nixon in the 1972 election where he carried every state except Massachusetts and the District of Columbia. Exactly what did media have to do with the unraveling of Nixon's presidency is a question that has been asked often in the years since Nixon left the White House to Gerald Ford, the only person never elected by the American people to serve as either the president or vice president to hold both offices. The answer is dependent upon the person asked, but there can be little doubt that media were able to keep the Watergate affair in the public spotlight. The *Washington Post* provided the bulk of the investigative journalism that led up to the resignation, and television's gavel-to-gavel coverage of the Watergate hearings gave the nation real-time access to the information that ultimately proved that Nixon knew all about the affair. TV coverage of the hearings drew in an estimated 85 percent of all U.S. households to either part or all of the testimony presented in congressional hearings.[117]

On June 17, 1972, five men were arrested trying to bug the Democratic National Committee offices in the Watergate complex in the nation's capital. No one at the time had any idea of the connections of the burglars with anyone within the Nixon administration.[118] The Nixon administration's first official comment on the arrests called it a "third-rate burglary." The Watergate break-in quickly shifted to the back burner of the nation's news, the election and Vietnam being the events that consumed most of media's attention. But at the *Washington Post*, the Watergate story never disappeared. In the first six months after the break-in, the *Post* published more than two hundred news articles about Watergate, more than twice as many stories as any other media outlet. Nearly twelve hundred articles about Watergate, written by wire services and other reporters, appeared in print sources during the same time period, many of them based on the *Post*'s coverage.[119] As those stories and the investigation by reporters continued, more questions were raised as to the implication of the burglary at the Watergate and just who may have ordered it and known about it. The *Post* stayed on the story even as other media outlets tended to shy away from it.

Carl Bernstein and Bob Woodward led the *Post*'s investigation. Because of their book, *All the President's Men*, and the subsequent movie that adapted it, Woodward and Bernstein created a special place for themselves in the Watergate saga. Though the Federal Bureau of Investigation insisted that it was always one step ahead of the reporters, the pair revealed to the nation in bits and pieces information about who knew about the break-in and who had ordered it one new detail after another. In October 1972, the *Post* announced that "FBI agents have established that the Watergate bugging incident stemmed from a massive campaign of political spying and sabotage conducted on behalf of President Nixon's re-election and directed by officials of the White House and the Committee for the Re-election of the President."[120]

Woodward and Bernstein, because the *Post* decided to follow the story so closely, had great advantages over other reporters. They were also privy to inside information from the source they called "Deep Throat," who in 2005 was revealed to be Mark Felt, the FBI's day-to-day manager of operations.[121] Woodward and Bernstein were the ones who kept the story in front of the nation.

In 1973 and 1974, television assumed a pivotal role in the coverage of Watergate, principally via congressional hearings. One of the key elements that undid Nixon in relation to Watergate occurred when the president decided to hold meetings on it in the Oval Office and record those conversations. The first of those recorded dialogs contained almost nineteen minutes that were erased, which led to speculation that the president and his chief of staff H. R. Haldeman discussed the break-in then. The tapes, which special Watergate prosecutor Leon Jaworski subpoenaed, provided the final piece of evidence against Nixon. He would resign six weeks after the court ruled that Nixon had to give the tapes up. With most Americans watching the hearings and learning of the tapes' revelations, Nixon's approval ratings fell to around 30 percent. As Louis Liebovich noted, the issues surrounding Watergate became a battle of public opinion between the Nixon White House and media; television was the battleground, and there was no way that the president could manipulate coverage.[122] Every day, Americans watched as the most influential men within the president's Oval Office inner circle testified. Each day, the hearings brought the Watergate affair closer to Nixon. Once the tapes were a part of the coverage and the so-called smoking gun, which revealed that the president had agreed to ask the FBI to stop its investigation into the Watergate break-in, citing it as a national security issue. That revelation would lead almost all of the president's supporters in Washington to abandon him and for him to decide to resign.

The role of the *Washington Post* and then network television in the Watergate affair turned into a media battle for Richard Nixon. In his memoirs, Nixon said that he found coverage of it all to be infuriating.[123] The *Post*, though, decided to make the break-in a moral issue and followed its leads with that in mind, according to Paul Johnson.[124] And, according to Michael Schudson, it really does not matter whether Woodward and Bernstein, the *Washington Post*, and the rest of media really unseated the president because the public continues to believe that is exactly what happened. "[T]wo young Washington Post reporters brought down the president of the United States," Schudson explains of how Americans viewed Watergate. "And the good guys win. The press, truth its only weapon, saves the day."[125] Ben Bagdikian, who was assistant managing editor of the *Post* in 1971 and 1972, called the work of Woodward and Bernstein "the most spectacular single act of serious journalism [in the twentieth] century."[126] Their work, David Greenberg says, "shaped the way Watergate unfolded."

CBS News anchor Dan Rather believed that both Congress and the courts "didn't have a clue" about Watergate. "[O]nly after repeated and constant coverage" did federal investigators grasp the significance of what media were reporting, Rather, who Nixon attempted to get fired for his aggressive reporting on Watergate, said, adding, "The record clearly shows that the cover-up would have worked if the press hadn't done its job."[127] According to Ben Bradlee of the *Washington Post*, Watergate changed the way media approached all stories in the years immediately following it, covering, for example "the most routine rural fires as if they were Watergate."[128] Because this happened, journalism changed, not just in the United States but globally. Silvio Waisbord explained in his study of South American journalism that Watergate transcended the borders of the United States. "Every investigative journalist has the dream of Watergate," one South American reporter said.[129] And, "gate" became the suffix for nearly all scandals, "Monicagate" with President Bill Clinton being but one example. Watergate and the work of Woodward and Bernstein specifically, helped change the journalism element of media. Just as Muckrakers had created change seventy years earlier, the investigative journalists of the Watergate era did the same.

✳✳✳

Media's interaction with the events of the 1950s through 1970s transcended nearly every aspect of American life. From information dispersal to entertainment, the growing media landscape provided the nation with the ability to read, hear, and see as never before. By the last half of the twentieth century, according to Stanley Kutler, media had become an essential component in the task of governing in the United States.[130] Bill Kurtis, who covered the 1968 Democratic National Convention for Chicago TV station WBBM, agreed. That convention, Kurtis said, changed reporting completely because all—protesters, politicians, everyone in media's spotlight—had an agenda and knew the whole world was watching. Media—television specifically—provided an input into nearly every American home.[131]

Media pushed the limits of the First Amendment with some of their journalism. The publication of the Pentagon Papers, which revealed a nearly three-decade U.S. involvement in Vietnam and how the Johnson administration had intentionally mislead the nation about that involvement and American casualties, was an intentional effort on the part of the *New York Times*, the *Washington Post*, and a few other newspapers. Publication was carefully considered, the *Post*'s Katherine Graham said, along with the implication of the revelations within the items that Daniel Ellsberg copied and removed from the Pentagon.[132] In February 1971 just before the Pentagon Papers were released, CBS news aired "The Selling of the Pentagon," which claimed that the military was freely spending taxpayer dollars in a public relations effort to bolster the military's image. This story and the Pentagon's use of media were perfect examples of the way that the media were being used as the conduit to effect public opinion. The CBS story led to a congressional investigation where the government demanded the network turn over its material to a special subcommittee of the Foreign Commerce Committee.[133] CBS refused and fought the order, just as the *Washington Post* and *New York Times* would fight the Nixon administration's injunctions against the publishing of the Pentagon Papers. In both cases, media triumphed, providing more impetus to produce pieces that uncovered malfeasance on the part of the government.

Other changes in media coverage and presentation of news developed, too. In September 1968, CBS launched the first news magazine, *60 Minutes*, though CBS considered it a news documentary along the lines of Edward R. Murrow's *See It Now*. Created by Don Hewitt, *60 Minutes* was the television precursor to the type of investigative reporting that would arise with Watergate in 1972, and the program became the most successful of its type in TV history. The program sought out corruption and presented it to an eager American audience. *60 Minutes*, according to Madsen, was able to focus its stories on issues that involved conflict surrounding issues and groups or individuals, which produced television that was entertaining but also newsworthy.[134]

Shortly after *60 Minutes* went on the air, a journalism with tinges of that created by Nellie Bly and others burst onto the scene. Dubbed "gonzo" as a reference to intense drinking, this new journalism was always subjective and was pioneered by Hunter S. Thompson. Just as Nellie Bly was always a character in a Nellie Bly story, Thompson did the same with gonzo journalism. "True gonzo reporting," Thompson would say, "needs the talents of a master journalist, the eye of an artist/photographer and the heavy balls of an actor. Because the writer *must* be a participant in the scene, while he's writing it—or at least taping it."[135]

Thompson's journalism—and those who used the same style—created information that bridged the information function of media and that of entertainment. The entertainment in the post-1950s United States introduced a world of cultural icons that changed the nation. Recorded music, as discussed, created massive change among young people, but television's ability to reach into nearly every household with identical programming had a greater impact because it transcended generational boundaries. From Desi Arnaz' Spanish tirades aimed at his red-haired wife, Lucille Ball, to the groundbreaking comedies of Norman Lear, TV entertainment grew to give the nation catch-phrases, clever jingles for advertisements and program introductions, and finally a way to use entertainment to comment on the national situation. Millions of Americans knew that the wiggle of a nose could bring about magic from *Bewitched*, and they could sing the bluegrass theme song of *The Beverly Hillbillies*. African Americans—specifically Nat King Cole, Flip Wilson, Bill Cosby, and Diahann Carroll—pioneered changes in the way the nation was entertained, which coincided with the civil rights movement. Their rise to prominence in television entertainment also coincided with the move from black and white programming to color, which seemed appropriate for all the events and programs that were viewed during these decades. And, Lear's creations, spearheaded by 1971's *All In the Family*, changed forever entertainment programming. Centered on a bigoted New Yorker named Archie Bunker, Lear used comedy—which was accompanied initially with a warning label—to comment on the nation's social ills of racism, war, and the cultural divide. His other programs, like *Maude* and *The Jeffersons* continued comedy as social commentary and paved the way for *M*A*S*H* in 1972, which used the Korean War as commentary on Vietnam. When it ended its eleven-year run, the comedy pulled in an estimated 125 million viewers in 50.2 million households that tuned in, the most watched program in television history.[136]

NOTES

1. U.S. Bureau of Census, *Statistical Abstract of the United States, 1970*, 91st ed. (Washington, D.C.: U.S. Department of Commerce, 1970), 496.
2. Numbers of televisions 1950, Edwin R. Bayley, *Joe McCarthy and the Press* (Madison: University of Wisconsin Press, 1981), 176. Costs of televisions taken from advertisements for televisions of the respective years found at *Television History–The First 75 Years*, http://www.tvhistory.tv/tv-prices.htm. Monetary conversions, *The Inflation Calculator*, http://www.westegg.com/inflation/infl.cgi. Other TV figures, U.S. Census Bureau, *Statistical Abstract of the United States 2003* (Washington, D.C.), 721.
3. See Burns W. Roper, *Trends in Public Attitudes Toward Television and Other Mass Media, 1959–1974* (New York: Television Information Office, 1975).
4. Maxwell McCombs and Donald Shaw, "The Agenda-Setting Function of Mass Media," *Public Opinion Quarterly* 36 (1972): 176–187.
5. U.S. Bureau of Census, *Statistical Abstract of the United States, 1970*, 498, 500, 495.
6. "No Easy Walk," *Eyes on the Prize: America's Civil Rights Years, 1954–1963*, produced by Henry Hampton, 120 minutes (Blackside, 1986), Videocassette.
7. Peter Braestrup, *Big Story: How the American Press and Television Reported and Interpreted the Crisis of Tet 1968 in Vietnam and Washington* (New Haven, Conn.: Yale University Press, 1977), xi–xii.
8. Samuel P. Huntington, "The United States," in Michael J. Crozier, Samuel P. Huntingon, and Joji Watanuki, *The Crisis of Democracy* (New York: New York University Press, 1975), 98–99.

9. In Daniel C. Hallin, *The "Uncensored War": The Media and Vietnam* (New York and Oxford: Oxford University Press, 1986), 5.
10. Information compiled from W. Horace Carter, interview by Jerry Lanier, tape recording, January 17, 1976, *Documenting the American South*, University of North Carolina at Chapel Hill, http://docsouth.unc.edu/sohp/B-0035/B-0035.html, and Center for the Study of the American South, "The Carter-Klan Documentary Project," University of North Carolina at Chapel Hill, 2006, http://www.carter-klan.org/index.html.
11. "The Carter-Klan Documentary Project."
12. Horace Carter, "No Excuse for KKK," *Tabor City (N.C.) Tribune*, July 26, 1950.
13. "Judge in Klan Robe Killed in Gun Battle," *New York Times*, August 29, 1950, 28.
14. "South Carolina Klan Chief Arrested After Riot Fatality," *Christian Science Monitor* (Boston), September 1, 1950, 11; Horace Carter, "True or False?" *Tabor City (N.C.) Tribune*, November 15, 1950.
15. See, for example, "Hooded Night Riders Flog 2 Men in New Carolina Terror Outbreak, *Washington Post*, January 21, 1951, M20; Horace Carter, "Bring the Charges to Court," *Mount Tabor (N.C.) Tribune*, Februay 7, 1951.
16. Carter, interview.
17. Roy J. Harris Jr. *Pulitzer's Gold: Behind the Prize for Public-Service Journalism* (Columbia: University of Missouri Press, 2007), 187.
18. Basic facts about the Greensboro sit-ins come from Thomas Adams Upchurch, *Race Relations in the United States, 1960–1980* (Westport, Conn.: Greenwood Press, 2008), 6–10.
19. Marvin Sykes, "Negro College Students Sit at Woolworth Lunch Counter," *Greensboro (N.C.) Record*, February 2, 1960.
20. Claude Sitton, "Negroes Extend Store Picketing," *New York Times*, February 11, 1960, 22.
21. "Counters To Remain Closed," *Greensboro (N.C.) Record*, February 8, 1960.
22. Howard Zinn, *SNCC: The New Abolitionists* (Boston: Beacon Press, 1964; reprint, Cambridge, Mass.: South End Press, 2002), 17.
23. Gene Roberts and Hank Klibanoff, *The Race Beat: The Press, The Civil Rights Struggle, and the Awakening of a Nation* (New York: Vintage Books, 2006), 226.
24. Roberts and Klibanoff, *The Race Beat*, 226.
25. Claude Sitton, "Negroes Extend Store Picketing," *New York Times*, February 11, 1960, 22; David Halberstam, "A Good City Gone Ugly," *Reporter* (Nashville, Tennessee), March 31, 1960, 17–19, reprinted in *Reporting Civil Rights: American Journalism 1941–1963* 2 vols. (New York: Library of America, 2003), 444.
26. Upchurch, *Race Relations in the United States*, 10.
27. "Integration Gain Listed by Stores," *New York Times*, October 18, 1960, 47.
28. "Rights in Conflict," *Washington Post*, April 17, 1960, B4.
29. Claude Sitton, "Stores in South Prosper with Integrated Counters," *New York Times*, June 6, 1960, 1.
30. "Quiet Progress from Sit-Ins," *Christian Science Monitor* (Boston), July 28, 1960, 20.
31. Roberts and Klibanoff, *The Race Beat*, 303–05.
32. In Tom Macklin, "25,000 Converge on State Capitol, End 5-Day Protest Trek from Selma," *Montgomery (Ala.) Advertiser*, March 26, 1965.

33. Quoted in Manning Marable, *Race, Reform, and Rebellion: The Second Reconstruction in Black America, 1945–1990*, 2nd ed. (Jackson: University Press of Mississippi, 1991), 70.
34. Jon Wiener, "Confrontation in Black and White," *Los Angeles Times*, November 12, 2006, http://articles.latimes.com/2006/nov/12/books/bk-wiener12.
35. Quoted in Roberts and Klibanoff, *The Race Beat*, 407.
36. Richard M. Nixon, September 23, 1952.
37. Kathleen Hall Jamieson and David S. Birdsell, *Presidential Debates: The Challenge of Creating an Informed Electorate* (New York and Oxford: Oxford University Press, 1990), 99.
38. Quoted in Joe Garner, *Stay Tuned: Television's Unforgottable Moments* (Kansas City, Mo.: Andrews McMeel, 2002), 67.
39. Erika Tyner Allen, "The Kennedy-Nixon Presidential Debates, 1960," *The Museum of Broadcast Communications*, http://www.museum.tv/archives/etv/K/htmlK/kennedy-nixon/kennedy-nixon.htm.
40. Quoted in Joe Garner, *Stay Tuned*, accompanying DVD.
41. Steve Michael Barkin, *American Television News: The Media Marketplace and Public Interest* (New York: M.E. Sharpe, 2003), 36.
42. Barkin, *American Television News*, 37.
43. Barbie Zelizer, *Covering the Body: The Kennedy Assassination, the Media, and the Shaping of Collective Memory* (Chicago: University of Chicago Press, 1992), 62.
44. Zelizer, *Covering the Body*, 61.
45. Quoted in William M. Hammond, *Media & Military at War* (Lawrence: University Press of Kansas, 1998), 293.
46. Lyndon Baines Johnson, *The Vantage Point: Perspectives on the Presidency, 1963–1969* (New York: Holt, Rinehart and Winston, 1971), 384.
47. Hammond, *Media & Military at War*, 126.
48. *NBC Evening News*, January 31, 1968, Jack Perkins reporting, "Tet 1968," *Vietnam: A Television History*, produced by WGBH, 660 minutes (Public Broadcasting Service, 1983), Digital video recording.
49. Harry McPherson, quoted in Herbert Y. Schandler, *The Unmaking of a President* (Princeton, N.J.: Princeton University Press, 1977), 81.
50. Max Frankel, interview by Todd Gitlin, March 23, 1979, quoted in Todd Gitlin, *The Whole World Is Watching: Mass Media in the Making & Unmaking of the New Left* (Berkeley: University of California Press, 1980), 205.
51. Quoted in Daniel C. Hallin, *The "Uncensored War": The Media and Vietnam* (New York and Oxford: Oxford University Press, 1986), 172.
52. John Chancellor, *NBC Nightly News*, February 1, 1968, in Carole Fink, Philipp Gassert, and Detlef Junker, eds., *1968, The World Transformed* (Cambridge: Cambridge University Press, 1998), 66.
53. Eddie Adams, "Eulogy," *Time*, July 27, 1998, http://www.time.com/time/magzine/article/0,9171,988783,00.html.
54. "Guerrilla Dies," *New York Times*, February 2, 1968, 1; "Death," *New York Times*, February 2, 1968, 12.
55. Quoted in Andrew Hoskins, *Televising War: From Vietnam to Iraq* (London and New York: Continuum, 2004), 19.

56. George A. Bailey and Lawrence W. Lichty, "Rough Justice on a Saigon Street: A Gatekeeper Study of NBC's Tet Execution Film," *Journalism Quarterly* 49 (summer 1972): 227–28.
57. Harry McPherson interview, "Tet 1968," *Vietnam: A Television History*, produced by WGBH, 660 minutes (Public Broadcasting Service, 1983), Digital video recording.
58. Hallin, *The "Uncensored War,"* 171.
59. Oscar Patterson III, "Television's Living Room War in Print: Vietnam in the Newsmagazines," *Journalism Quarterly* 61 (spring 1984): 35–40.
60. Hammond, *Reporting Vietnam*, 122, 126.
61. Hammond, *Reporting Vietnam*, 122; Braestrup, *Big Story*, xiii.
62. See Chester J. Pach Jr., "And That's the Way It Was: The Vietnam War on the Network Nightly News," in David Farber, ed., *The Sixties: From Memory to History* (Chapel Hill: University of North Carolina Press, 1994), 94–101.
63. Fink, Gassert, and Junker, *1968, The World Transformed*, 58.
64. Walter Cronkite, "Who, What, When, Where, Why: Report from Vietnam," *CBS News*, February 27, 1968, quoted in Gitlin, *The Whole World Is Watching*, 206.
65. Tom Johnson interview, *Walter Cronkite: Eyewitness to History*, produced by Paul F. Ryan and CBS News Productions, 100 minutes (A&E Network, 1998), Video recording.
66. Hammond, *Reporting Vietnam*, 122.
67. Larry Berman, *Lyndon Johnson's War: The Road to Stalemate in Vietnam* (New York: XX, 1989), 84–85.
68. Quoted in Michael S. Sweeney, *The Military and the Press: An Uneasy Truce* (Evanston, Ill.: Northwestern University Press, 2006), 137.
69. Kathleen J. Turner, *Lyndon Johnson's Dual War: Vietnam and the Press* (Chicago and London: University of Chicago Press, 1985), 186.
70. Jerry Rubin, *Do It! Scenarios of the Revolution* (New York: Ballantine, 1970), 106.
71. Steve Cone, *Powerlines: Words That Sell Brands, Grip Fans, & Sometimes Change History* (New York: Bloomberg Press, 2008), 135–36.
72. Harry Mills, *Artful Persuasion: How to Command Attention, Change Minds, and Influence People* (New York: AMACOM, 2000), 20.
73. James T. Patterson, *Grand Expectations: The United States, 1945–1974* (New York: Oxford University Press, 1996), 315.
74. David M. Potter, *People of Plenty* (Chicago: University of Chicago Press, 1954), 167.
75. Thomas Frank, *The Conquest of Cool: Business Culture, Counterculture, and the Rise of Hip Consumerism* (Chicago: University of Chicago Press, 1997), 136–37.
76. Christopher H. Sterling and John M. Kittross, *Stay Tuned: A History of American Broadcasting*, 3rd ed. (Mahwah, N.J.: Lawrence Erlbaum, 2002), 428.
77. Barry Richards, Iain Mac Rury, and Jackie Botterill, *Dynamics of Advertising* (New York: Routledge, 2000), 59–61.
78. William Lazer, *Handbook of Demographics for Marketing & Advertising* (New York: Lexington Books, 1994), 125.
79. Alan Wells and Ernest A. Hakanen, *Mass Media & Society* (Greenwich, Conn.: Ablex Publishing, 1997), 550.
80. Caroline Knapp, *Appetites: Why Women Want* (New York: Counterpoint, 2003), 147.
81. Vickie Rutledge Shields and Dawn Heinecken, *Measuring Up: How Advertising Affects Self-Image* (Philadelphia: University of Pennsylvania Press, 2002), 49.

82. John E. Richardson, *Marketing*, 21st ed. (New York: McGraw-Hill,1999), 120.
83. "Bridgeport Ban Stops the Music: Halts 'Rock 'n Roll' Teen Dances," *New York Times*, March 26, 1955, 17.
84. William H. Young and Nancy K. Young, *The 1950s* (Westport, Conn.: Greenwood Press, 2004), 179.
85. Quoted in Cameron Crowe, "Liner Notes," Bob Dylan, *Biograph*, Columbia Records, 38830LP.
86. Bob Dylan, *The Times They Are A-Changin'*, Columbia Records, CL 2105.
87. Bob Spitz, *Dylan: A Biography* (New York: W. W. Norton, 1991), 248.
88. Robert Shelton, "Guthrie's Heirs," *New York Times*, June 14, 1964, X16.
89. Robert Shelton, "Pop Singers and Song Writers Racing Down Bob Dylan's Road," *New York Times*, August 27, 1965, 17.
90. Barry McQuire, *Eve of Destruction*, Dunhill Records, DS-50003.
91. Jefferson Airplane, *Volunteers*, RCA Records, LSP-4238.
92. "Study Shows Sub-Par Yule Sale of LP Disks," *Billboard*, February 6, 1961, 3.
93. "LP and Singles Business Booms in Latest Sales Statistics," *Billboard*, November 14, 1960, 3.
94. Andre Millard, *America on Record: A History of Recorded Sound*, 2nd ed. (Cambridge: Cambridge University Press, 2005), 331. The $1.6 billion figure from Simon Frith and Andrew Goodwin, *On Record: Rock, Pop, & the Written Word* (Routledge, 1990), 127.
95. Frith and Andrew Goodwin, *On Record*, 127.
96. Quoted in Millard, *America on Record* 335.
97. Bob Spitz, *The Beatles: The Biography* (New York and Boston: Little, Brown and Company, 2005), 473; Val Adams, "Garry Moore Suggests Program To Replace His Canceled Show," *New York Times*, February 11, 1964, 79.
98. Walter Everett, *The Beatles as Musicians* (Oxford: Oxford University Press, 1999), 91–92.
99. David Kronemyer, "Deconstructing Pop Culture: How Many Records Did the Beatles Actually Sell?" *Music Industry Newswire*, April 29, 2009, http://musicindustrynewswire.com/2009/04/29/min1592_195858.php.
100. See Michael R. Frontani, *The Beatles: Image and the Media* (Jackson: University Press of Mississippi, 2007).
101. Richard Goldstein, "Why Do The Kids Dig Rock," *New York Times*, November 24, 1968, H1, 14.
102. Quoted in Tim Furniss, *A History of Space Exploration and Its Future* (Guilford, Conn.: Lyons Press, 2003), 42.
103. National Aeronautics and Space Administration, *Sputnik and The Dawn of the Space Age*, October 10, 2007, http://history.nasa.gov/sputnik/.
104. John F. Kennedy, speech, Rice University (Houston, Texas), September 12, 1962.
105. Harlen Makemson, *Media, NASA, and America's Quest for the Moon* (New York: Peter Lang, 2009), 59.
106. Lawrence Laurent, "Networks' 'Pool' Ready for Astronaut Launching," *Washington Post*, May 2, 1961, A21.
107. Quoted in Makemson, *Media, NASA, and America's Quest for the Moon*, 71.
108. Garner, *Stay Tuned*, 78.
109. Jack Gould, "The Whole World Sat Front-Row Center," *New York Times*, July 27, 1969, D15.

110. Makemson, *Media, NASA, and America's Quest for the Moon*, 78.
111. Jack Gould, "TV: Lunar Scenes Top Admirable Apollo Coverage," *New York Times*, July 22, 1969, 79.
112. Louise Sweeney, "The Week We Went to the Moon," *Christian Science Monitor* (Boston), January 3, 1969, 4.
113. Fred Ferretti, "Cronkite on Endurance: 'You Don't Think of That,'" *New York Times*, July 24, 1969, 75.
114. Makemson, *Media, NASA, and America's Quest for the Moon*, 178.
115. Lawrence Laurent, "Greatest Show Of Earth," *Washington Post*, July 22, 1969, C1.
116. Laurence Laurent, "Moon Missions Ratings Race," *Washington Post*, July 25, 1969, C7.
117. Garner, *Stay Tuned*, 86.
118. Alfred E. Lewis, "5 Held in Plot to Bug Democrats' Office Here," *Washington Post*, June 18, 1972, 1.
119. Louis W. Liebovich, *Richard Nixon, Watergate, and the Press* (Westport, Conn.: Praeger, 2003), 67.
120. Carl Bernstein and Bob Woodward, "FBI Finds Nixon Aides Sabotaged Democrats," *Washington Post*, October 10, 1972, 1.
121. Bob Woodward, "How Mark Felt Became 'Deep Throat,'" *Washington Post*, June 2, 2005, 1.
122. Liebovich, *Richard Nixon, Watergate, and the Press*, 82.
123. Richard Nixon, *In the Arena* (New York: Simon and Schuster, 1990), 934, in Anthony R. Fellow, *American Media History* (Belmont, Calif.: Tomson Wadsworth, 2005), 352.
124. Paul Johnson, *Modern Times: The World from the Twenties to the Eighties* (New York: Harper & Row, 1983), 651.
125. Michael Schudson, *Watergate in American Memory: How We Remember, Forget, and Reconstruct the Past* (New York: Basic, 1992), 104. This is the argument made by Mark Feldstein, "Watergate Revisited," *American Journalism Review*, August/September 2004, http://www.ajr.org/Article.asp?id=3735.
126. Quoted in David Greenberg, *Nixon's Shadow: The History of an Image* (New York: W. W. Norton, 2003), 162.
127. Quoted in Feldstein, "Watergate Revisited."
128. Quoted in Schudson, *Watergate in American Memory*, 119.
129. Silvio Waisbord, *Watchdog Journalism in South America: News, Accountability, and Democracy* (New York: Columbia University Press, 2000), 170.
130. Stanley I. Kutler, *The Wars of Watergate* (New York: W. W. Norton 1990), 162.
131. Bill Kurtis, "Forty Years Later: How the Tumultuous 1968 Democratic National Convention in Chicago Impacted American Institutions and Individuals," panel presentation, Association for Education in Journalism and Mass Communication, Chicago, August 7, 2008.
132. Quoted in Schudson, *Watergate in American Memory*, 105.
133. Liebovich, *Richard Nixon, Watergate, and the Press*, 31–32.
134. Axel Madsen, *60 Minutes: The Power and the Politics of America's Most Popular TV News Show* (New York: Dodd, Mead, 1984), 14.

135. Quoted in John Peter Sugden and Alan Tomlinson, eds., *Power Games: A Critical Sociology of Sport* (New York: Routledge, 2002), 14.
136. James H. Wittebols, *Watching M*A*S*H, Watching America: A Social History of the 1972–1983 Television Series* (Jefferson, N.C.: McFarland, 2003), 138.

10. *The Scramble to Fill 24 Hours of Air Time: The End of the Twentieth Century*

In 1980, the face of news presentation in the United States altered dramatically. That is when a burgeoning media mogul from Atlanta, Georgia, introduced the Cable News Network. CNN, rather than confining news to the early morning news broadcasts, multi-second updates during the day, and the evening news, promised to devote all day, every day to the nation's and the world's events. Initially, doing so was difficult. As one CNN executive said, "In the beginning, we had to scramble to fill 24 hours of air time."[1] Soon, though, CNN was providing the nation with a live, ninety-minute news summary every evening, and in January 1991, the cable network provided sounds and then images from Baghdad as the United States sent Patriot missiles into Iraq at the start of the Persian Gulf War. Doing so, one observer noted, changed American television news coverage by providing people with real-time access to a wealth of information. As a result, the relationship between newsmakers, news carriers, and the news audience was altered forever.[2]

CNN was not the only new network introduced to the American public in the early 1980s. On August 1, 1981, Music Television started its cable broadcast life. Playing music videos hosted and presented by VJ's, MTV soon became the most profitable twenty-four-hour network, reaching 30.8 million households within five years of the first broadcast. Just as CNN became a transformative agent for media news, MTV breathed renewed life into the medium of recorded music and affected social norms almost immediately. Its programming impacted film, fashion, radio, and, eventually, other TV programming. Perhaps nothing in the 1980s became more ubiquitous in youth fashion than red and black leather jackets, as seen on MTV in the Michael Jackson 1983 video "Thriller." "With video, it's an era of instant communication," one department store executive said, "Kids see what the stars are wearing tonight, and they want to wear it tomorrow."[3] "I want my MTV" became a literal, figurative, and cultural component of American society for teens, young adults, and even children.

CNN and MTV were simply a part of the media expansion of the 1980s and 1990s. Driven by cable broadcasts and later via satellite systems, the public quickly discovered that its options for news and entertainment were not limited to traditional print and through-the-air broadcast. In the late 1970s, nearly 90 percent of all those who watched television nightly tuned in to their source—the major TV networks and a few independent stations via an antenna. By the end of the 1980s, more than half of all U.S. households subscribed to a cable TV system, and cable programming—like that of CNN and MTV—had sucked away a full 20 percent of the network audience. On top of that, technological advancements were beginning to change everything that Americans knew about media usage. In 1990, with about 70 percent of all homes owning a video cassette recorder, people began to "time-shift" their broadcast viewing habits.[4] Rather than watch a program at its presented time, they would record it, watch it later, and almost always fast forward through the commercials.

The VCR was but one media-associated invention that would change the nation forever. Portable devices like Sony's Walkman, the compact disc, and the personal computer would alter the nation, information dispersal, and even information itself before the new century. The widespread use of the Internet that occurred in the last five years of the century and the twenty-four-hour news cycle would produce what has been called a "frenzy of feeding sharks,"[5] where media outlets—in the quest to be the first to break a story—turned almost anything into headline news.

For some clever orators, media—specifically broadcast—became what President Theodore Roosevelt had called a "bully pulpit." President Ronald Reagan turned television into his own personal tool to persuade the nation to adopt his policies, while commentator Rush Limbaugh used radio to promote his agenda of conservatism for any number of issues but especially for electing politicians. Their ability to evoke change via media was some of the most powerful by individuals that the nation had seen in decades. Despite the ability of media to influence at the national and even international level, local media were still great influencers. Effecting change via media on the local level was given new names—civic, community, or public journalism—and print and broadcast outlets adopted slogans such as "If it matters to (insert community's name), it matter to (insert media outlet's name)." But, civic journalism, where "citizen-journalists" were "intimately involving themselves in the welfare of the place, in the civic life of their towns, participating as active members of the very community they're covering,"[6] was really just a new name for the participatory journalism practiced by Joseph Pulitzer and William Randolph Hearst, the Muckrakers, even the journalism of Samuel Adams and Benjamin

Franklin. By the end of the twentieth century, however, the crusading journalist had many more media tools with which to effect change on any societal level—even the ability to do so in real time—something they sometimes did.

POLITICS OF PERSUASION

A number of U.S. presidents knew how to use media to push their agendas and effectively change the course of the nation. Only a few since the introduction of broadcast media, though, were able to do so. Franklin Roosevelt used radio; John Kennedy charmed via television. But Ronald Reagan used his earlier career as an actor to make his on-air appearances have all the polish of entertainment television. Timing, gestures, how to stand or look at cameras, the ability to control emotions—Reagan knew all the tricks of the actor's trade, and they made him charming to the America public and liked almost universally by reporters.[7] His ability to establish what seemed to be a dialog with individuals, though he spoke to millions via television, earned him the title "the great communicator."

Reagan used television to influence the nation, but TV reporters used the medium to influence the president, too. It was a symbiotic relationship. According to Robert Donovan and Ray Scherer, Reagan and wife Nancy spent hours watching television, especially news programming. Reporters and other government officials soon learned this, and they would then produce or pitch stories that would affect the national agenda. The result—regardless the origin of it—was that Reagan would then take the issue and persuasively present it to the American people and to Washington politicians. When one administration official planned to eliminate a subsidy program that provided heating assistance to the poor, NBC produced a story on it, focusing on one Pennsylvania couple. The heart-wrenching story, which included a conversation with the retired worker and his disabled wife, was seen by Reagan who then vetoed the proposed cut. Reagan, who had promised to excise the nation's tax burden in his landslide 1984 election, refused to cut a program that would have surely been eradicated had NBC not provided the story on the *Nightly News*, knowing the president would be watching and have a soft spot for such personal testimony.[8]

According to columnist David S. Broder, "The one clear American policy revolution of [the 1980s] was engineered by Ronald Reagan when he came to power intent on reversing the almost half-century growth of the welfare state. He succeeded to a significant degree in realizing that ambition." Through television, Reagan identified himself with the values and traditions of millions of American voters. Day in and day out, on a scale never before seen in the White House, Reagan and his assistants meticulously planned how they could use television to build support for the president's programs in Congress.⁹ And, it worked.

Sax appeal

On June 3, 1992, Arkansas governor and Democratic presidential candidate Bill Clinton did the unthinkable. He appeared on a late-night television program. Wearing sunglasses with a saxophone around his neck, the governor joined the house band on *The Arsenio Hall Show* and jammed on the Elvis Presley tune "Heartbreak Hotel." Afterward, Clinton joined Hall for a conversation, one where the future president could do what he did best: talk casually to people all the while being witty and charming. While news commentators such as the *Washington Post*'s William Raspberry dismissed such ploys by candidates as unworthy of news coverage, Clinton's appearance on an entertainment program changed campaigning forever. Besides being charming, Clinton was able to discuss issues that the United States faced. He discussed racism in America, a topic that worked perfectly with Hall, an African American. Clinton realized something about media that was critical. A medium such as television offered a candidate a way to have direct access to millions of Americans without having the conversation filtered by the gatekeepers of news media. Clinton was able to have an extended conversation with the voting public that was neither edited nor directed by reporters. From this point until he defeated incumbent George H.W. Bush in November, Clinton employed this tactic, using news outlets as little as possible.[10]

The Clinton appearance on *The Arsenio Hall Show* created a firestorm of controversy in media in the midst of a campaign that was already unconventional. Independent candidate Ross Perot announced his candidacy on CNN's *Larry King Live*. Clinton had already appeared—while running for the Democratic nomination—on *The Phil Donohue Show*. But, now, he was the Democratic contender for the presidency. Somehow, playing a saxophone and talking about marijuana use—a big part of the controversy—did not seem to be what a potential president should be doing, many media commentators said, especially since the talk-show circuit allowed candidates to bypass the centuries-old policy of turning to traditional new media outlets as the means of reaching the masses. One of Clinton's advisers, Frank Greer, however, pointed out that Americans spent more time with entertainment programming than with TV news. "These types of programs have become much more important as a source of information for people. Ten years ago, these programs didn't have the audience or the impact that they have now." Another observer, Darrell West, a professor of political science at Brown University, explained that doing what Clinton did on *The Arsenio Hall Show* was "generate stories" about himself that went beyond the actual appearance.[11]

Next, Clinton turned to MTV where he held a ninety-minute forum with voters aged 18 to 24. As the cable network with the largest viewing audience, Clinton acquired a direct link to millions, especially when you realize that MTV rebroadcast the appearance at least six times during a three-day period.[12] Clinton's ploys to get the ear of voters did not play well with traditional journalists. Barbara Walters, on ABC's *This Week with David Brinkley*, said "there is something about a presidential candidate with his shades on, playing the saxophone that is endearing on the one hand, but not very dignified." Clinton, of course, bet the presidency on the former, and he was correct. Tom Wicker said that for Clinton "to appear on television, playing jazz, with dark shades on *The Arsenio Hall Show* after midnight, I don't think that enhances his standing on family values."[13] But, Clinton's comments, his talents, and his ability to make himself appear as an everyday person, worked, and it forced a change in the way news media—television particularly—approached those running for elected office.

Clinton and Perot realized that there was a better way to campaign. They did not bypass the networks or the medium that had elected Kennedy and Reagan and ended the presidency of Nixon; they simply used the entertainment element rather than the news one because it provided the greatest access to the populous. The Center for Media and Public Affairs at Harvard University discovered that in 1968 a presidential candidate received an average of 42.3 seconds of air time for a sound bite. In 1992, that time on TV news had shrunk to a mere 7.3 seconds.[14] Martin Luther King, Jr. Robert Kennedy, and others in the 1960s tailored their remarks to sound bites, but they had more than six times the amount of time to say something as did Clinton, Perot, and Bush in 1992. Clinton—and Perot to a lesser extent—discovered a way to speak unimpaired for extended times. The results were successful—Clinton defeated Bush by nearly 6 million popular votes and 370 to 168 in electoral votes. Perot captured 19.7 million votes, the most of any third-party candidate since George Wallace and more than six times the votes that Ralph Nader received in the 2000 presidential race.

By 2000, entertainment programming was essential to reaching the electorate and led late-night host David Letterman to tell George W. Bush somewhat tongue-in-cheek, "The road to Washington runs through me." Where Tom Wicker had questioned the dignity of candidates appearing in such forums, *Los Angeles Times*' writer Howard Rosenberg wondered, after seeing Bush's performance on *Late Night with David Letterman*, "Do we want a Chief Executive who, when the pressure tightens like a vice, is ad-libless? With not one wisecrack to deploy on behalf of the nation or world peace?"[15] While Rosenberg was trying to make a point that comedic wit was not necessarily what the nation needed in its president, being able to think quickly and being able to reach people in this new way when running for office were. By the end of the century, the lines between entertainment and news had blurred, and candidates are now indebted to appear on entertainment television. Arnold Schwarzenegger appeared on *The Tonight Show with Jay Leno* in 2003 to promote his new motion picture *Terminator 3*. What the nation got, instead, was Schwarzenegger's announcement of his bid to unseat California Governor Gray Davis. Schwarzenegger did not need a press conference to make the announcement. He already had an audience of millions, and the uniqueness of the forum produced even more commentary for news analysts, who, by this time, knew that the entertainment element of media had established itself as a pivotal component of American politics.

DITTO HEADS AND RADIO PERSUASION

In 1988, a Sacramento, California, radio talk show host signed a contract with WABC radio in New York City to host a national talk show. Rush Limbaugh quickly became a voice of conservatism in the United States. By January 1990, about 177 stations carried Limbaugh's program.[16] Before year's end, that number had jumped to about three hundred stations with approximately 5 million regular listeners, more than any other person on the growing and increasingly powerful radio of opinion. "Limbaugh sounds like he's on a soapbox. He's intoxicated by words, especially those flowing from his own lips," the *New York Times* said. "Alone with the multitudes in his soundproof room, he sits chanting his right-wing rap, joyful in his work."[17] Multitudes agreed with Limbaugh's right leanings and would call in to his three-hour program. They began to say "ditto" when they agreed with Limbaugh's comments and soon millions adopted the name "ditto heads" to mean that they were among those who followed Limbaugh and his opinions on women, homosexuals, Democrats, and on the direction the nation should take.

Limbaugh's popularity on radio continued to grow in the 1990s. Conservative estimates put the number of listeners by the end of the decade at 10 million with other estimates saying that between 15 million and 20 million listeners tuned in to *The Rush Limbaugh Show* each day.[18] But Limbaugh's ability to influence the 1994 election and the direction of the country may have been unprecedented in media history, especially when you consider the multiplicity of information sources that the nation had in 1994. As election day approached, the *New York Times* referred to Limbaugh as the "national precinct captain for the Republican insurgency." Limbaugh constantly reminded his millions of listeners, "This is not the time to be depressed. This is the time to remember the weapon that you have, and that is the vote."[19]

Limbaugh and a growing number of right-wing commentators helped the Republicans regain control of Congress, something that had not occurred in decades. One newly elected member of Congress told Limbaugh that "talk radio, with you in the lead, is what turned the tide," while another Congressman acknowledged that "Rush is as responsible for what happened here as much as anyone." Limbaugh, responded in an uncharacteristically modest way by saying, "I'm just a media guy."[20] But Limbaugh's ability to reach millions weekly made him more than a "media guy." He was able, as many in Washington and across the nation rightly observed, to affect the outcome of many of the political contests in 1994. Mary Matalin, a Republican strategist who turned to TV political commentary for the election, estimated that 44 percent of voters reported that they received their news principally from radio. Matalin summed that commentary up by saying, "That's Rush."[21] And, some estimates suggested that more than 90 percent of those who listened to Limbaugh voted.[22] President Clinton referred to Limbaugh's media clout in his State of the Union Address. Though he did not refer directly to the radio commentator, all who heard Clinton knew to whom the president was referring when he said that "more and more of our citizens now get most of their information in very negative ways."[23]

Limbaugh, though certainly the most influential of the nation's radio commentators, was not the only one actively involved in working to influence politics and the nation's direction in the 1990s. A New York Times/CBS News Poll taken before the 1994 elections discovered that half of all Americans said that they listened—at least occasionally—to political call-in programs.[24] For many, however, Limbaugh was not vitriolic enough, nor vocal enough on homosexuals, feminists, and minorities. An estimated 150 radio stations in the 1990s carried what has been called hate radio, talk programs that promote white supremacy and attacks on minority groups. Some of these programs, which overtly advocated white supremacy, have been referred to as a "Ku Klux Klan rally of the airwaves—cruel, racist, with hints of violence."[25] As the 1990s continued and access and availability to the Internet grew, many of these groups expanded their reach online with at least two thousand sites operating by the beginning of the twenty-first century.[26]

Perhaps one of the best examples of Americans who turned to hate radio in the 1990s was Timothy McVeigh, the person behind the bombing of the Alfred P. Murrah building in Oklahoma City in 1995. McVeigh was an avid listener to talk radio that advocated violence as a way to keep the country from losing its white dominance. Another was Francisco Duran, who was a regular follower of hate-radio commentator Chuck Baker and Limbaugh. Baker openly advocated taking out the "slimeballs" in Washington, that is, to kill Congressmen and wage an armed revolution against the government. In 1994, Duran did exactly what Baker advocated, went to the nation's capital and fired dozens of bullets at the White House before being arrested.[27]

Talk radio, which tends to lean much more toward conservatism, provided people with an immediate way to voice opinions and to talk and discuss issues with like-minded individuals. But, talk radio, especially in the case of Rush Limbaugh and other big-name commentators was also entertainment, and the material presented often was never vetted in the way information presented in more traditional news formats was. Innuendo and blatant falsehoods on talk shows and hate radio became commonplace. This did not and continues not to matter to the millions who listen daily. Talk radio commentary has altered the political scene, as witnessed in the 1994 election. It is not alone, though, in terms of influence. Talk shows on television carry tremendous power; consider *The Oprah Winfrey Show*, which has a weekly audience of approximately 28 million. Winfrey's ability to influence is easily seen and has been called "the greatest media influence on the adult population" of the United States. Unlike Limbaugh, though, her influence tends to be in areas other than politics,[28] though she did weigh in on the 2008 presidential election. The Internet, however, began to usurp radio and television as a source for vindictive commentary by century's end, and what is said there operates under the protection of the First Amendment.[29] How many people visit those sites is not known, but the Internet provides a means to reach people who may not be able to tune in similar radio programming. Ultimately, the power of conversation via media has altered the way people obtain, comprehend, and react to information about the direction of the nation.

LIVE FROM BAGHDAD, VIDEO FROM LA

On January 16, 1991, the United States opened an aerial assault on Iraq in what would be known as the Persian Gulf War. When the United States became involved in Vietnam, years passed before any consistent media coverage occurred. When Japan bombed Pearl Harbor on December 7, 1941, the nation learned of the attack after the fact and only with terse radio comment. But, when bombs began falling in Baghdad, millions of Americans listened as Bernard Shaw announced, "The skies over Baghdad have been illuminated. We're seeing bright flashes going off all over the sky." Technology and the twenty-four-hour news cycle ushered the nation into what has been called "war-as-it happens TV reporting," and it led Bill Clinton to say later, "When some emergency happens somewhere in the world, there's a fifty-fifty chance I can look at it on CNN quicker than I can get a report from the State Department."[30]

Soon after CNN's live audio reporting from Baghdad, the broadcast networks joined in with live visuals. Americans were able to watch night bombings and missile strikes from the sanitary confines of their homes. What they watched resembled video games as missiles moved in on targets and provided a glow when the target was hit. CNN had changed the face of information presentation. It was the first network with the ability to send a signal from and to anywhere in the world via satellite coverage. CNN globalized television imagery, and that meant that all other players in visual communication and information dispersal had to do the same. This ability was soon labeled the "CNN effect," and as Stephen Hess observed, this coverage "meant that the world was now wired, open to instantaneous coverage, and that the coverage affected everyone and everything, including world leaders and their tactics and strategy."[31] CNN made sure that the Persian Gulf War was *the story* of that period. The network placed 150 broadcasters, technicians, and support staff in the Gulf, and to ensure its war stories would be related to all other news it presented, CNN had another fifteen hundred people working on the story worldwide.[32]

But CNN's twenty-four hours of news has transformed the nation and its information in other ways. The live broadcast ability, the desire to be first with news, and the fact that the network (as well as others) is always "on" to break the latest happenings for people means that everything that's being reported for the first time somehow becomes more significant than if there were no race to present information first or the ability to let people know what happens in Washington, New York, London, or anywhere that a network reporter is stationed or there is a person with ties to a news organization with satellite uplink capabilities. But the coverage of the first Iraq war, with its live images, slick visuals, and round-the-clock availability, also resembled what people were watching as entertainment. Networks created logos that looked more like those that were used with television mini-series. As one commentator observed during this period when flipping channels on television, it became impossible to distinguish between the high-tech graphics that showed planes soaring in the sky over Iraq and those in trailers for the movie *Top Gun*. "It was only when Tom Cruise came on the screen that I knew which was the movie," the commentator said. "Even then, he didn't look much different from the pilots whose faces I saw on CNN."[33]

The fact that Persian Gulf War coverage and media entertainment looked so much alike was new. World War II newsreels were in black and white. Though what was viewed was real, it did not look like the Technicolor real world. Images from Vietnam were much more graphic than those of World War II and real, but they did not look like any entertainment programming of the era. In the 1990s, coverage of the Persian Gulf War looked like games on Nintendo, like motion pictures, and like broadcast entertainment programming. Graphics, instant videotape, logos, theme music—all were a part of entertainment programming and a part of the coverage of the war. Reality and the imaginary had the same look. The Persian Gulf War was instant war, but it was presented as instant amusement, too.[34] The death and destruction of a video game or TV show were presented in the same manner as that from the real war, and that was a major change in media information presentation that continues to affect the nation.

Videotape from LA

On March 3, 1991, at 12:45 a.m., Los Angeles police arrested a suspect in a highway pursuit who was suspected of being high on phencyclidine (PCP). This fact may have been all that would have been known of Rodney King's arrest had not George William Holliday, who lived in an apartment complex adjacent to where King was arrested not turned on his camcorder and filmed the arrest from his second-floor balcony. In what would become a perfect example of one big story replacing another in the twenty-four hour news cycle, the video of King's arrest two days later assumed lead role over the Persian Gulf War, which, ironically, ended the same day of the arrest. The violence of the bombings in Iraq was replaced by a home video of an African-American male being beaten into submission by a host of Los Angeles police and California Highway Patrol officers. Ed Turner, executive vice president of CNN noted, "Television used the tape like wallpaper."[35] In less than a week from the time of his arrest, the video turned Rodney King into someone nearly every American knew.

The King videotape was edited down to sixty-eight seconds from its original eighty-one. What was left showed King on the ground being bludgeoned fifty-six times by the officers surrounding him. Even Daryl Gates, the chief of the Los Angeles Police Department, was shocked at what he saw. So were Americans because the tape ran hundreds, if not thousands, of times on the news in the days after the arrest occurred. It remained in the repeating loop on Headline News, on CNN, and on the major broadcast networks. ABC, CBS, and NBC produced eighty-seven stories on the beating for their nightly news programs during the next year.[36] In litigation that took place for five years after the beating, charges were brought against officers, but they were acquitted, sparking race riots in Los Angeles and more video footage of a white truck driver being pulled from his cab and beaten by blacks. By the time the riots ended, fifty-three people had been killed and more than seven thousand arrested. Property loss was estimated at more than $1 billion. In a civil suit against Los Angeles, King won $3.8 million, but in a similar case against the policemen, he won nothing. President George H. W. Bush ordered a federal inquiry into the Los Angeles Police Department, and that produced a federal grand jury that charged four of the policemen with violating King's civil rights. Two were convicted and sent to prison. Elements of the case against the officers reached the Supreme Court in June 1996, and that effectively put an end to issues surrounding the officers' imprisonment.[37]

What made the King episode so significant was not the brutality that it portrayed nor the subsequent riots—though neither are trivial. The video, shot by a novice, altered the way media were to operate. Suddenly, because of technological advancements, anyone with a video camera, a recorder, or any other device created for personal use, now turned all citizens into potential journalists. Media outlets were soon asking for input from citizens, and as the quality of the product created by personal technology improved, so, too, did its use. By the end of the 1990s and the rapid rise of the Internet and its usage in the United States, information dispersal was changing. The twenty-four hour news cycle was now not confined solely to professional journalists. Anyone at any time could add information to the public sphere, and gradually, tens of thousands did. With the Rodney King event, the way that Americans could obtain information changed. And, as we will see in the next chapter, it did more than change media; it changed the way the nation viewed and understood what was taking place.

COMMUNITY JOURNALISM

In the 1990s, a group of journalists concluded that media had grown so large that they no longer adequately allowed people to have a voice in the democratic process. Though media were constantly being used to effect change in the nation—as witnessed by Limbaugh's Ditto Heads, the journalists felt that media were not giving people enough opportunities to do this on the local level. As a result, they began to push community or public journalism where media outlets did more than report on events and issues. Instead—in conjunction with those the outlet served—media needed to promote and improve the quality of life for a community.[38] Media were to provide a means for the people to discuss issues and even guide and direct what the media outlet discussed. The rationale of community journalism was that when local media listened to and allowed the public to speak, community leaders could hear them and react. If the people did not like the reactions, they could vote people out of office or push for their removal if they were in an appointed position. Or, as one of the pioneers of public journalism Jay Rosen has said, media should allow the public to shape the news agenda. The result would be to effect change in local policies or attitudes.[39]

This is exactly what the *Akron (Ohio) Beacon Journal* did.[40] In 1993, reacting to the Rodney King incident in California, the *Beacon Journal* decided to look at race relations in the city in an effort to improve them. The multi-year effort by the newspaper was called "Question of Color," and the campaign produced thirty different articles. From the beginning, the *Beacon Journal* turned to the citizens of Akron to help create ways to improve race relations. The newspaper also created a civic organization that it called the "Coming Together Project." In order to allow public input into the news stories and consequently have citizens' voices heard by city leaders, the paper conducted a telephone survey, established focus groups, and conducted numerous in-depth interviews with black and white Akron residents. After completing these information-gathering practices, the *Beacon Journal* used it to set the agenda for the direction of its reporting. As a result, the paper's news stories were filled with commentary from "ordinary" citizens much more often than quotes from community leaders. This meant, among other things, that the newspaper's offerings were filled with multiple personal examples of racism in Akron along with community suggestions on how to solve racial issues.

The *Beacon Journal* went beyond the "Coming Together Project" to involve itself in the effort to improve race relations. It hired facilitators—white and black—to increase public involvement in the "Question of Color" campaign. They, in turn, were able to get more than two hundred local civic organizations in contact with the newspaper. After all this input, the *Beacon Journal* began offering solutions to racial issues in Akron based on community input. The project quickly gained national attention, and in 1997 President Bill Clinton chose Akron as the site of his first town hall meeting on race. Clinton said he picked Akron because of the "Coming Together Project," a direct result of a media outlet's efforts to enlighten and use its audience to change the community.

FEEDING FRENZY, 24-7

While the rise of the twenty-four/seven news cycle intensified media scrutiny, especially if politicians were involved, any number of issues and events from the mid-1970s on dominated news cycles—even those prior to the twenty-four-hour one. Often, the events, when analyzed from a later date, seemed tremendously overblown in relation to media attention versus the significance of the actual issue. Though the events often did not alter the direction of the nation, the overblown coverage of them did take media attention away from other issues. The nation focused its conversation on this "breaking news." Dozens of examples exist. Gary Hart, running as a Democratic presidential candidate, became embroiled in a media frenzy surrounding an affair with Donna Rice, a model, actress, and marketer. President Gerald Ford, whose clumsiness became a parody feature on *Saturday Night Live* during his presidency, was followed by cameras when he went skiing to record any falls that he might have. President Jimmy Carter's reference to himself as a born-again Christian led to the president's admission that though he tried to live a sin-free life that he had "lust in my heart," which sent media into a frenetic conversation about evangelical Christianity among other subjects. Vice President Dan Quayle became a media focus for any number of issues, included his misspelling of potato and for his feud with mythical TV character Murphy Brown about what Quayle called "family values" after the TV character decided to have a child as a single mother.[41]

As the twentieth century drew to a close, the intensity and competition within media led led one observer to note that media "go after a wounded politician like sharks in a feeding frenzy." He also noted that, just as in the days of yellow journalism, "the journalists now take center stage in the process, creating the news as much as reporting it, changing both the shape of election-year politics and the contours of government."[42] Media outlets—in the quest to be the first to break a story—turned almost anything into headline news, and it did not have to revolve solely around politicians. In fact, some of the media's most prominent subjects were those in the public spotlight, though not all politicians. The "perfect storm" for media and the American public occurred in 1994 and 1995 surrounding National Football League Hall of Fame running back O.J. Simpson. For months, the murder of Simpson's wife, Nicole, and her friend Ronald Goldman—coupled with speculation that Simpson killed them, his subsequent arrest for their murder, followed by the murder trial, captured front-page space in newspapers, considerable amounts of time of nightly TV news broadcasts, and talk radio and Internet conversation. Once Simpson's trial began, TV networks provided nightly recaps while some followed the case gavel to gavel. The trial ensured that a little-watched cable network, Court TV, would soon be a household name. It was truly what has been called a mega-media spectacle, where media outlets—in competition for audience and revenue—allowed information about the spectacle to supersede regular programming and to dominate the media conversation.[43] Even Russian president Boris Yeltsin, when he met President Clinton in 1995 asked him, "Do you think O.J. did it?"[44]

On June 12, 1994, media first reported on the murder of Nicole Simpson and Goldman. Because of the connection to O.J. Simpson, who after gaining fame as an NFL star became an actor and celebrity pitchman and was one of the nation's more recognizable figures, the murder garnered national attention. Simpson's home in Brentwood, California, immediately became the location where dozens of media outlets sent reporters and cameras. On June 17, police believed they had enough evidence to issue a warrant for Simpson's arrest. More than one thousand reporters waited at police headquarters for Simpson's arrival. Instead of showing up, however, Simpson, with friend Al Cowlings driving a white Ford Bronco, led police and nearly two dozen TV helicopters on a slow-speed chase down a California interstate. The chase ended at Simpson's mansion. The major broadcast networks interrupted regular programming to join the chase in progress with approximately 95 million Americans watching, 10 million more than watched the 1969 moon walk.[45] Perhaps the most interesting comment of the evening came from Tom Brokaw, news anchor for NBC. After interrupting an NBA playoff game with the chase, Brokaw said, "For people who have known the public O.J. Simpson or even the private, it is inconceivable that it could come to this, a man who has spent most of his adult life in the public eye, almost always in adulation. Now, the dark side of his role in public life." Brokaw's pronouncement, no doubt an ad lib statement, declared Simpson to be guilty, and media conversations immediately tended to make this assumption.

During Simpson's trial, CNN provided 631 hours of coverage with an average audience of 2.2 million, a five-fold increase in the network's regular daytime audience. Analysts discussed the trial in terms of ratings, of who won or lost—the prosecution or defense—that day in court.[46] American workers' productivity fell by an estimated $25 billion from the beginning of the trial in January 1995 through the announcement of the verdict in October. At least two thousand reporters covered the trial that cost California an estimated $20 million. When the verdict was aired early in the afternoon Eastern time, an estimated 150 million Americans followed it, roughly 91 percent of all who were watching television.[47] Larry King announced, "If we had God booked and O.J. was available, we'd move God."[48] The verdict spurred radical reaction and created a racial riff in the United States, all of which was mediated by media, and it became impossible at times to distinguish between the tabloid journals of the supermarket checkout line and more traditional outlets because of the rush to be first in reporting information or to break some new aspect of the Simpson saga. The three major television networks—ABC, CBS, and NBC—gave more air time to the coverage of the Simpson trial than they gave to the Oklahoma City bombing and the war in Bosnia, both of which occurred during the time of the trial, combined in what analysts dubbed "the crime of the century."[49]

Coverage of Simpson did not end, however; a civil suit brought by the families of those slain followed. That trial, where jurors held Simpson responsible for the deaths, awarded $25 million to both families. But, the media spectacle that surrounded Simpson was only the beginning. In 1997, Diana, princess of Wales, died in a car crash in Paris. Though an international incident not involving an American, Diana had become a darling of the world, her wedding to Prince Charles having been watched worldwide. The reports surrounding Diana's death blamed the throngs of media who constantly sought her and other celebrities out for photographs. More than 33 million Americans watched the funeral, just 4 million fewer than watched President Barack Obama's inauguration and 2 million more than watched Michael Jackson's funeral in 2009.[50] In 1996, media affixed their attention on the murder of JonBenet Ramsey, a 6-year-old Colorado girl who disappeared on Christmas day. Ramsey's disappearance and death filled media for days. Because her parents had pushed her into numerous child beauty pageants, video of Ramsey, made up to look as if she were years older than she was, ran constantly on television. Media claimed that the avalanche of stories on Princess Di and

JonBenet were what the public wanted. But one news analyst said that the reverse was true, explaining that "media drag their readers and viewers into it, but not kicking and screaming."[51]

Breaking news, Internet style

In 1995, President Bill Clinton told the nation that "more and more of our citizens now get most of their information in very negative and aggressive ways that are hardly conducive to honest and open conversations."[52] Clinton and controversy had been closely related even before he was elected via an alleged twelve-year affair with Gennifer Flowers and through Whitewater, where the president and wife Hillary were accused of questionable land dealings in Arkansas. In 1998, however, the president and media scandal became synonymous. The terms negative and aggressive barely describe the way the media discussed allegations of an affair in the Oval Office between Clinton and White House intern Monica Lewinsky. What was new and radical about the Lewinsky scandal was the means in which it was "broken" to the U.S. public.

On January 17, 1998, Matt Drudge, who operated a Web site called the *Drudge Report*, posted a story that claimed, "A White House intern carried on a sexual affair with the President of the United States!" Drudge's account went on to claim that the woman "was a frequent visitor to the White House after midnight," had "intimate phone conversations" that were recorded and saved, and that the "young woman was bragging about the affair to others."[53] Although everything the story claimed was false other than the affair part, it did not matter. The story was in the public sphere, a source that could hardly be considered reputable in comparison with traditional American media had broken it, and now all other media would need to play catch-up. And catch up they did. From January 21 to April 20, one study of network news discovered, 46 percent of the stories on the nightly news on ABC, CBS, and NBC focused on the Clinton sex scandal, more than two times as many stories than the closest other subject—foreign affairs—that aired nightly. The study also found that the networks used the Lewinsky scandal as a means to introduce non-scandal related stories, too. CBS anchor Dan Rather on January 21 opened a story about Palestinian Liberation Organization by saying, "Amidst all the turmoil surrounding the White House today, PLO chief Yasir Arafat arrived for a meeting tomorrow with President Clinton." The *CBS Evening News* simply put Arafat's visit into a frame that kept viewers focused on the unfolding scandal. In fact, more than 29 percent of the stories that appeared on the network nightly news programs during the first three months of the scandal contained at least one reference to the Clinton sex scandal.[54]

The irony of the feeding frenzy surrounding the sex scandal was the fact that Clinton's approval ratings climbed during this period to 72 percent. As the media focused on the scandal, the nation was enjoying some of its most prosperous times ever, and Clinton introduced a balanced budget and a budget surplus of $9.5 billion, which garnered only four news stories to 411 about the sex scandal.[55] The media quickly turned their attention to the investigations led by Kenneth Starr. Media rushed to be the first to break a story, even if a retraction was required later. Some media outlets chose to run unedited transcripts of testimony even though much of it was considered offensive and problematic for younger listeners or readers. But, most importantly, the Clinton-Lewinsky sex scandal validated the possibilities of the Internet, something that would change the media landscape in the twenty-first century, and it reinforced the idea that media, with twenty-four hours of time to fill and a multiplicity of ways to reach the public, often operated as a frenzy of feeding sharks.

***

In the last two decades of the twentieth century, media's move into twenty-four-hour mode and the melding of news and entertainment greatly affected the nation. Though it may be overly simplifying the era by focusing upon CNN and MTV as prime examples of the way the American populous was reached and affected, the programming content and styles of the two were able to garner the attention of the nation. But, these sources of news and entertainment were not all that made a difference. In fact, it could be argued that technological advancements related to media were of more importance. The introduction of the personal computer in 1984 by Apple set in motion a means for sharing and creating content that would find a ready outlet in the Internet in the 1990s. As seen with the case of Rodney King, personal devices such as the video recorder, though they may have been sold as a means to keep a record of the family, were useful tools for media, and they quickly gave anyone who owned one the ability to capture information that might be of interest to others. As we will see, this type of information sharing was set to change everything from entertainment to presidential campaigning in the twenty-first century via social networking and the Internet.

Because of the multiplicity of media outlets toward the end of the twentieth century, clever politicians, journalists, and even ordinary citizens were able to direct national and local conversation through mass-mediated public conversation. Media rediscovered ways to make change on the local level by using print and broadcast as the public sounding board. Media outlets served as the moderator of the conversation and often its director. By the end of the century, the lines that had distinguished media elements for decades were disappearing as was the dominance of the traditional power of print and broadcast. Though newspapers and television specifically still garnered more of the advertising dollar than other traditional media, there were signs that all was about to change. Matt Drudge's ability to break the Monica Lewinsky scandal was only a first sign. By the beginning of 2000, there were approximately 75 million Web site hosts on the Internet where there had been fewer than 160,000 at the beginning of 1990.[56] Information dispersal was about to have a new element. While the older media forms would not disappear nor become uninfluential, in the twenty-first century, being online would change and affect the nation in radical ways.

NOTES

1. Quoted in Edwin Diamond, *The Media Show: The Changing Face of News, 1985–1990* (Cambridge, Mass.: MIT Press, 1991), 55–56.
2. Diamond, *The Media Show*, 57.
3. R. Serge Denisoff, *Inside MTV* (New Brunswick, N.J.: Transaction Publishers, 1988), 1, 258.
4. James L. Baughman, *The Republic of Mass Culture: Journalism, Filmmaking, and Broadcasting in America Since 1941*, 3rd ed. (Baltimore: Johns Hopkins University Press, 2006), 213–14.
5. Quoted in Diamond, *The Media Show*, 173.
6. Jock Lauterer, *Community Journalism: Relentlessly Local*, 3rd ed. (Chapel Hill: University of North Carolina Press, 2005), preface.
7. Robert J. Donovan and Ray Scherer, *Unsilent Revolution: Television News and American Public Life* (Cambridge: Cambridge University Press, 1992), 177–78.
8. Donovan and Scherer, *Unsilent Revolution*, 178–79.
9. Donovan and Scherer, *Unsilent Revolution*, 184–85.
10. Joseph Hayden, *Covering Clinton: The President and the Press in the 1990s* (Westport, Conn.: Greenwood Press, 2002), 17–18.
11. Quoted in Elizabeth Kolbert, "Whistle-Stops à la 1992: Arsenio, Larry and Phil," *New York Times*, June 5, 1992, 1.
12. Maralee Schwartz and Howard Kurtz, Clinton Gets His MTV, Wants Its Viewers' Votes," *Washington Post*, June 17, 1992, A16.
13. Quoted in Hayden, *Covering Clinton*, 19.
14. Kiku Adatto, Center for Media and Public Affairs, in "The Incredible Shrinking Sound Bite," *New York Times*, June 21, 1992, A19.
15. Howard Rosenberg, "If You Can't Be Funny at Least You Can Run for President," *Los Angeles Times*, March 6, 2000, 1, http://articles.latimes.com/2000/mar/06/entertainment/ca-5802?pg=1.
16. Jeffrey Yorke, "Limbaugh's Play on Words," *Washington Post*, January 2, 1990, C7.
17. Lewis Grossberger, "The Rush Hour," *New York Times*, December 16, 1990, SM58, 92.
18. Thomas A. Greenfield, "Rush Limbaugh," in *American Icons: An Encyclopedia of the People, Places, and Things That Have Shaped Our Culture*, eds. Dennis R. Hall and Susan Grove Hall (Westport, Conn.: Greenwood Press, 2006), 390.
19. Robin Toner, "The 1994 Campaign," *New York Times*, November 3, 1994, A29.

20. Katharine Q. Seelye, "Republicans Get a Pep Talk from Rush Limbaugh," *New York Times*, December 12, 1994, A16.
21. Greenfield, "Rush Limbaugh," 391.
22. Robert L. Hilliard and Michael C. Keith, *Waves of Rancor: Tuning in the Radical Right* (Armonk, N.Y.:M. E. Sharpe, 1999), 6.
23. William Jefferson Clinton, "State of the Union Address, January 24, 1995, http://govinfo.library.unt.edu/npr/library/speeches/22e2.html.
24. Toner, "The 1994 Campaign," *New York Times*, November 3, 1994, A29.
25. Jeff Cohen and Norman Solomon, "Spotlight Finally Shines on White Hate Radio," *Media Beat* 43 (November 1994): 138.
26. Robert L. Hilliard, "Band of Hate: Rancor on the Radio," in *Radio Cultures: The Sound Medium in American Life*, ed. Michael C. Keith (New York: Peter Lang, 2008), 203.
27. Hilliard and Keith, *Waves of Rancor*, 7.
28. Janet Lowe, *Oprah Winfrey Speaks: Insights from the World's Most Influential Voice* (New York: Wiley, 1998), 1.
29. Hilliard and Keith, *Waves of Rancor*, 28.
30. Quoted in Baughman, *The Republic of Mass Culture*, 221–22.
31. Stephen Hess and Marvin L. Kalb, eds., *The Media and the War on Terrorism* (Washington, D.C.: The Brookings Institution, 2003), 23, 63.
32. Edward Wakin, *How TV Changed America's Mind* (Lincoln, Neb.: iUniverse, 1996), 207.
33. Caryl Rivers, "It's Tough to Tell a Hawk From 'Lonesome Dove,'" *New York Times*, February 10, 1991, H29.
34. Rivers, "It's Tough to Tell a Hawk From 'Lonesome Dove,'" H32.
35. Quoted in Lou Cannon, *Official Negligence: How Rodney King and the Riots Changed Los Angeles and the LAPD* (Boulder, Col.: Westview Press, 1999), 21.
36. Cannon, *Official Negligence*, 23–24,
37. Douglas O. Linder, "The LAPD Officers' Trials: A Chronology," *Famous American Trials*, http://www.law.umkc.edu/faculty/projects/ftrials/lapd/kingchronology.html.
38. Theodore Glasser and Francis L. F. Lee," "Repositioning the Newsroom: The American Experience with Public Journalism," in *Political Journalism: New Challenges, New Practices*, eds. Raymond Kuhn and Eric Neveu (London: Routledge, 2002), 203.
39. Jay Rosen, *Getting the Connections Right: Public Journalism and the Troubles in the Press* (New York: Twentieth Century Fund Press,1996), 71.
40. Discussion of the *Akron Beacon Journal*'s public journalism campaign is adapted from Tanni Haas, *The Pursuit of Public Journalism: Theory, Practice, and Criticism* (New York: Routledge, 2007), 84–101.
41. Peter J. Wallison, *Ronald Reagan: The Power of Conviction and the Success of His Presidency* (Boulder, Col.: Westview Press, 2004), 115.
42. Larry L. Sabato, *Feeding Frenzy: How Attack Journalism Has Transformed American Politics* (New York: Free Press, 1991), 1.
43. Douglas Kellner, *Media Spectacle* (New York: Routledge, 2003), 94.
44. Quoted in Richard Thompson Ford, *The Race Card: How Bluffing About Bias Makes Race Relations Worse* (New York: Farrar, Straus and Giroux, 2008), 312.

45. Jim Moret, "Simpson Trial and TV: New High, or Low?" *Cable News Network*, December 19, 1995, http://www.cnn.com/US/OJ/daily/9512/12-20/index.html?iref=newssearch; Steve Michael Barkin, *Television News: The Media Marketplace and the Public Interest* (New York: M.E. Sharpe, 2003), 70.
46. Barkin, *Television News*, 71–72.
47. Thomas L. Jones, "The O.J. Simpson Murder Trial: Prologue," *Notorious Murders*, http://www.trutv.com/library/crime/notorious_murders/famous/simpson/index_1.html.
48. Quoted in Maggie Wykes and Kirsty Welsh, *Violence, Gender and Justice* (Thousand Oaks, Calif.: Sage, 2009), 15.
49. Kellner, *Media Spectacle*, 94.
50. Associated Press, July 9, 2009, http://www.thestar.com/news/world/article/663231.
51. Joe Sheibly, *News-Sentinel* (Fort Wayne, Indiana), quoted in Cathy Coleman, "The Junk Food News Stories of 1997," in *Censored, 1998: The News that Didn't Make the News*, ed. Peter Phillips (New York: Seven Stories Press, 1998), 138.
52. Clinton, "State of the Union Address," January 24, 1995.
53. "Newsweek Kills Story on White House Intern," *Drudge Report*, http://www.drudgereportarchives.com/data/2002/01/17/20020117_175502_ml.htm.
54. Kate M. Kenski, "The Framing of Network News Coverage During the First Three Months of the Clinton-Lewinsky Scandal," in *Images, Scandal, and Communication Strategies of the Clinton Presidency*, eds. Robert E. Denton Jr. and Rachel L. Holloway (Westport, Conn.: Greenwood Press, 2003), 254–56.
55. Kenski, "The Framing of Network News Coverage," 257.
56. Robert H. Zakon, *Hobbes' Internet Timeline*, http://www.zakon.org/robert/intenet/timeline/.

11. Show-Me Journalism: Media Transformation in the Twenty-First Century

In 2005, the Project for Excellence in Journalism released its annual report. Framing it was this headline: "Technology is transforming citizens from passive consumers of news produced by professionals into active participants who can assemble their own journalism." The report also declared that "the era of '*Show Me*' journalism has begun.¹ What started at the end of the twentieth century in terms of information dispersal via the Internet exploded in the post-9/11 world. By July 2005, more than 350 million hosts—computer systems with registered Internet protocol addresses—existed. That number had been less than 160,000 in 1990.² The ease in use and growing ubiquity of personal computers turned the nation's populous into a country of information sharers, and it seemed as if it did not matter what information was being disclosed—personal, professional, or that which mattered to the welfare of communities or the nation. While millions turned to a form of voyeurism as they watched personal videos that could be about anything, the real strength of the Internet in terms of shaping society was predicted in the 1990s by Howard Rheingold, one of the first people to explore the use of the Internet as a social tool. "The ability of groups of citizens to debate political issues is amplified enormously by instant, widespread access to facts that could support or refute assertions made in those debates," he wrote in *The Virtual Community*. "This kind of citizen-to-citizen discussion, backed up by facts available to all, could grow into the real basis for a possible electronic democracy of the future."³

Technology and information sharing in the twenty-first century has become one of the greatest definers of the nation, and it is similar to the way media were used early in the nation's history. Individuals, writing to newspapers or creating broadsides, shared information or presented ideas. People everywhere discussed these ideas out loud in taverns and coffeehouse, on the streets, and in homes. Letters and essays produced more letters and essays along with more discussion. As this more personal style of information sharing was subsumed by growing media in the late nineteenth and twentieth century, people tended to lose in some cases the individual ability to effect change. As we have seen, though, this was not always the case, but larger media tended to drown out many smaller voices. That is why public opinion polling as initiated by George Gallup in the late 1920s was so important, and no doubt why people responded so positively to talk radio in the 1990s, where they were given the opportunity to voice opinions on air. When the personal computer and Internet provided people the opportunity to expand their dialog in the public sphere, they grasped it, literally, with two hands strapped to the keyboard. Now, instead of public discussions, they debated in a virtual marketplace.

Traditional media still played a critical role in the twenty-first century, and that could be seen on all levels. Lu Ann Cahn in 2005 used television to expose corrupt practices in Delaware County, Pennsylvania, with her story "Dirty Little Secret" and kept alive the idea that a crusading journalist could make a difference for and in a community. ABC, the same year, decided that Peter Jennings' fight with lung cancer afforded the network the opportunity to lead a crusade against cigarette smoking. ABC used all available outlets—television, radio, and online—to do this. Major traditional media made a massive impact on the nation with two events, those surrounding the terrorist attacks of September 11, 2001, and the catastrophe of Hurricane Katrina. Historian David McCullough called coverage of 9/11 "media's finest hour."[4]

But individuals via the Internet could and did make a huge impact. Philip DeFranco, who uses YouTube videos as a means to comment on issues, recorded more than 1 million hits on his August 29, 2008, commentary on Republican vice presidential candidate Sarah Palin in less than two weeks. While the video mixes humor with political commentary and had approached 2 million views a year after it was uploadeded, the fact that more than 1 million people watched and listened to DeFranco's commentary on a free Web site while the nation was focused on the Republican and Democratic national conventions helps explain how pivotal the individual has become to information dispersal and opinion shaping in the United States. With "show-me" journalism, people assemble the news they want from as many sources as they choose. Some are traditional sources, managed by professional journalists. Others are like DeFranco's videos, put together by individuals who have access to technology. What happens, according to the State of Journalism report, is that people "google" together information to satisfy their needs. As a result, the world of information is more fractured than it has ever been, but it is no less influential.

THE 2000 ELECTION

Coverage of presidential elections has always been a central role of media, and that was no different in 2000 than before, and presidential candidates through the first two elections of the 2000s spent a majority of their budget on media publicity, especially on TV advertisements. This was not problematic. Media did tend in the months leading up to the 2000 election to ask questions of the candidates that probably would not have been asked had it not been for the media relationship and frenzy that took place following Bill Clinton's talk-show campaigning and with the Monica Lewinsky scandal. As a result, the media pried into the lives of George W. Bush and Al Gore with questions about sexual affairs, alcoholism, and drug use.

As a result, media did not greatly alter voting on election day the way that the social networking aspect of the Internet did for the 2008 election, but post-election studies of the presidential debates did show that the major television networks tended to provide Bush with more positive comments, more negative ones for Gore, and, in general, spoke of Bush more positively in election coverage.[5] The media did become a seed of controversy and ultimately affected the nation with coverage of election results the night of the election. Because television networks—and all mainstream national media—had developed a mindset of being first to report information, being the first to predict a state in the Bush or Gore camp became paramount. The practice of announcing a state as being one won by a candidate as soon as polling data revealed who would win it had been an issue of contention before 2000, but in this election, the announcements would be pivotal. And, it turned out, the desire to be first led the networks to sacrifice accuracy for immediacy.[6]

As polls closed in the Eastern time zone at 7 p.m., the networks began to call states for the candidates. States along the Atlantic seaboard tended to fall into line as expected with Bush winning Southern states and Gore taking those in New England. Still, two pivotal states because of their electoral votes and because the race was so close in them became critical in media's call of states for Bush or Gore. Around 8 p.m., the networks put Florida into the Democratic fold, and Pennsylvania quickly followed. These predictions, if accurate, almost assured Gore the election, and the networks approached their coverage as if that were true. Around the country, though, Republican commentators chimed in that Florida might not be so easily won. Soon, Florida swang to Bush, and chaos in coverage ensued. Americans heard that Florida went Democratic, then Republican, and finally, too close to call. The early calls of states in the East would later be blamed for some of the voter turnout in the West, the other states in the nation followed their pre-election pattern, which meant that the 2000 election hinged on Florida.

Coverage of elections, Mark Brewin says in his book about media and election days, is supposed to bring the nation together. The loser concedes, and the winner accepts. The differences in parties and candidates become secondary to the process. But in 2000, the ineptitude in commentary and coverage left the nation divided, and according to Brewin, created the rise of blue and red America.[7] No clear winner was decided until the Supreme Court on December 12, 2000, ruled that all recounts of votes must halt and that Bush, who had been ultimately declared the Florida winner before any recount, be given the state's electoral votes. The subsequent issues surrounding counting ballots in Florida also led the media to follow the recount story with great scrutiny. That, in turn, led to voting reform in a number of states.[8]

SEPTEMBER 11, 2001

Certain events stand as defining moments of generations. In its history, Americans of assorted generations could tell you where they were and what they were doing when they learned that the Japanese had bombed Pearl Harbor, that John F. Kennedy had been assassinated, when Neil Armstrong stepped onto the Moon. The same holds true for the September 11, 2001, terror attacks that brought down the World Trade Center in New York City; damaged the Pentagon in Washington, D.C.; and sent another airliner hurtling to the ground in a rural area of Pennsylvania. What else holds true for these events is that almost universally among Americans, the revelation of these events was closely tied to media. Easy access by millions of Americans to radio, then television, and then both plus online for the 9/11 tragedy meant that an event, that was not media driven, became dependent almost completely upon the media by the public. And, from that point of revelation on, the nation became dependent upon media to keep everyone apprised. The desire for more news meant that media drove much of what happened from that point on.

With 9/11 though, another chilling aspect of media and its ability to affect the nation was revealed. The terrorists who planned the hijacking of planes and their use as weapons of mass destruction were depending upon media coverage to enhance the awfulness of their actions. Breaking news in the United States became the vehicle not only to inform the American public but to spread the terrorist message. Those who planned the September 11 attacks were counting on the unfolding events to become something that media would cover live. Media coverage became an essential part of the terrorists' plans. As Brooke Barnett and Amy Reynolds pointed out in their study of terrorism and the press, terrorist attacks are rarely about the victims; they are about gaining media attention and through media the attention of the public and government.[9]

When media realized what was happening on September 11, coverage changed. Not since the assassination of John Kennedy in 1963 had media—specifically television—preempted all programming including commercials for so long. For four days, round the clock, the networks followed the story. Even cable networks that never provided news, like MTV, either ran what the major broadcast networks that their media conglomerate owned or news from local stations in New York City or Washington, D.C. Others stations simply ran patriotic images, like that of the American flag. The unknown that surrounded the first hours following the attacks was reflected in coverage. The Bush administration framed attacks as war declared on the United States, and media, almost universally, presented information to the nation in the same manner. CNN's reporting created the feeling that the United States declaring war on the perpetrators should be expected and was justified.[10] Universally, on-air personalities affixed American flags to their lapels as a sign of support for the nation. Not wearing a flag was questioned. "George Bush is the president," CBS anchor Dan Rather even told David Letterman on *Late Night with David Letterman*. "Wherever he wants me to line up, just tell me where."[11]

Media became avid supporters of all that the government proposed, including the subsequent invasions of Afghanistan and Iraq. Doing so was the natural response, though Lisa Flannagan says that in by being so, media failed to do their job of asking questions that answer the basic who, what, when, where, why, and how questions and in questioning how government frames any question, especially the war in Iraq.[12] That claim is a moot point in terms of media defining the nation. What is significant is that President Bush's approval ratings according to a Gallup poll rose to 90 percent, the highest approval rating ever recorded by the poll, which began operation in the 1920s, and it remained high for at least a year—at 70 percent on September 16, 2002.[13] These numbers can be attributed at least in part to media's overwhelming support of the administration's policies and the use of media officials for almost all of media comments about the events surrounding 9/11.

Just as Dan Rather promised to follow the president unequivocally, so, too, did other media. After the United States attacked Afghanistan in October 2001, CNN almost always ran images of the burning World Trade Center during some segment of coverage of bombings and destruction in Afghanistan because, as CNN chairman Walter Isaacson said in a memo, the network had to remind Americans that the Taliban provide a safe haven for terrorists. "We must talk about how the Taliban are using civilian shields and how the Taliban have harbored the terrorists responsible for killing close to five thousand innocent people," Isaacson wrote in a reminder to correspondents who were supplying positive reports about the good the Taliban had done in some parts of Afghanistan.[14] Whenever someone in media questioned policies, pressure was brought against them. Bill Maher had his ABC late-night program *Politically Incorrect* eventually cancelled and immediately dropped by several affiliates after he raised questions about the president's characterization of the terrorists as cowards. "We have been the cowards, lobbing cruise missiles from 2,000 miles away. That's cowardly," Maher opined. "Staying in the airplane when it hits the building, say what you want about it, it's not cowardly."[15] The administration's response to Maher was that we all "need to watch what we say."[16] That statement helped justify Maher's dismissal and helped reinforce media's support of the government. The media adopted wholly the Bush administration's framing of all that happened in the post-9/11 world as the "war on terror." As a writer for the *New Yorker* later said, thinking of the United States' response to the September 11 terrorist attacks as anything other than the nation waging war "just isn't in circulation."[17]

So why did David McCullough call the coverage of September 11 "media's finest hour"? It was because media did all within their power to keep the nation current on everything that was happening. "This is the most important story of my lifetime," NBC President David Shapiro said. "I think it's our job to stay on the air." An estimated 79.5 million Americans watched that coverage on the night of September 11.[18] More than 170 newspapers issued extras after the attacks. Media made sure the nation understood the devastation and amount of death that had occurred but were sensitive to viewers in terms of what they showed. All had access to video of the nearly two hundred people who jumped from the World Trade Center, but none showed those people's impact with the ground. And, that footage almost immediately was removed from the news cycle. At CNN, the images were blurred once it was realized that what was falling was a human and then eliminated as they were at NBC, CBS, and Fox where the images were also presented. HBO even worried that its miniseries *Band of Brothers* might be offensive because of its images of gear-clad World War II soldiers and despite the fact that the series had a patriotic tenor.[19] Media simply placed the welfare of the nation above the bottom line of the ledger. Media companies lost millions of dollars in advertising revenue, and none complained. Even if media for the most part promoted the government's directives and did little questioning of any U.S. responses to the attacks, media allowed the nation to focus in upon the event, to grieve together, and to unite. As NBC news anchor Tom Brokaw said, "I can't remember when our profession in all its dimensions performed so well under such difficult circumstances."[20]

KATRINA

The way media affected the nation after Hurricane Katrina struck the Gulf coast on August 28, 2005, has been viewed as either an outstanding job of reporting and watchdog journalism or as coverage that exacerbated race relations in some cases or sensationalized news in others. No matter which way coverage is viewed, media's attack on the Bush administration's response to Katrina sent the president's approval ratings on a downward spiral from which they never recovered. Two weeks after Katrina with relentless media coverage of New Orleans and the Mississippi coast, Bush's approval rating stood at 35 percent, dropping twelve points in just one week.[21]

New Orleans' only newspaper, the *Times-Picayune*, lost its printing presses but turned to the city's Web site, NOLA.com, in order to publish and to provide information. A blog associated with NOLA.com turned into a means of locating people stranded in the city. Because the *Times-Picayune* was online and available to people everywhere, people throughout the nation, who had heard from someone in New Orleans through text messages, contacted the paper and rescue authorities through the blog. Those people were then found and moved to safety.[22] Other media outlets assumed an active role in the efforts to find survivors and to rebuild the region. ABC's *Good Morning America*, for example, announced an effort to rebuild Pass Christian, Mississippi. The hallmarks of community journalism were applied to localities, but outlets such as ABC used their national reach to do the work.

Media were also locked into their watchdog function during Katrina. Perhaps no better example exists than the work on Anderson Cooper of CNN. Though he was not the only reporter on the scene who confronted how the government handled the situation, he certainly became the most prominent. On September 1, Cooper interviewed Louisiana Senator Mary Landrieu, who during her appearance on *Anderson Cooper 360* thanked the federal government for all the aid it was providing. "I haven't heard that because for the last four days I've been seeing dead bodies in the streets here," Cooper said, interrupting the senator. "And to listen to politicians thanking each other and complimenting each other, you know, I gotta tell you there are a lotta people here who are very upset and very angry and very frustrated and when they hear politicians, you know, thanking one another it just, you know, kind of cuts them the wrong way." Cooper then told of seeing a body lying in the street being eaten by rats and asked Landrieu, "Do you get the anger that is out here?"[23]

Cooper did not hide his frustration and emotions by trying to be objective and stoic on air. Instead, he cried on air outside the Superdome and then displayed righteous indignation at what he considered appalling conduct by politicians and the government in its efforts after the hurricane. Other reporters, such as Fox News Channel's Shepard Smith, simply called government officials liars and allowed disgust to come through, just as Cooper did. "The government said, 'You go here, and you'll get help,' or 'You go in that Superdome and you'll get help,'" Smith ranted in one of his stories. "And they didn't get help. They got locked in there. And they watched people being killed around them. And they watched people starving. And they watched elderly people not get any medicine." Reports such as these helped turn public opinion against the Bush administration as much as any other element of the Katrina disaster because, as one observer noted, media provided "aggressive, in-your-face reporting" by shining a light on those in greatest need and keeping the story focused there until help actually arrived.[24]

One of the most widely discussed elements of media coverage of Katrina centered around residents of New Orleans who went into abandoned stores to obtain food and other living necessities. This reporting quickly turned much of the Katrina coverage into a racial issue that focused attention on the social and economic divides present in New Orleans. It also caused people to look closely at media and to question whether stereotyping, which had long been used in entertainment programming (though not necessarily grounded in racism), was also a part of news presentation. Because African Americans comprised 67 percent of New Orleans' citizens before Katrina, they should have naturally been the principal people in images seen from the city. When blacks and whites were compared finding food, however, people quickly discerned that something different was being said about the races. In one image from Agence France-Presse, two white New Orleans residents are shown carrying a bag of food. The photo's caption said, "Two residents wade through chest-deep water after finding bread and soda from a local grocery store in New Orleans." A similar photograph from the Associated Press with black residents said, "Looters carry bags of groceries through floodwaters after taking the merchandise away from a wind damaged convenience story in New Orleans." *Time* magazine credited the AP caption to the rush to present news and the sensationalizing of information.[25]

The captions and the questions surrounding race raised larger issues for media in terms of how they portray African Americans. Studies have shown that media and Americans in general are more likely to label questionable activity by blacks as criminal than when the same acts are viewed and being done by whites.[26] Media's missteps, which were probably innocent but still troubling nonetheless, pointed out something about the nation. Though media began to self-police themselves more closely after this, the debate probably did not change anything about race relations in the nation. It did, however, stimulate discussion and debate. And, media self-policing of its word use—referring to those in New Orleans as victims versus refugees being a problem along with looters versus residents—no doubt helped to change some attitudes nationally.

WORLD WIDE WEB, 2.0

In 2006, *Time* magazine named "you," that is, everyone, its "Person of the Year." *Time* declared the highlight of the year was "a story about community and collaboration on a scale never seen before. It's about the cosmic compendium of knowledge Wikipedia and the million-channel people's network YouTube and the online metropolis MySpace. It's about the many wresting power from the few and helping one another for nothing and how that will not only change the world, but also change the way the world changes." The Web had existed and was used by multitudes in the late 1990s, *Time* noted but then explained, "The new Web is a very different thing. It's a tool for bringing together the small contributions of millions of people and making them matter. Silicon Valley consultants call it Web 2.0, as if it were a new version of some old software. But it's really a revolution." The Web, *Time* concluded, provides people once again the ability to seize the reins of media and frame a new digital democracy—all from their chairs in front of a computer screen.[27]

The use of Web 2.0 tools is the latest and most extensive manifestation of community journalism. Instead of allowing mainstream media to serve as the gatekeepers of information, that is media outlets managing information by deciding which stories to pass along to the public and which to filter out, community journalism allows the public to determine the subject and focus of some of media's offerings. Web 2.0 tools expand those possibilities. The public becomes a group of "gatewatchers," who decide for themselves the news that is important.[28] They then frame that information through Web logs, Twitter, Face Book, MySpace, and dozens of other online tools that are being developed daily. People filter information and then publish it themselves. Columnist Kathleen Parker said the use of Web 2.0 tools as a means to bypass traditional media represented a funeral "for traditional journalism as the omnipotent gatekeeper of information."[29] What makes this shift in the control of sharing information unique in the world of media is that the filter process is not being handled solely by a professional editor, but by an individual who then shapes that information to meet personal or community needs. Threads to the stories and links to others are often added, so that the information takes on a life of its own,[30] exactly the way information presented by the press did more than two centuries ago. As Saul Cornell explained about the early nineteenth century, once an essay, letter, or comment was published, its interpretation and further comment were open to any who read it.[31] The same holds true with Internet information posted by individuals today. Now, though, the conversation is not limited within a spatial frame. Comments do not have to come from the community in which the writer lives unless the issue being discussed is central to a certain community.

Consider the situation that led to Kathleen Parker's comment on the funeral of traditional media's gatekeeping function. In 2004, CBS News anchor Dan Rather, on *60 Minutes*, provided the country with a report on President George W. Bush's Vietnam-era military service with the Air National Guard. Using a letter that was largely unsubstantiated and a source that probably had issues with the president, Rather and CBS presented their damning report just ahead of the presidential election. Immediately, the Internet blogosphere came to life. There, the authenticity of the letter was easily called into question and discredited. The attacks from the world of Internet writers was too great even for Rather, who, in 2005, agreed to leave his position at CBS, one he had held since the retirement of Walter Cronkite in 1981.

At the time of the CBS exposé, about 57 million Americans said they read Web logs with millions admitting that they followed specific blogs daily. Another 12 million adults reported that they kept a blog.[32] While all of those blogs are not political in nature, consider the power of 12 million different voices within the public sphere on a somewhat regular basis as a means of shaping conversation. By 2007, the number of regular visitors to blogs had increased to more than 94 million, according to eMarketer, an increase of 65 percent. This figure represents half of all people who use the Internet. Facebook drew in another 41 million viewers and MySpace, 75.1 million, according to an August 2008 tally by comScore MediaMetrix.[33] Within the blogosphere individuals like Philip DeFranco and Ana Marie Cox possess powerful voices. Cox, who posts a political blog called *Wonkette* and also works for *Time* magazine, pulls in about a half-million regular readers, but she also has more than 1.1 million followers on Twitter, the 140-character-per-post site that many Congressmen began using during sessions in 2008. This means that *Wonkette* has a larger regular readership than almost all American daily newspapers. It pulls in more viewers than most local TV news broadcasts, and it competes well with viewership of most cable news channels.

When the numbers of people who use the Internet are compared with figures from other traditional media sources, one can see how Americans turning to these outlets can change national conversation. Traditional broadcast networks average between 4 million and 6 million viewers per night. When totaled, that means that about 25 million Americans tuned into ABC, CBS, and NBC each night in 2008, but in 1980, 16 million Americans regularly followed the *CBS Evening News* with anchor Walter Cronkite alone, and total viewership has fallen by 53 percent since Cronkite left the air.[34] CNN in the last month of 2008 had an average prime-time audience of fewer than 1 million, though that figure stood at more than 2 million just before the election. Fox News pulls in the largest number of viewers on Cable, and Bill O'Reilly with more than 3.7 million regular viewers leads all TV commentators.[35] This does tell us that Americans like the individual, powerful voice, which explains listeners to Rush Limbaugh and followers of Ana Marie Cox, but in each case, those who follow have the ability to comment, and that has grown in importance to the public and helps to explain the continued growth of Internet tools such as WordPress and Twitter. Traditional media in the twenty-first century have also made a concerted effort to provide more of their information online. In 2008, unique visitors to the top three online news sites topped 35 million each, with about 39 million visiting MSNBC's Web site and about 37 million visiting both Yahoo! News and CNN. In the case of MSNBC, this was more than a 34 percent increase in traffic.[36]

While the power of the Internet in terms of providing a means for people to gain information and to be a part of public dialog cannot be denied, the numbers tell us something else about media and its role within U.S. society. Americans are using every available medium to which they have access. They are dependent upon media as never before. While network television figures are down, the ability to reach 25 million Americans each night with nightly news casts is powerful. The same is true of cable news programming. With online sites, traditional media outlets are finding new ways to connect with the populace. Even though the audience for newspapers has declined, "The State of the News Media" report for 2009 discovered that the number of daily newspapers in the United States actually grew in 2008, something most Americans would never have believed, and about 48 million papers are sold every day, meaning print journalism's reach is considerably greater than the number of products sold. Also, people are increasing readership online.[37]

All of the above figures apply to news. The same elements could be applied to entertainment as well. While traditional broadcast television has lost audience, much of it has been consumed on the Web, and just as individuals through blogs, Twitter, YouTube, and other networking sites are making their presence felt in the political and social arenas, so, too, are individuals via entertainment. After Michael Jackson's death in 2009, a YouTube video sponsored by Evian bottled water of baby's skating and dancing to Jackson's "Billie Jean" recorded nearly 3 million views in two months, while Jackson's own video of "Billie Jean" captured more than 35.5 million YouTube views and another 9 million views of the song from a CBS special on Jackson during the same two-month period.

THE 2008 ELECTION

Don Hewitt, the creator *60 Minutes* and the director of the first presidential debates in 1960, noted that after those debates it was impossible for a person to run for president, much less be elected, without running advertising on television.[38] In 2008, the John McCain and Barack Obama campaigns spent more than $361 million on television advertising,[39] but the deciding factor for the election may well have been the use of online tools, specifically, Web 2.0 tools. And Obama understood this better than anyone ever. During the campaign, Obama raised approximately $600 million, with an estimated $500 million coming from online donors.[40] But, Obama did more online than raise money; he used all of the online tools possible to reach voters, and his campaign achieved what it did because it used the Internet as a means to allow voters to share in what was transpiring, not just through donations, but through conversation. The Obama campaign was, according to one commentary, as much a twenty-first century media operation as it was a political campaign. And, because most of the Internet campaigning that the Obama campaign did cost almost nothing to produce, Obama was able to use his $600 million to bombard the nation via traditional media at four times the amount that the McCain campaign could.[41]

The Obama campaign established Facebook and MySpace sites where Obama supporters could contact other Obama supporters. The campaign encouraged supporters to create their own Web sites and Web logs to raise money and to discuss issues. This allowed Obama supporters to build grassroots, virtual communities, that often established real-world communities and action for the candidate. The campaign also created MyBarackObama.com, which encouraged more participation and involvement by citizens. By the time of the election, about 2 million individual user profiles had been created on the site along with more than 35,000 volunteer groups. Through MyBarackObama.com, about 200,000 offline events were planned for the candidate, few of them involving an appearance by Obama himself. The campaign used email and text messaging to create what looked like individualized messages and sent out more than 1 billion emails. To ensure that no one missed an email because they were away from a computer, the campaign used mobile texting and Twitter to keep in constant contact with supporters, and more than 1 million people signed up to receive these messages. The campaign posted 1,650 videos to YouTube that were viewed for more than 14.5 million hours. Obama's speech on race, which was thirty-seven minutes long, produced more than 5 million views by Sepember. The campaign even created its own online TV network, BarackTV at www.barackobama.com/tv.[42]

What the Obama campaign did was as unique as what Bill Clinton did in the 1992 campaign. By establishing a controlled presence on the Internet, Obama was able to sidestep traditional media and speak directly to people about issues. If a smear tactic was used against him, the campaign could issue a YouTube video to set the record straight. The seven-second sound bite of television news essentially became null and void for the Obama campaign, which could speak at length personally to people via the Web. And, interestingly, the YouTube videos were watched most by those in the 45- to 55-year-old age group. As president-elect, Obama continued to use social networking via something his transition team created called the Citizen Briefing Book and via his Web site, Change.gov.[43] That site changed to Whitehouse.gov after the inauguration.

✻✻✻

As the nation moved into the twenty-first century, the use of online media continued to expand, and no better way exists to see that power than the 2008 election. Traditional media turned to the Web as did individuals. Consequently, the power of Internet connection and commentary has made it possible for the voices of thousands to be heard and not be ignored. When President Obama, in September 2009, announced that he would deliver a speech to all the nation's school children, which would be broadcast live into the nation's classrooms, school administrators okayed the idea. But then, some people charged that the president would be using the speech as a means of pushing a political agenda and of indoctrination of the young. Kathleen Parker pointed out that "a protest of one (or a few) can instantly morph into a babble of thousands,"[44] amplified by online social networking—through Twitter and blogs, through emails and Web sites. As a result, numerous school systems changed their minds about wholesale viewings of the president's speech, opting to let parents decide whether their children would watch or skip the speech. The media-driven backlash to the speech no doubt altered some of the things Obama planned to discuss.

Personal technology, from mobile to iPod, is also driving information dispersal. From information updates sent to phones to podcasts available on digital players, advancements in technology keep making it easier to reach individuals with information designed to inform and shape opinion. Whether the message is sent from a media conglomerate, an interest group, or an individual, the message has a means of reaching people in new ways. This does not mean that the more traditional media and their messages are not still viable. The role that television played in the 2000 election and coverage of Hurricane Katrina are prime examples. Though not discussed, radio played an essential role in Katrina and in other deadly storms in the century. Radio stations, an Arbitron study discovered, were the principal source of information during emergencies, especially once a storm hit and immediately afterwards. Radio, unlike television, could be battery powered and portable. Radio, people said, tended to have a consistent signal. This fact will play out even more now that the nation has made the switch to all-digital television. In addition, radio provided much more local information, which was essential in a time of crisis or disaster.[45] Online connectivity also plays a large part in information dispersal, too, as the efforts of the *Times-Picayune* demonstrated during Katrina. Americans were simply looking to gather information from whatever source was readily and most easily available, and individuals and media looked to dissemination in the same way.

Elements of the online world have also provided people with a place to work and disseminate information that resided in the realm of science fiction in the 1990s. Second Life, a virtual online world that launched in 2003, provides a place where people can interact with others. This three-dimensional world, where people create their virtual representation called an avatar, is housed by real-world businesses, educational units, media—anything that exists in the real world. Second Life has a currency, the Linden, named for Second Life's founder Philip Rosedale but Philip Linden in Second Life. It is even possible to make money in the real world via the virtual site. In the first quarter of 2009, Second Life logged an amazing 124 million user hours with an average of more than 88,000 concurrent users operating in Second Life at any one time. The number of user hours had jumped from 7 million and the number of concurrent users from five thousand at the beginning of 2006.[46] Second Life has become a place where real-world people introduce real-world items for sale. The virtual world turned into the real world for many.[47] Like other elements of the online world, Second Life is not something that affects all Americans directly, but its reach is large. Because most major corporations operate there, the effect is greater on society than can be realized.

iPods, likewise, have transformed methods of information dispersal. In 2006, Duke University shared the technology with its students who then used the portable digital players—besides listening to music—to download lectures, class notes, and other programming available online. Duke even experimented with students uploading their assignments in different formats rather than writing traditional research papers.[48] The portable MP3 device has also transformed the world of recorded music. The Recording Industry Association of America reported nearly a 50 percent drop in physical sales of recorded music from 2000 to 2006, and in 2008. Sales converted to digital downloads.

Communication technology will continue to expand and with it the public sphere in which voices share and debate information. With every new development, people will work to shape public opinion in an effort to direct the nation. One of the more unsavory lessons from the twenty-first century centered on the ability to fabricate information. Venerable journalist Walter Cronkite, in a 2004 address to the Society of Professional Journalists, called the evolution of journalism in the twenty-first century, the journalism of scandal mongers.[49] No doubt Cronkite was thinking of the amalgamation of events from the 1990s and the 2004 scandal at the *New York Times* surrounding Jayson Blair, who copied facts and quotes from others' stories and fabricated information. While the most prominent of reporters caught up in unethical activity, Blair was certainly not the only media professional who raised questions about the quality of journalism at the beginning of the century. But Cronkite was probably also thinking about the blatant way information on many of the ever-expanding number of sites on the World Wide Web appeared to have an almost universal disregard for providing sources for information, of how plagiarism seemed irrelevant, how verification of facts appeared to be unimportant. Vigilance will be required, and that may be one of the most important lessons of the use of media to shape the direction of the nation in the twenty-first century.

NOTES

1. *The State of the New Media: An Annual Report on American Journalism*, Project for Excellence in Journalism, http://www.stateofthemedia.org/2005/execsum.pdf, 2, 6.
2. Robert H. Zakon, *Hobbes' Internet Timeline*, http://www.zakon.org/robert/internet/timeline/.
3. Quoted in Janna Quitney Anderson, *Imagining the Internet: Personalities, Predictions, Perspectives* (Lanham, Md.: Rowman & Littlefield, 2005), 189.
4. David McCullough, "Baird Pulitzer Prize Lecture," Elon University, September 19, 2001.
5. Douglas Kellner, *Grand Theft 2000: Media Spectacle and a Stolen Election* (Lanham, Md.: Rowman & Littlefield, 2001), 8.
6. Mary E. Stuckey and Kristina E. Curry, "Presidential Elections and the Media," in *Media Power, Media Politics*, eds. Mark J. Rozell and Jeremy D. Mayer (Lanham, Md.: Rowman & Littlefield, 2008), 186.
7. Mark W. Brewin, *Celebrating Democracy: The Mass-Mediated Ritual of Election Day* (New York: Peter Lang, 2008), 205.
8. Bernd-Peter Lange and David Ward, eds. *The Media and Elections: A Handbook and Comparative Study* (Mahwah, N.J.: Lawrence Erlbaum, 2004), 51.
9. Brooke Barnett and Amy Reynolds, *Terrorism and the Press: An Uneasy Relationship* (New York: Peter Lang, 2009), 1, 3.
10. See Amy Reynolds and Brooke Barnett, "'American under Attack': CNN's Verbal and Visual Framing of September 11," in *Media Representations of September 11*, eds. Steven Chermak, Frankie Y. Bailey, and Michelle Brown (Westport, Conn.: Praeger, 2003), 85–101.
11. Quoted in Lisa Finnegan, *No Questions Asked: News Coverage Since 9/11* (Westport, Conn.: Praeger, 2007), xix.
12. Finnegan, *No Questions Asked*, 2.
13. Daron R. Shaw, *Race to 270: The Electoral College and the Campaign Strategies of 2000* (Chicago: University of Chicago Press, 2006), 163.
14. Barnett and Reynolds, *Terrorism and the Press*, 80.
15. Bill Maher, *Politically Incorrect*, September 17, 2001, quoted in Mary L. Dudziak, *September 11 in History: A Watershed Moment?* (Durham, N.C.: Duke University Press, 2003), 49–50.
16. Cynthia Billhartz, "Maher's Comments Lead to Show's Suspension," *St. Louis Post Dispatch*, September 22, 2001, 15.
17. Quoted in Nancy Snow, *Information War: American Propaganda, Free Speech and Opinion Control Since 9-11* (Toronto, Canada: Seven Stories Press, 2003), 78.
18. David Bauder, "New Executives Believe Terrorist Attack Coverage Important to Continue," *Pittsburgh Post-Gazette*, September 15, 2001, B-9.
19. Bauder, "New Executives Believe Terrorist Attack Coverage Important to Continue"; Barnett and Reynolds, *Terrorism and the Press*, 81.
20. Quoted in Michael D. Murray, "The Contemporary Media," *The Media in America: A History*, ed. Wm. David Sloan, 5th ed. (Northport, Ala.: Vision Press, 2002), 485.
21. William James Willis, *The Media Effect: How the News Influences Politics and Government* (Westport, Conn.: Praeger, 2007), 67.

22. Anthony R. Fellow, "The Information Function: Mediating Reality," in *Mass Communication in a Global Age*, eds. David Copeland and Anthony Hatcher, 2nd ed. (Northport, Ala.: Vision Press, 2007), 53–54.
23. Quoted in Robert Phillip Kolker, *Media Studies: An Introduction* (Malden, Mass.: John Wiley, 2009), 82–83.
24. Quoted in Peter Johnson, "Katrina Rekindles Adversarial Media," *USA Today Online*, September 5, 2005, http://www.usatoday.com/life/columnist/mediamix/2005-09-05-media-mix_x.htm.
25. Madison Gray, "The Press, Race and Katrina," *Time*, August 30, 2006, http://www.time.com/time/nation/article/0,8599,1471224,00.html.
26. Samuel R. Sommers, Evan P. Apfelbaum, Kristin N. Dukes, Negin Toosi, and Elsie J. Wang, "Race and Media Coverage of Hurricane Katrina: Analysis, Implications, and Future Research Questions," *Analyses of Social Issues and Public Policy* 6, 1 (2006): 6.
27. Lev Grossman, "Time's Person of the Year: You" *Time*, December 13, 2006, http://www.time.com/time/magazine/article/0,9171,1569514,00.html.
28. See Axel Bruns, *Gatewatching: Collaborative Online News Production* (New York: Peter Lang, 2005).
29. Kathleen Parker, "Speak Now and Forever Wish You Hadn't," *Orlando Sentinel*, February 16, 2005.
30. Axel Bruns, *Blogs, Wikipedia, Second Life, and Beyond: From Production to Produsage* (New York: Peter Lang, 2008), 74–75.
31. Saul Cornell, *The Other Founders: Anti-Federalism & the Dissenting Tradition in America, 1788–1828* (Chapel Hill: University of North Carolina Press, 1999), 22.
32. Joseph Graf, "The Audience for Political Blogs," *Institute for Politics, Democracy & the Internet*, http://www.ipdi.org/uploadedfiles/the%20audience%20for%20political%20blogs.pdf, 3.
33. "State of the Blogosphere/2008," *Technorati*, http://technorati.com/blogging/state-of-the-blogosphere.
34. Figures taken from Nielson Media Research, in "The State of the News Media 2009," *Pew Project for Excellence in Journalism*, http://www.stateofthemedia.org/2009/narrative_networktv_audience.php?media=6&cat=2#NetAud1.
35. "The State of the News Media 2009," http://www.stateofthemedia.org/2009/narrative_cabletv_intro.php?media=7.
36. Figures taken from Nielsen Online in "The State of the News Media 2009," http://www.stateofthemedia.org/2009/narrative_online_audience.php?media=5&cat=2#topnewssites.
37. "The State of the News Media 2009," http://www.stateofthemedia.org/2009/narrative_newspapers_intro.php?media=4.
38. Joe Garner, *Stay Tuned: Television's Unforgottable Moments* (Kansas City, Mo.: Andrews McMeel, 2002), accompanying DVD.
39. Andrei Scheinkman, Xaquín G.V., Alan Mclean and Stephan Weitberg, "The Ad Wars," *New York Times*, August 11, 2009, http://elections.nytimes.com/2008/president/advertising/index.html.
40. Jose Antonio Vargas, "The Clickocracy. Obama Raised Half a Billion Online," *Washington Post Online*, November 20, 2008, http://voices.washingtonpost.com/44/2008/11/20/obama_raised_half_a_billion_on.html.

41. Andrew Rasiej and Micah L. Sifry, "With New Media, Obama Camp Takes Stage," *Politico.com*, September 11, 2008.
42. Laura Gordon-Murnane, "The 51st State: The State of Online," *Searcher* Medford Vol. 17, Iss. 5 (May 2009): http://proquest-.umi-.com/pqdweb-?did=1732390471-&sid=2-&Fmt=3-&clientId=15031-&RQT=309-&VName=PQD; Rasiej and Sifry, "With New Media, Obama Camp Takes Stage."
43. Gordon-Murnane, "The 51st State."
44. Kathleen Parker, "What Was the Fuss About?" *Burlington (N.C.) Times News*, September 9, 2009, 4.
45. See *Riding Out the Storm: The Vital Role of Local Radio in Times of Crisis* (New York: Arbitron, 2005).
46. T Linden, "The Second Life Economy–First Quarter 2009 in Detail," *Second Life*, https://blogs.secondlife.com/community/features/blog/2009/04/16/the-second-life-economy—first-quarter-2009-in-detail, April 16, 2009.
47. See Peter Ludlow and Mark Wallace, *The Second Life Herald: The Virtual Tabloid that Witnessed the Dawn of the Metaverse* (Cambridge, Mass.: MIT Press, 2007).
48. Ken Fuson, "iPods Now Double as Study Aids," *USA Today*, March 14, 2006, http://www.usatoday.com/tech/products/2006-03-14-ipod-university_x.htm.
49. Walter Cronkite, Address, Society of Professional Journalists, 2004 SPJ Convention, http://www.ojr.org/ojr/workplace/1096589178.php.

Epilogue

The idea of using the media of the times to shape opinion has been central to the American experience. From the moment that Europeans decided to inhabit the land that would become the United States, they formulated ways to shape thought. Colonization literature of the 1500s promised that America was "the goodliest and most pleasing Territorie of the world."[1] That, of course, depended upon perspective. For those who made a fortune off of sending people to the New World, this was true. For the thousands who died from disease, at sea, or attack from Native Americans, the statement was blatantly false. In the first decade of the twenty-first century, the multiplicity of media has made it possible for people to share anything. Technology has made it possible to portray anything as "the goodliest and most pleasing" via manipulation of images and the copying, pasting, and altering of text.

But dishonesty has not been at the root of the use of media as a means of shaping and defining the nation. It is only one aspect that requires vigilance among the people. What we have seen with media and the defining of the nation is a means of injecting ideas into the public sphere where they often take on a life of their own, and they inform, direct, and reveal so that people can make decisions. It can seem idyllic for the optimist. For the pessimist, media use is manipulation. One thing is certain; throughout American history people have "cooked up paragraphs," as John Adams said, but they have also uncovered and exposed all sorts of malfeasance that has changed the lives of people and changed the direction of the country.

Technology is the vehicle for change, and it has come full circle. In the colonial era, individual voices tended to dominate the conversation. Information disseminated via the printing press was read in homes, taverns, coffeehouses, on the streets, and it was debated. As the country grew, so did the size of media outlets. By the end of the 1800s, the nation had dozens of media outlets with circulations in excess of 100,000 each day, and New York

City had two papers alone that surpassed the 1 million mark in daily readership. With the introduction of broadcast, the individual voice seemed as though it had been consumed. It was never completely drowned out, though, because media operate on so many different levels. Where a few could make no difference on the national level, they could on the local level. And other individuals, working through giant media, did the same for the nation. Muckrakers did it at the beginning of the twentieth century. Edward R. Murrow and others did the same in the 1950s, and these are just two well-known examples. The ability to effect change is the defining feature of media in the shaping of the nation. With technological advancements at the end of the twentieth century, the multiplicity of voices that could operate on a national level again has returned, and media operate on a variety of levels that could not be predicted just a generation ago except by the most imaginative of thinkers.

Media have had the power to change the course of the nation. It has been the active voice of individuals, journalists, advertisers, anyone who has had access to media who have shaped and defined the nation. Would the nation have developed without media to disseminate information and provide a place of debate and discussion? Of course. But what would have been the result?

NOTE

1. Richard Lane, "Letter to M. Richard Hakluyt Esquire, and another Gentleman of the Middle Temple, from Virginia," in Richard Hakluyt, *The Principal Navigations Voyages Traffiques & Discoveries of the English Nation*, 12 vols. (London, 1598; reprint, Glasgow, 1903–1905), 8:299.

Bibliography

Media Sources

American Minerva (New York).
American Watchman and Delaware Republican (Wilmington).
American Weekly Mercury (Philadelphia).
American Whig Review (New York).
Arkansas Gazette (Little Rock).
Atlanta Daily Constitution.
Aurora General Advertiser (Philadelphia).
Balance and Columbian Repository (Hudson, New York), *The.*
Bee (Earlington, Ky.), *The.*
Billboard.
Boston American.
Boston Courier.
Boston Evening Transcript.
Boston Evening-Post.
Boston Gazette, and Country Journal.
Boston Gazette, or Weekly Advertiser.
Boston Gazette.
Boston News-Letter.
Boston Post.
Cable News Network.
CBS News.
Centinel of Freedom (Newark).
Chicago Defender.
Chicago Tribune.
Christian Science Monitor (Boston).
Cleveland Press.
Collier's.
Colored American (New York), *The.*
Columbian Centinel (Boston).

Columbian Herald or the Patriotic Courier of North-America (Charleston).
Connecticut Courant (Hartford).
Connecticut Gazette (New Haven).
Cosmopolitan.
Courier and Enquirer (New York).
Crisis, The.
Cumberland Gazette (Portland, Maine).
Daily Advertiser (Boston).
Daily Advertiser (New York).
Daily Arkansas Gazette (Little Rock).
Daily Evening Bulletin, (San Francisco).
Daily Inter Ocean, (Chicago).
Daily National Intelligencer (Washington, D.C.).
De Bow's Review (New Orleans).
Denver Evening Post.
Drudge Report.
Dunlap's Pennsylvania Packet or the General Advertiser (Philadelphia).
Emancipator (New York City), *The.*
Emporia (Kan.) Gazette.
Enquirer (Richmond).
Everybody's Magazine.
Family Visitor (Richmond).
Federal Republican (Baltimore).
Freedom's Journal (New York).
Freeman's Journal: or, The North-American Intelligencer (Philadelphia), *The.*
Gazette (Pittsburgh, Pa.).
Gazette of the United States (Philadelphia).
General Advertiser (Philadelphia).
Greensboro (N.C.) Daily News.
Greensboro (N.C.) Record.
Harper's Monthly Magazine.
Harper's Weekly (New York).
Independent Chronicle and the Universal Advertiser (Boston).
Independent Gazetteer, or, the Chronicle of Freedom (Philadelphia).
Independent Journal: or, the General Advertiser (New York).
Independent Reflector (New York).
Indiana State Journal (Indianapolis).
Journal of the Telegraph.
Kentucky Irish American (Louisville).
Ladies' Home Journal.
Liberator (Boston).
Los Angeles Herald.
Macon (Ga.) Telegraph.
Maryland Chronicle or the Universal Advertiser (Fredericktown).
Maryland Gazette (Annapolis).
Massachusetts Centinel (Boston).
Massachusetts Gazette (Boston).
Massachusetts Mercury (Boston).

Massachusetts Spy Or, Thomas's Boston Journal.
Massachusetts Spy: Or, American Oracle of Liberty (Worcester, Mass.).
Mercury (Charleston, S.C.).
Milwaukee Daily Sentinel.
Milwaukee Journal.
Montgomery (Ala.) Advertiser.
Morning Oregonian (Portland).
Motion Picture Herald.
Musical Leader.
National Advocate (New York).
National Era (New York), *The.*
National Era (Washington, D.C.).
National Intelligencer (Washington, D.C.).
National Reformer (Rochester, N.Y.).
NBC Evening News.
New-England Courant (Boston).
New-England Weekly Journal (Boston).
New-Hampshire Gazette (Portsmouth).
New-Hampshire Gazette; or State Journal, and General Advertiser (Portsmouth).
New-Hampshire Mercury and the General Advertiser (Portsmouth).
New-Hampshire Patriot (Concord).
New-Haven Gazette, and the Connecticut Magazine.
New-Jersey Journal (Elizabethtown).
New-London Gazette.
Newport Herald.
Newport Mercury.
News and Observer (Raleigh, N.C.).
New York Call.
New-York Evening Post.
New York Freeman.
New-York Gazette Revived in the Weekly Post-Boy.
New-York Gazette.
New-York Gazette: or the Weekly Post-Boy.
New York Herald.
New York Journal.
New-York Journal; or the General Advertiser.
New-York Mercury.
New-York Packet.
New York Spectator.
New-York Times.
New-York Tribune.
New-York Weekly Journal.
New-York Weekly Post-Boy.
New York World.
Niles' Weekly Register (Baltimore).
Norfolk (Va.) Gazette and Publick Ledger.
North American (Philadelphia).
Ogden (Utah) Morning Examiner.

Omaha (Neb.) Daily Bee.
Orlando Sentinel.
Paducah (Ky.) Sun.
Pearson's Magazine.
Pennsylvania Chronicle, and Universal Advertiser (Philadelphia).
Pennsylvania Gazette (Philadelphia).
Pennsylvania Herald, and General Advertiser (Philadelphia).
Pennsylvania Journal and Weekly Advertiser (Philadelphia).
Pennsylvania Packet or the General Advertiser (Philadelphia).
Philadelphia Public Ledger.
Philanthropist (Cincinnati).
Pittsburgh Courier.
Pittsburgh Post-Gazette.
Providence Gazette; and Country Journal.
Raleigh Register, and North-Carolina Weekly Advertiser.
Richmond Dispatch.
Rivington's New-York Gazetteer.
Rocky Mountain News (Denver).
Salt Lake Herald (Salt Lake City).
San Francisco Call.
San Francisco Chronicle.
Saturday Evening Post.
Scientific American.
South-Carolina Gazette (Charleston).
Southern Literary Messenger (Richmond).
St. Louis Enquirer (Missouri Territory).
St. Louis Globe-Democrat.
St. Louis Post Dispatch.
Sun (Baltimore).
Tabor City (N.C.) Tribune.
Time Piece (New York).
Time.
Times-Picayune (New Orleans).
Trenton (N.J.) Federalist.
Union (Washington, D.C.).
United States Magazine and Democratic Review.
USA Today.
Vermont Chronicle (Bellows Falls).
Village Record (West Chester, Pa.).
Virginia Gazette (Williamburg, Purdie and Dixon).
Virginia Gazette (Williamsburg, Pinkney).
Wall Street Journal (New York).
Washington (D.C.) Federalist.
Washington Herald (Washington, D.C.).
Washington Post.
Weekly Raleigh Register, and North Carolina Gazette.
Weekly Rocky Mountain News (Denver).

Wheeling (West Virginia) Intelligencer.
Whig (Boston).
Yank.

Books, Articles, and Web Sites

Adam, Thomas, Ed. *Germany and the Americas: Culture, Politics, and History.* Santa Barbara, Calif.: ABC Clio, 2005.
Adams, Hopkins. *The Great American Fraud.* Chicago: Journal of the American Medical Association, 1905.
Adams, John. *The Works of John Adams, with a Life of the Author, Notes and Illustrations.* Ed. C.F. Adams, 10 vols. Boston, 1850–1856.
Adler, Selig. *The Isolationist Impulse: Its Twentieth Century Reaction.* London: Abelard-Schuman, 1957.
Alien Registration Act of 1940 18 U.S.C. § 2385.
Allen, Erika Tyner. "The Kennedy-Nixon Presidential Debates, 1960." *The Museum of Broadcast Communications.* http://www.museum.tv/archives/etv/K/htmlK/kennedy-nixon/kennedy-nixon.htm.
Anderson, Fred. *Crucible of War: The Seven Years' War and the Fate of Empire in British North America, 1754–1766.* New York: Alfred A. Knopf, 2000.
Anderson, Janna Quitney. *Imagining the Internet: Personalities, Predictions, Perspectives.* Lanham, Md.: Rowman & Littlefield, 2005.
Anderson, Laird B. and Wm. David Sloan, Eds. *Pulitzer Prize Editorials: America's Best Writing, 1917–2003.* 3rd Ed. Ames: Iowa State Press, 2003.
Atomic Energy Act of 1954, 42 U.S.C. § 2011.
Audoin-Rouzeau, Stéphane and Annette Becker. *14–18: Understanding the Great War.* New York: Hill and Wang, 2002.
Axtell, James. *Beyond 1492: Encounters in Colonial America.* New York and Oxford: Oxford University Press, 1992.
Bailey, Edwin R. *Joe McCarthy and the Press.* Madison: University of Wisconsin Press, 1981.
Bailey, Thomas A. "Congressional Opposition to Pure Food Legislation, 1879–1906." *American Journal of Sociology*, 36, 1 (July 1930): 52–64.
Bailyn, Bernard and John B. Hench, Eds. *The Press and the American Revolution.* Boston: Northeastern University Press, 1981.
Baldasty, Gerald J. "The Press and Politics in the Age of Jackson." *Journalism Monographs* 89 (1984).
Balio, Tino. *Grand Design: Hollywood as a Modern Business Enterprise, 1930–1939.* Berkeley: University of California Press, 1995.
Barkin, Steve Michael. *American Television News: The Media Marketplace and Public Interest.* New York: M.E. Sharpe, 2003.
Barnett, Brooke and Amy Reynolds. *Terrorism and the Press: An Uneasy Relationship.* New York: Peter Lang, 2009.
Bartlett, Kenneth G. "Social Impact of the Radio," *Annals of the American Academy of Political and Social Science* 250 (March 1947).
Baughman, James L. *The Republic of Mass Culture: Journalism, Filmmaking, and Broadcasting in America Since 1941.* 3rd Ed. Baltimore: Johns Hopkins University Press, 2006.

Bayley, Edwin R. *Joe McCarthy and the Press.* Madison: University of Wisconsin Press, 1981.
Beasley, Maurine H. and Sheila J. Gibbons. *Taking Their Place: A Documentary History of Women and Journalism* 2nd Ed. State College, Pa.: Strata Press, 2003.
Berlin, Edward A. *King of Ragtime: Scott Joplin and His Era.* New York and Oxford: Oxford University Press, 1995.
Berman, Larry. *Lyndon Johnson's War: The Road to Stalemate in Vietnam.* New York: W.W. Norton, 1989.
Bernays, Edward. *Propaganda.* New York: Horace Liveright, 1928.
Bernays, Edward L. *Biography of an Idea: Memoirs of Public Relations Counsel Edward L. Bernays.* New York: Simon & Schuster, 1965.
Bernstorff, Count. *My Three Years in America.* New York: Charles Scribner's Sons, 1920.
Black Press: Soldiers Without Swords, The. Produced and directed by Stanley Nelson. 86 min. California Newsreel, 1999.
Blanchard, Margaret A. *Revolutionary Sparks: Freedom of Expression in Modern America.* New York and Oxford: Oxford University Press, 1992.
Blum, John M., et al., *The National Experience.* 6th Ed. 2 Vols. San Diego: Harcourt Brace Jovanovich, 1985.
Bly, Nellie. *Nellie Bly's Book. Around the World in Seventy-Two Days.* New York: The Pictorial Weeklies Co., 1890.
———*Ten Days In a Mad-House.* New York: Ian L. Munro, 1887.
Bohn, Thomas W. and Richard L. Stromgren. *Light and Shadows: A History of Motion Pictures.* Mountain View, Calif.: Mayfield Publishing, 1987.
Braestrup, Peter. *Big Story: How the American Press and Television Reported and Interpreted the Crisis of Tet 1968 in Vietnam and Washington.* New Haven, Conn.: Yale University Press, 1977.
Breen, T. H. *The Marketplace of Revolution: How Consumer Politics Shaped American Independence.* Oxford: Oxford University Press, 2004.
Brewin, Mark W. *Celebrating Democracy: The Mass-Mediated Ritual of Election Day.* New York: Peter Lang, 2008.
Brian, Denis. *Pulitzer: A Life.* New York: John Wiley & Sons, 2001.
Bristol, Roger P. *Supplement to Charles Evans' American Bibliography.* Charlottesville: University Press of Virginia, 1970.
Brooks, Tim and Richard Keith Spottswood. *Lost Sounds: Blacks and the Birth of the Recording Industry, 1890–1919.* Urbana: University of Illinois Press, 2004.
Brown, Richard D. *Knowledge Is Power: The Diffusion of Information in Early America, 1700–1865.* New York and London: Oxford University Press, 1989.
Browne, Benjamin P. *Christian Journalism for Today.* Philadelphia: The Judson Press, 1952.
Bruns, Axel. *Blogs, Wikipedia, Second Life, and Beyond: From Production to Produsage.* New York: Peter Lang, 2008.
———. *Gatewatching: Collaborative Online News Production.* New York: Peter Lang, 2005.
Burnham, Walter Dean. *The Current Crisis in American Politics.* New York: Oxford University Press, 1982.
Burt, Elizabeth. *The Progressive Era.* Westport, Conn.: Greenwood Press, 2004.
Calhoun, Craig, Ed. *Habermas and the Public Sphere.* Cambridge, Mass.: MIT Press, 1992.
Callow, Alexander B., Jr. *The Tweed Ring.* New York: Oxford University Press, 1966.

Campbell, W. Joseph. "1897: Journalism's Exceptional Year." *Journalism History* 29, 4 (2004).
———. *The Year That Defined American Journalism: 1897 and the Clash of Paradigms.* New York: Routledge, 2006.
———. *Yellow Journalism: Puncturing the Myths, Defining the Legacies.* Westport, Conn.: Praeger, 2001.
Cannon, Lou. *Official Negligence: How Rodney King and the Riots Changed Los Angeles and the LAPD.* Boulder, Col.: Westview Press, 1999.
Carter, W. Horace. Interview by Jerry Lanier. Tape recording. January 17, 1976. *Documenting the American South.* University of North Carolina at Chapel Hill. http://docsouth.unc.edu/sohp/B-0035/B-0035.html.
Cassedy, James H. *Medicine in America: A Short History.* Baltimore, Md.: Johns Hopkins University Press, 1991.
Chadbourn, James Harmon. *Lynching and the Law.* Chapel Hill: University of North Carolina Press, 1933.
Chandler, Alfred D, Jr. and James W. Cortada, Eds. *A Nation Transformed by Information.* Oxford and New York: Oxford University Press, 2000.
Cherny, Robert. "*The Jungle* and the Progressive Era," *History Now* 16 (June 2008).
Clark, Charles E. *The Public Prints: The Newspaper in Anglo-American Culture, 1665–1740.* New York and Oxford: Oxford University Press, 1994.
Clark, Thomas D. *The Southern Country Editor.* Indianapolis, Ind.: Bobbs-Merrill, 1948; reprint, Columbia: University of South Carolina Press, 1991.
Clinton, William Jefferson. "State of the Union Address. January 24, 1995. http://govinfo.library.unt.edu/npr/library/speeches/22e2.html.
Collins, Ross F. *World War I.* Westport, Conn.: Greenwood Press, 2003.
Commanger, Henry Steele, Ed. *Documents of American History.* 8th ed. New York: Appleton-Century-Crofts, 1968.
Cone, Steve. *Powerlines: Words That Sell Brands, Grip Fans, & Sometimes Change History.* New York: Bloomberg Press, 2008.
Cook, Fred J. *The Unfinished Story of Alger Hiss.* New York: William Morrow, 1958.
Copeland, David A. "'Join, or Die': America's Press during the French and Indian War," *Journalism History* 24:3 (1998).
———. "A Series of Fortunate Events: Why People Believed Richard Adams Locke's 'Moon Hoax,'" *Journalism History* 33 No. 3, (Fall 2007): 140–150.
———. *Colonial American Newspapers: Character and Content.* Newark: University of Delaware Press, 1997.
———. *Debating the Issues in Colonial Newspapers.* Westport, Conn.: Greenwood, 2000.
———. *The Antebellum Era.* Westport, Conn.: Greenwood Press, 2003.
———. *The Idea of a Free Press: The Enlightenment and Its Unruly Legacy.* Evanston, Ill.: Northwestern University Press, 2006.
Copeland, David A. and Anthony Hatcher. *Mass Communication in a Global Age.* 2nd Ed. Northport, Ala.: Vision Press, 2007.
Copeland, David A. and Carol Sue Humphrey. *The War of 1812.* In *Greenwood Library of American War Reporting.* Ed. David A. Copeland. 8 Vols. Westport, Conn.: Greenwood Press, 2005.
Cornell, Saul. *The Other Founders: Anti-Federalism & the Dissenting Tradition in America, 1788–1828.* Chapel Hill: University of North Carolina Press, 1999.

Creel, George. *How We Advertised America*. New York and London: Harper & Brothers, 1920.
Crozier, Michael J., Samuel P. Huntingon, and Joji Watanuki. *The Crisis of Democracy*. New York: New York University Press, 1975.
Cumings, Bruce. *The Origins of the Korean War*. 2 Vols. Princeton, N.J.: Princeton University Press, 1990.
Daniel, Douglas K. *The Korean War. The Greenwood Library of American War Reporting*. Ed. David A. Copeland. 8 Vols. Westport, Conn.: Greenwood Press, 2005.
Daniels, George H. *American Science in the Age of Jackson*. New York: Columbia University Press, 1968.
Davidson, James West Davidson, et al. *Nation of Nations*. New York: McGraw-Hill, 1990.
Davidson, Philip. *Propaganda and the American Revolution, 1763–1783*. Chapel Hill: University of North Carolina Press, 1941.
Dawley, Thomas Robinson Jr. *The Child That Toileth Not: The Story of a Government Investigation*. New York, Gracia, 1912.
DeArmond, Anna Janney. *Andrew Bradford: Colonial Journalist*. Newark: University of Delaware Press, 1949.
Decosta-Willis, Miriam, Ed. *The Memphis Diary of Ida B. Wells: An Intimate Portrait of the Activist as a Young Woman*. Boston: Beacon Press, 1995.
DeKay, James E. *Anniversary Address on the Progress of the Natural Sciences in the United States: Delivered before the Lyceum of Natural History of New York, February, 1826*. New York, 1826.
Denisoff, R. Serge. *Inside MTV*. New Brunswick, N.J.: Transaction Publishers, 1988.
Denton, Robert E. Jr. And Rachel L. Holloway, Eds. *Images, Scandal, and Communication Strategies of the Clinton Presidency*. Westport, Conn.: Greenwood Press, 2003.
Dewey, John. *Democracy and Education*. New York: Henry Holt, 1916.
Diamond, Edwin. *The Media Show: The Changing Face of News, 1985–1990*. Cambridge, Mass.: MIT Press, 1991.
Dickens, A. G. *Reformation and Society in Sixteenth-Century Europe*. London: Thames and Hudson, 1966.
Dickerson, Donna L. *The Reconstruction Era*. Westport, Conn.: Greenwood Press, 2003.
Dill, William A. *Growth of Newspapers in the United States*. Lawrence: University of Kansas, 1928.
Dizard, Wilson, Jr. *Old Media New Media: Mass Communications in the Information Age*. 3rd Ed. New York: Addison Wesley Longman, 2000.
Doherty, Thomas. *Hollywood, American Culture, and World War II*. New York: Columbia University Press, 1993.
Donovan, Robert J. and Ray Scherer. *Unsilent Revolution: Television News and American Public Life*. Cambridge: Cambridge University Press, 1992.
Dooley, Patricia L. *The Early Republic*. Westport, Conn.: Greenwood Press, 2004.
Douglas, Susan J. *Inventing American Broadcasting, 1899–1922*. Baltimore and London: The Johns Hopkins University Press, 1987.
Douglass, Frederick. *Life and Times of Frederick Douglass Written by Himself*. New York: Citadel Press, 1983.
Drehle, David Von. *Triangle: The Fire That Changed America*. New York: Grove/Atlantic, 2003.
Dudziak, Mary L. *September 11 in History: A Watershed Moment?* Durham, N.C.: Duke University Press, 2003.

Dylan, Bob. *The Times They Are A-Changin'*. Columbia Records, CL 2105.
Edwards, Bob. *Edward R. Murrow and the Birth of Broadcast Journalism*. Hoboken, N.J.: John Wiley, 2004.
Elkins, Stanley and Eric McKitrick. *The Age of Federalism*. New York and Oxford: Oxford University Press, 1993.
Ellis, Jack C. *A History of Film*. Englewood Cliffs, N. J.: Prentice-Hall, 1990.
Emery, Michael and Edwin Emery. *The Press and America: An Interpretive History of the Mass Media*. 6th Ed. Englewood Cliffs, N.J.: Prentice Hall, 1988.
Evans, Charles. *American Bibliography*. 14 vols. Chicago, 1904.
Everett, Walter. *The Beatles as Musicians*. Oxford: Oxford University Press, 1999.
Eyes on the Prize: America's Civil Rights Years, 1954–1963. Produced by Henry Hampton. 120 minutes. Blackside, 1986.
Farber, David, Ed. *The Sixties: From Memory to History*. Chapel Hill: University of North Carolina Press, 1994.
Feldstein, Mark. "Watergate Revisited." *American Journalism Review*. August/September 2004. http://www.ajr.org/Article.asp?id=3735.
Fellow, Anthony R. *American Media History*. Belmont, Calif.: Thomson, 2005.
Filler, Louis. *The Muckrakers*. 2nd Ed. University Park: Pennsylvania State University Press, 1976.
Fink, Carole, Philipp Gassert, and Detlef Junker, Eds. *1968, The World Transformed*. Cambridge: Cambridge University Press, 1998.
Finnegan, Lisa. *No Questions Asked: News Coverage Since 9/11*. Westport, Conn.: Praeger, 2007.
Fischer, David Hackett. *Albion's Seed: Four British Folkways in America*. New York and Oxford: Oxford University Press, 1989.
Fisher, David E. and Marshall Jon Fisher. *Tube: The Invention of Television*. San Diego: A Harvest Book, 1997.
Ford, Richard Thompson. *The Race Card: How Bluffing About Bias Makes Race Relations Worse*. New York: Farrar, Straus and Giroux, 2008.
Fousek, John. *To Lead the Free World: American Nationalism and the Cultural Roots of the Cold War*. Chapel Hill: University of North Carolina Press, 2000.
Frank, Thomas. *The Conquest of Cool: Business Culture, Counterculture, and the Rise of Hip Consumerism*. Chicago: University of Chicago Press, 1997.
Franklin, Benjamin. *Writings*. New York: Library of America, 1987.
Freedman, Jo. *We Will Be Heard: Women's Struggles for Political Power in the United States*. Lanham, Md.: Rowman & Littlefield, 2008.
Friendly, Fred W. *Due to Circumstances Beyond Our Control*. New York: Random House, 1967.
Frith, Simon and Andrew Goodwin. *On Record: Rock, Pop, & the Written Word*. New York: Routledge, 1990.
Frontani, Michael R. *The Beatles: Image and the Media*. Jackson: University Press of Mississippi, 2007.
Furniss, Tim. *A History of Space Exploration and Its Future*. Guilford, Conn.: Lyons Press, 2003.
Ganley, Gladys D. "Power to the People via Personal Electronic Media," *Washington Quarterly* 10, 3 (1992).
Gardner, Alexander. *Gardner's Photographic Sketch Book of the War*. 2 Vols. Washington, D.C.: Philip & Solomons, 1865–1866.

Garner, Joe. *Stay Tuned: Television's Unforgettable Moments*. Kansas City, Mo.: Andrews McMeel, 2002.
Gatewood, Willard B., Jr., Ed., *Free Man of Color: The Autobiography of Willis Augustus Hodges*. Knoxville: University of Tennessee Press, 1982.
Gibbons, Edward. *Floyd Gibbons–Your Headline Hunter*. New York: Exposition Press, 1953.
Giddings, Paula J. *Ida: A Sword Among Lions*. New York: HarperCollins, 2008.
Gilmore, William J. *Reading Becomes a Necessity of Life: Material and Cultural Life in Rural New Britain, 1780–1835*. Knoxville: University of Tennessee Press, 1989.
Gitelman, H. M. *Legacy of the Ludlow Massacre: A Chapter in American Industrial Relations*. Philadelphia: University of Pennsylvania Press, 1988.
Gitlin, Todd. *The Whole World Is Watching: Mass Media in the Making & Unmaking of the New Left*. Berkeley: University of California Press, 1980.
Glaeser, Edward L. and Claudia Goldin, Eds. *Corruption and Reform: Lessons from America's Economic History*. Chicago: University of Chicago Press, 2006.
Golden Age of Radio 1935–50. March 2007. http://history.sandiego.edu/gen/recording/radio2.html.
Grady, Henry W. *The New South*. New York: Robert Bonner's Sons, 1890.
Gravlee, G. Jack and James R. Irvine, Eds. *Pamphlets and the American Revolution: Rhetoric, Politics, Literature, and the Popular Press*. Delmar, N.Y.: Scholars' Facsimilies & Reprints, 1976.
Greenberg, David. *Nixon's Shadow: The History of an Image*. New York: W. W. Norton, 2003.
Greene, Jack. *Pursuits of Happiness: The Social Development of Early Modern British Colonies and the Formation of American Culture*. Chapel Hill: University of North Carolina Press, 1988.
Grieveson, Lee and Peter Krämer, Eds. *The Silent Cinema Reader*. New York: Routledge, 2004.
Griffith, Robert. *The Politics of Fear: Joseph R. McCarthy and the Senate*. Lexington: University of Kentucky Press, 1970.
Griffith, Sally Foreman. *Home Town News: William Allen White and the* Emporia Gazette. New York and Oxford: Oxford University Press, 1989.
Haas, Tanni. *The Pursuit of Public Journalism: Theory, Practice, and Criticism*. New York: Routledge, 2007.
Habermas, Jürgen *The Structural Transformation of the Public Sphere: an Inquiry into a Category of Bourgeois Society*. Trans. Thomas Burger. Cambridge, Mass.: Harvard University Press, 1989.
Hakluyt, Richard. *Divers Voyages Touching the Discovery of America and the Islands Adjacent*. 1598; reprint, London: The Hakluyt Society, 1850.
———. *The Principal Navigations Voyages Traffiques & Discoveries of the English Nation*. 12 vols. London, 1598; reprint, Glasgow, 1903–1905.
Hall, Dennis R. and Susan Grove Hall. Eds. *American Icons: An Encyclopedia of the People, Places, and Things That Have Shaped Our Culture*. Westport, Conn.: Greenwood Press, 2006.
Hallin, Daniel C. *The "Uncensored War": The Media and Vietnam*. New York and Oxford: Oxford University Press, 1986.

Hamm, Bradley and Donald L. Shaw. *World War II, The Asian Theater. Greenwood Library of American War Reporting.* Ed. David A. Copeland. 8 Vols. Westport, Conn.: Greenwood Press, 2005.

Hammond, William M. *Media & Military at War.* Lawrence: University Press of Kansas, 1998.

Harris, Roy J. Jr. *Pulitzer's Gold: Behind the Prize for Public-Service Journalism.* Columbia: University of Missouri Press, 2007.

Hayden, Joseph. *Covering Clinton: The President and the Press in the 1990s.* Westport, Conn.: Greenwood Press, 2002.

Hays, Will H. *See and Hear.* New York: Motion Picture Producers and Distributors of America, 1929.

Herman, Edward S. and Noam Chomsky. *Manufacturing Consent: The Political Economy of the Mass Media.* New York: Pantheon Books, 2002.

Hess, Stephen and Marvin L. Kalb, Eds. *The Media and the War on Terrorism.* Washington, D.C.: The Brookings Institution, 2003.

Hickey, Donald R. *The War of 1812: The Forgotten War.* Urbana: University of Illinois Press, 1989.

Hilliard, Robert L. and Michael C. Keith. *Waves of Rancor: Tuning in the Radical Right.* Armonk, N.Y.: M. E. Sharpe, 1999.

Historical Statistics of the United States, Colonial Times to 1970, Part 1. Washington, D.C., U.S. Bureau of Census, 1975.

Hodges, Graham Russell. *Root and Branch: African Americans in New York and East Jersey, 1613–1863.* Chapel Hill: University of North Carolina Press, 1999.

Hohenberg, John. *Foreign Correspondence: The Great Reporters and Their Times.* New York: Columbia University Press, 1964.

Holmes, Oliver Wendell. *The Works of Oliver Wendell Holmes.* 13 Vols. Boston and New York: Houghton, Mifflin & Co., 1892.

Horten, Gerd. *Radio Goes to War: The Cultural Politics of Propaganda During World War II.* Berkeley: University of California Press, 2003.

Hoskins, Andrew. *Televising War: From Vietnam to Iraq.* London and New York: Continuum, 2004.

Hudson, Linda S. *Mistress of Manifest Destiny: A Biography of Jane McManus Storm Cazneau, 1807–1878.* Austin: Texas State Historical Association, 2001.

Humphrey, Carol Sue *The Revolutionary Era.* Westport, Conn.: Greenwood Press, 2003.

———. *The Press of the Young Republic, 1783–1833.* Westport, Conn.: Greenwood Press, 1996.

Huntzicker, William E. *The Popular Press, 1833–1865* (Westport, Conn.: Greenwood Press, 1999.

Hutton, Frankie. *The Early Black Press in America, 1827–1860.* Westport, Conn.: Greenwood Press, 1993.

Jackson, Andrew. *The Correspondence of Andrew Jackson.* Ed. John Spencer Bassett. 7 vols. Washington, D. C., 1926–35.

Jamieson, Kathleen Hall and David S. Birdsell. *Presidential Debates: The Challenge of Creating an Informed Electorate.* New York and Oxford: Oxford University Press, 1990.

Jefferson Airplane. *Volunteers.* RCA Records. LSP-4238.

Jefferson, Thomas. *The Writings of Thomas Jefferson.* Ed. Andrew A. Lipscomb and Albert Ellery Bergh. 20 vols. Washington, D.C.: Thomas Jefferson Memorial Association, 1905.

Johnson, Lyndon Baines. *The Vantage Point: Perspectives on the Presidency, 1963–1969.* New York: Holt, Rinehart and Winston, 1971.

Johnson, Paul. *Modern Times: The World from the Twenties to the Eighties.* New York: Harper & Row, 1983.

Keith, Michael C., Ed. *Radio Cultures: The Sound Medium in American Life.* New York: Peter Lang, 2008.

Kellner, Douglas. *Grand Theft 2000: Media Spectacle and a Stolen Election.* Lanham, Md.: Rowman & Littlefield, 2001.

———. *Media Spectacle.* New York: Routledge, 2003.

Kessler, Lauren. *The Dissident Press: Alternative Journalism in American History.* Beverly Hills, Calif.: Sage, 1984.

Kielbowicz, Richard B. *News in the Mail: The Press, Post Office, and Public Information, 1700–1860s.* Westport, Conn.: Greenwood Press, 1989.

Klein, Milton M., Ed. *The Independent Reflector.* Cambridge, Mass.: Harvard University Press, 1963.

Kline, Ronald R. *Consumers in the Country: Technology and Social Change in Rural America.* Baltimore: Johnson Hopkins University Press, 2000.

Knapp, Caroline. *Appetites: Why Women Want.* New York: Counterpoint, 2003.

Knudson, Jerry W. *In the News: American Journalists View Their Craft.* Wilmington, Del.: Scholarly Resources, 2000.

Kobre, Sidney. *Development of American Journalism.* Dubuque, Iowa: Wm. C. Brown, 1969.

Kolker, Robert Phillip. *Media Studies: An Introduction.* Malden, Mass.: John Wiley, 2009.

Koppes, Clayton R. and Gregory D. Black. *Hollywood Goes to War: How Politics, Profits, and Propaganda Shaped World War II Movies.* New York: Free Press, 1987.

Kroger, Brooke. *Nellie Bly: Daredevil, Reporter, Feminist.* New York: Crown, 1994.

Kuhn, Raymond and Eric Neveu, Eds. *Political Journalism: New Challenges, New Practices.* London: Routledge, 2002.

Kunczik, Michael. *Images of Nations and International Public Relations.* Mahwah, N.J.: Lawrence Erlbaum Associates, 1997.

Kurtis, Bill. "Forty Years Later: How the Tumultuous 1968 Democratic National Convention in Chicago Impacted American Institutions and Individuals." Panel presentation, Association for Education in Journalism and Mass Communication. Chicago, August 7, 2008.

Kutler, Stanley I., *The Wars of Watergate.* New York: W. W. Norton 1990.

Lambert, Frank. "'Pedlar of Divinity': George Whitefield and the Great Awakening," *Journal of American History* 77 (1990): 812–37.

———. *The Founding Fathers and the Place of Religion in America.* Princeton and Oxford: Princeton University Press, 2003.

Lange, Bernd-Peter and David Ward, Eds. *The Media and Elections: A Handbook and Comparative Study.* Mahwah, N.J.: Lawrence Erlbaum, 2004.

Lanham, Edward Connery, Comp. *Chronological Tables of American Newspapers, 1690–1820.* Barre, Mass.: American Antiquarian Society & Barre Publishers, 1972.

Lasswell, Harold D. *Propaganda Technique in the World War.* New York: Alfred A. Knopf, 1927.

Lauterer, Jock. *Community Journalism: Relentlessly Local.* 3rd ed. Chapel Hill: University of North Carolina Press, 2005.
Lazer, William. *Handbook of Demographics for Marketing & Advertising.* New York: Lexington Books, 1994.
Lee, Alfred McClung. *The Daily Newspaper in America.* New York: Macmillan, 1937.
Leonard, Thomas C. *News for All: America's Coming-of-Age with the Press.* New York and Oxford: Oxford University Press, 1995.
Levy, Leonard W. *Emergence of a Free Press.* New York and Oxford: Oxford University Press, 1985.
Liebovich, Louis W. *Richard Nixon, Watergate, and the Press.* Westport, Conn.: Praeger, 2003.
Linder, Douglas O. "The LAPD Officers' Trials: A Chronology." *Famous American Trials.* http://www.law.umkc.edu/faculty/projects/ftrials/lapd/kingchronology.html.
Lloyd, Chester S. *The Young Man and Journalism.* New York: Macmillan, 1922.
Long, Priscilla. *Where the Sun Never Shines: A History of America's Bloody Coal Industry.* St. Paul, Minn.: Paragon House, 1989.
Lorenz, Alfred Lawrence. *Hugh Gaine: A Colonial Printer-Editor's Odyssey to Loyalism.* Carbondale: Southern Illinois University Press, 1972.
Lowe, Janet. *Oprah Winfrey Speaks: Insights from the World's Most Influential Voice.* New York: Wiley, 1998.
Ludlow, Peter and Mark Wallace. *The Second Life Herald: The Virtual Tabloid that Witnessed the Dawn of the Metaverse.* Cambridge, Mass.: MIT Press, 2007.
Lynn, Kenneth S. *Charlie Chaplin and His Times.* New York: Simon and Schuster, 1997.
Lynn, Naomi B., Ed. *Women, Politics, and the Constitution.* New York: Haworth Press, 1990.
Madsen, Axel. *60 Minutes: The Power and the Politics of America's Most Popular TV News Show.* New York: Dodd, Mead, 1984.
Makemson, Harlen. *Media, NASA, and America's Quest for the Moon.* New York: Peter Lang, 2009.
Mandelbaum, Seymour J. *Boss Tweed's New York.* New York: John Wiley, 1965.
Marable, Manning. *Race, Reform, and Rebellion: The Second Reconstruction in Black America, 1945–1990.* 2nd Ed. Jackson: University Press of Mississippi, 1991.
Martelle, Scott. *Blood Passion: The Ludlow Massacre and Class War in the American West.* New Brunswick: Rutgers University Press, 2007.
Martin, Robert W. T. *The Free and Open Press: The Founding of American Democratic Press Liberty, 1640–1800.* New York: New York University Press, 2001.
Martineau, Harriet. *The Martyr Age of the United States.* Boston: Weeks, Jordan & Co., 1839.
Marvin, Carolyn. *When Old Technologies Were New: Thinking About Electric Communication in the Late Nineteenth Century.* New York and Oxford: Oxford University Press, 1988.
McClellan, Scott. *What Happened: Inside the Bush White House and Washington's Culture of Deception.* New York: Public Affairs, 2008.
McClure, S. S. *My Autobiography.* New York: Frederick A. Stokes, 1914.
McCombs, Maxwell and Donald Shaw "The Agenda-Setting Function of Mass Media." *Public Opinion Quarterly* 36 (1972): 176–187.
McLuhan, Marshall. *Understanding Media.* London: Routledge and Kegan Paul, 1964.

McMurry, Linda O. *To Keep the Waters Troubled: The Life of Ida B. Wells.* New York and Oxford: Oxford University Press, 1998.

McQuire, Barry. *Eve of Destruction.* Dunhill Records. DS-50003.

Merrens, H. Roy. "The Physical Environment of Early America: Images and Image Makers in Colonial South Carolina." *Geographical Review* 59, 4 (October, 1969).

Messer, Peter C. *Stories of Independence: Identity, Ideology, and History in Eighteenth-Century America.* Dekalb: Northern Illinois University Press, 2005.

Millard, Andre. *America on Record: A History of Recorded Sound.* 2nd Ed. Cambridge: Cambridge University Press, 2005.

Mills, Harry. *Artful Persuasion: How to Command Attention, Change Minds, and Influence People.* New York: AMACOM, 2000.

Milton, John. *The Works of John Milton.* 20 vols. New York: Columbia University Press, 1931.

Mindich, David T. Z. *Just the Facts: How "Objectivity" Came to Define American Journalism.* New York: New York University Press, 1998.

Miner, Ward L. *William Goddard, Newspaperman.* Durham, N.C.: Duke University Press, 1962.

Morrison, Champlain W. *Democratic Politics and Sectionalism: The Wilmot Proviso Controversy.* Chapel Hill: University of North Carolina Press, 1967.

Morse, Sherman. "An Awakening in Wall Street: How the Trusts, After Years of Silence, Now Speak Through Authorized and Acknowledged Press Agents," *The American Magazine* 62 (September 1906): 457–63.

Morton, David. *Sound Recording: The Life Story of a Technology.* Baltimore: Johns Hopkins University Press, 2006.

Mott, Frank Luther. *A History of American Magazines, 1741–1850.* Cambridge: Harvard University Press, 1966.

———. *American Journalism, A History: 1690–1960.* 3rd ed. New York: MacMillan, 1962.

Murrow, Edward R. *This Is London.* New York: Simon and Shuster, 1941.

Musser, Charles. *Before the Nickelodeon: Edwin S. Porter and the Edison Manufacturing Company.* Berkeley: University of California Press, 1991.

Muybridge, Eadweard. *Muybridge's Complete Human and Animal Locomotion. All 781 Plates from the 1887 "Animal Locomotion."* 3 Vols. Mineola, N.Y.: Dover Publications, Inc. 1979.

Negrine, Ralph. *The Communication of Politics.* London: Sage, 1996.

Nevins, Allan. *American Press Opinion, Washington to Coolidge: A Documentary Record of Editorial Leadership and Criticism, 1785–1927.* Boston: D.C. Heath, 1928.

Newhall, Beaumont. *The History of Photography.* New York: Museum of Modern Art, 1982.

Nixon, Richard. *In the Arena.* New York: Simon and Schuster, 1990.

Nord, David Paul. *Faith in Reading: Religious Publications and the Birth of Mass Media in America.* Oxford: Oxford University Press, 2004.

Oderman, Stuart. *Roscoe "Fatty" Arbuckle: A Biography Of The Silent Film Comedian, 1887–1933.* Jefferson, N.C.: McFarland, 2005.

Pasley, Jeffrey L. *"The Tyranny of Printers": Newspaper Politics in the Early American Republic.* Charlottesville: University Press of Virginia, 2001.

Patee, Fred Lewis. *The First Century of American Literature:1770–1870.* New York: Cooper Square, 1966.

Patterson, James T. *Grand Expectations: The United States, 1945–1974.* New York: Oxford University Press, 1996.
Penn, Garland. *The Afro-American Press and Its Editors.* Springfield, Mass.: Wiley and Co., 1891.
Peters, John Durham and Peter Simonson *Mass Communication and American Social Thought.* Lanham, Md.: Rowman & Littlefield, 2004.
Phillips, Peter, Ed. *Censored, 1998: The News that Didn't Make the News.* New York: Seven Stories Press, 1998.
Potter, David M. *People of Plenty.* Chicago: University of Chicago Press, 1954.
———. *The Impending Crisis, 1848–1861.* New York: Harper & Row, 1976.
Procter, Ben. *William Randolph Hearst: The Early Years, 1863–1910.* New York: Oxford University Press, 1998.
Pyle, Ernie. *Here Is Your War.* New York: Henry Holt, 1943.
Radio Families–USA–1946. New York: Broadcast Measurement Bureau, 1946.
Ramsey, David. *The History of the American Revolution.* 2 Vols. Philadelphia, 1789.
Ratner, Lorman and Dwight L. Teeter, Jr. *Fanatics and Fire-eaters: Newspapers and the Coming of the Civil War.* Urbana: University of Illinois Press, 2003.
Reporting Civil Rights: American Journalism 1941–1963. 2 Vols. New York: Library of America.
Reynolds, Amy and Brooke Barnett. "'America under Attack': CNN's Verbal and Visual Framing of September 11." In *Media Representations of September 11.* Eds. Steven Chermak, Frankie Y. Bailey, and Michelle Brown. Westport, Conn.: Praeger, 2003.
Rhodes, James Ford. *Historical Essays.* New York: Macmillan, 1909.
Richards, Barry, Iain Mac Rury, and Jackie Botterill. *Dynamics of Advertising.* New York: Routledge, 2000.
Richardson, John E. *Marketing.* 21st Ed. New York: McGraw-Hill, 1999.
Riding Out the Storm: The Vital Role of Local Radio in Times of Crisis. New York: Arbitron, 2005.
Riegel, Robert E. "Standard Time in the United States," *American Historical Review* 33, 1 (October 1927): 84–89.
Risley, Ford. *Abolition and the Press: The Moral Struggle Against Slavery.* Evanston, Ill.: Northwestern University Press, 2008.
Robbins, Hollis. "*Uncle Tom's Cabin* and the Matter of Influence," *History Now* 16 (June 2008).
Roberts, Gene and Hank Klibanoff. *The Race Beat: The Press, The Civil Rights Struggle, and the Awakening of a Nation.* New York: Vintage Books, 2006.
Roeder, George H. Jr. *The Censored War: American Visual Experience During World War II.* New Haven, Conn.: Yale University Press, 1993.
Roper, Burns W. *Trends in Public Attitudes Toward Television and Other Mass Media, 1959–1974.* New York: Television Information Office, 1975.
Rosen, Jay. *Getting the Connections Right: Public Journalism and the Troubles in the Press.* New York: Twentieth Century Fund Press, 1996.
Rosenfeld, Richard N. *American Aurora: A Democratic-Republican Returns.* New York: St. Martin's, 1997.
Ross, Stewart Halsey. *Propaganda for War: How the United States Was Conditioned to Fight the Great War of 1914–1918.* Jefferson, N.C.: McFarland, 1996.
Rozell, Mark J. and Jeremy D. Mayer, Eds. *Media Power, Media Politics.* Lanham, Md.: Rowman & Littlefield, 2008.

Rubin, Jerry. *Do It! Scenarios of the Revolution.* New York: Ballantine, 1970.
Sabato, Larry L. *Feeding Frenzy: How Attack Journalism Has Transformed American Politics.* New York: Free Press, 1991.
Schandler, Herbert Y. *The Unmaking of a President.* Princeton, N.J.: Princeton University Press, 1977.
Schecter, Patricia A. *Ida B. Wells-Barnett and American Reform, 1880–1930.* Chapel Hill: University of North Carolina Press, 2001.
Schenck v. United States, 249 U.S. 47 (1919).
Schlesinger, Arthur M. *Prelude to Independence: The Newspaper War on Britain 1764–1776.* New York: Random House, 1958.
Schneirov, Matthew. *The Dream of a New Social Order: Popular Magazines in America, 1893–1914.* New York: Columbia University Press, 1994.
Schudson, Michael. *The Good Citizen: A History of American Civic Life.* New York and London: Oxford University Press, 1998.
———. *Watergate in American Memory: How We Remember, Forget, and Reconstruct the Past.* New York: Basic, 1992.
Seib, Philip. *Broadcasts from the Blitz: How Edward R. Murrow Helped Lead America Into War.* Washington, D.C.: Potomac Books, 2006.
Serrin, Judith and William Serrin, Eds. *The Muckrakers: The Journalism That Changed America.* New York: The New Press, 2002.
Sharpe, Joanne P. *Condensing the Cold War: Reader's Digest and American Identity.* Minneapolis and London: University of Minnesota Press, 2000.
Shaw, Daron R. *Race to 270: The Electoral College and the Campaign Strategies of 2000.* Chicago: University of Chicago Press, 2006.
Shaw, Donald L., Bradley J. Hamm and Diana L. Knott, "Social Melding: Mass Media Studies and the Four Ages of Place, Class, Mass and Space," *Journalism Studies* 1, 1 (2000): 57–79.
Shields, Vickie Rutledge and Dawn Heinecken. *Measuring Up: How Advertising Affects Self-Image.* Philadelphia: University of Pennsylvania Press, 2002.
Shoemaker, Pam and Stephen D. Reese. *Mediating the Message: Theories of Influence on Mass Media Content.* New York: Longman, 1996.
Siegel, Robert. "Anti-Lynching Law in U.S. History, *All Things Considered,* NPR (June 13, 2005), http://www.npr.org/templates/story/story.php?storyId=4701576.
Sinclair, Upton. *The Jungle.* New York: Doubleday, 1905; reprint by the author, 1920.
Sloan, Wm. David and Julie Hedgepeth Williams. *The Early American Press, 1690–1783.* Westport, Conn.: Greenwood, 1994.
Sloan, Wm. David Sloan, Cheryl S. Wray, and C. Joanne Sloan. *Great Editorials.* 2nd ed. North Port, Ala.: Vision Press, 1997.
Smillie, George W. *Public Health, Its Promise for the Future.* New York, Macmillan Co., 1955.
Smith, Culver H. *The Press, Politics, and Patronage: The American Government's Use of Newspapers 1789–1875.* Athens: University of Georgia Press, 1977.
Smith, James Morton. *Freedom's Fetters: The Alien and Sedition Laws and American Civil Liberties.* Ithaca, N.Y., 1956.
Smith, John. *Travels and Works of Captain John Smith.* Ed. Edward Arber. Edinburgh: John Grant, 1910.
Smith, John. *Writings with Other Narratives of Roanoke, Jamestown, and the First English Settlement of America.* New York: Library of America, 2007.

Smith, Starr. *Jimmy Stewart: Bomber Pilot*. St. Paul, Minn.: Zenith Press, 2005.
Smythe, Ted Curtis. *The Gilded Age Press, 1865–1900*. Westport, Conn.: Greenwood Press, 2003.
Snow, Nancy. *Information War: American Propaganda, Free Speech and Opinion Control Since 9-11*. Toronto, Canada: Seven Stories Press, 2003.
Sperber, A. M. *Murrow: His Life and Times*. New York: Freundlich Books, 1986.
Spitz, Bob. *Dylan: A Biography*. New York: W. W. Norton, 1991.
———. *The Beatles: The Biography*. New York and Boston: Little, Brown and Company, 2005.
Sproule, J. Michael. *Propaganda and Democracy: The American Experience of Media and Mass Persuasion*, Cambridge: Cambridge University Press, 1997.
Stanton, Elizabeth Cady. *A History of Woman Suffrage*. 2 Vols. Rochester, N.Y.: Fowler and Wells, 1889.
Steeples, Douglas and David Whitten. *Democracy in Desperation: The Depression of 1893*. Westport, Conn.: Greenwood Press, 1998.
Stein, Leon. *The Triangle Fire*. New York: J. B. Lippincott, 1962.
Sterling, Christopher H. and John M. Kittross. *Stay Tuned: A History of American Broadcasting*. 3rd. Ed. Mahwah, N.J.: Lawrence Erlbaum, 2002.
Sterling, Christopher. *Electronic Media: A Guide to Trends in Broadcasting and Newer Technologies 1920–1983*. New York: Praeger, 1984.
Stewart, Donald H. *The Opposition Press of the Federalist Period*. Albany: State University of New York Press, 1969.
Stoddard, Henry Luther. *Horace Greeley: Printer, Editor, Crusader*. New York: G.P. Putnam's Sons, 1946.
Stokes, Melvyn. *D. W. Griffith's The Birth of a Nation: A History of the Most Controversial Motion Picture of All Time*. Oxford: Oxford University Press, 2007.
Stout, Harry S. *The New England Soul: Preaching and Religious Culture in Colonial New England*. New York: Oxford University Press, 1986.
Stowe, Charles Edward. *Harriet Beecher Stowe: The Story of Her Life*. Boston: Houghton Mifflin.
Stowe, Harriet Beecher. *A Key to Uncle Tom's Cabin; Presenting the Original Facts and Documents upon Which the Story Is Founded. Together with Corroborative Statements Verifying the Truth of the Work*. Boston: John P. Jewett, 1853.
Strauven, Wanda, Ed. *The Cinema of Attractions Reloaded*. Amsterdam, Netherlands: University of Amsterdam Press, 2006.
Sugden, Peter and Alan Tomlinson, Eds. *Power Games: A Critical Sociology of Sport*. New York: Routledge, 2002.
Summers, Mark Wahlgren. *Rum, Romanism, & Rebellion: Making of a President*. Chapel Hill: University of North Carolina Press, 2000.
Swanberg, W. A. *Pulitzer*. New York: Charles Scribner's Sons, 1967.
Sweeney, Michael S. *From the Front: The Story of War Featuring Correspondents' Chronicles*. Washington, D.C.: National Geographic Press, 2002.
———. *The Military and the Press: An Uneasy Truce*. Evanston, Ill.: Northwestern University Press, 2006.
Tarbell, Ida M. *The History of the Standard Oil Company*. 2 Vols. Ed. David M. Chalmers. New York: McClure, Phillips & Co., 1904; reprint, Portland, Ore.: Dover, 2003.
Tebbel, John. *The Media in America*. New York: Thomas Y. Crowell, 1974.

Teel, Leonard Ray. *The Public Press, 1900–1945.* Westport, Conn.: Greenwood Press, 2006.
Television History–The First 75 Years. http://www.tvhistory.tv/tv-prices.htm.
The State of the New Media: An Annual Report on American Journalism. Project for Excellence in Journalism. http://www.stateofthemedia.org/2005/execsum.pdf, 2, 6.
Thomas, Isaiah. *The History of Printing in America.* 1810; reprint edition, New York: Weathervane Books, 1970.
Thompson, Luther. *Wiring a Continent: The History of the Telegraph Industry in the United States, 1832–1866.* Princeton, N.J.: Princeton University Press, 1947.
Thompson, Susan. *The Penny Press: The Origins of the Modern News Media, 1833–1861.* Northport, Ala.: Vision Press, 2004.
Tindall, George Brown and David Emory Shi. *America: A Narrative History.* 5th Ed. 2 Vols. New York: W. W. Norton, 1999.
Tobin, James. *Ernie Pyle's War: America's Eyewitness to World War II.* New York: Free Press, 1997.
Tocqueville, Alexis de. *Democracy in America.* 2 vols. New York: Alfred A. Knopf, 1946.
Trask, David F. *The War with Spain in 1898.* New York: The Free Press, 1981; reprint, Lincoln: University of Nebraska Press, 1996.
Tripp, Bernell. *Origins of the Black Press, New York, 1827–1847.* Northport, Ala.: Vision Press, 1992.
Turner, Kathleen J. *Lyndon Johnson's Dual War: Vietnam and the Press.* Chicago and London: University of Chicago Press, 1985.
U.S. Bureau of Census, *Statistical Abstract of the United States, 1970.* 91st Ed. Washington, D.C.: U.S. Department of Commerce, 1970.
U.S. Census Bureau, *Statistical Abstract of the United States 2003.* Washington, D.C.
Ullery, Jacob G., Ed. *History of Vermonters and Sons of Vermont.* Holyoke, Mass.: Transcript Publishing, 1894.
Upchurch, Adams. *Race Relations in the United States, 1960–1980.* Westport, Conn.: Greenwood Press, 2008.
Vietnam: A Television History. Produced by WGBH. 660 minutes. Public Broadcasting Service, 1983.
Von Schilling, James. *The Magic Window: American Television, 1939–1953.* Binghamton, N.Y.: Haworth Press, 2002.
Voss, Frederick S. *Reporting the War: The Journalistic Coverage of World War II.* Washington, D.C.: Smithsonian Institution, 1994.
Wade, Wyn Craig. *The Fiery Cross: The Ku Klux Klan in America.* New York and Oxford: Oxford University Press, 1998.
Waisbord, Silvio. *Watchdog Journalism in South America: News, Accountability, and Democracy.* New York: Columbia University Press, 2000.
Wakin, Edward. *How TV Changed America's Mind.* Lincoln, Neb.: iUniverse, 1996.
Wallison, Peter J. *Ronald Reagan: The Power of Conviction and the Success of His Presidency.* Boulder, Col.: Westview Press, 2004.
Walstreicher, David. *In the Midst of Perpetual Fetes: The Making of American Nationalism, 1776–1820.* Chapel Hill: University of North Carolina Press, 1997.
Walter Cronkite: Eyewitness to History. Produced by Paul F. Ryan and CBS News Productions. 100 minutes. A&E Network, 1998.
Warner, Michael *The Letters of the Republic: Publication and the Public Sphere in Eighteenth-Century America.* Cambridge, Mass.: Harvard University Press, 1990.

Washburn, Patrick S. *A Question of Sedition: The Federal Government's Investigation of the Black Press During World War II*. New York: Oxford University Press, 1986.

———. *World War II, The European Theater. Greenwood Library of American War Reporting*. Ed. David A. Copeland. 8 Vols. Westport, Conn.: Greenwood Press, 2005.

Weinberg, Arthur and Lila Weinberg, Eds. *The Muckrakers*. New York: Simon and Schuster, 1961; reprint, Urbana: University of Illinois Press, 2001.

Weinstein, Allen and Alexander Vassiliev. *The Haunted Wood: Soviet Espionage in America—The Stalin Era*. New York: Modern Library, 1999.

Weinstein, Israel. "Eighty Years of Public Health in New York City," *Bulletin Of The New York Academy Of Medicine*, 23, No. 2 (1947): 221–37.

Wells, Alan and Ernest A. Hakanen. *Mass Media & Society*. Greenwich, Conn.: Ablex Publishing, 1997.

Wells, Ida B. *Crusade for Justice: The Autobiography of Ida B. Wells*. Ed. Alfreda M. Duster. Chicago and London: University of Chicago Press, 1970.

———. *Southern Horrors. Lynch Law in All Its Phases*. New York: New York Age Print, 1892.

Wells-Barnett, Ida. *The Red Record: Tabulated Statistics and Alleged Causes of Lynching in the United States*. 1895).

White, Trumbull, Ed. *Our Wonderful Progress: The World's Triumphant Knowledge and Works*. 1902.

White, William Allen. *The Autobiography of William Allen White*. New York: Macmillan, 1946.

Williams, Julie Hedgepeth. *The Significance of the Printed Word in Early America: Colonists' Thoughts on the Role of the Press*. Westport, Conn.: Greenwood, 1999.

Willis, William James. *The Media Effect: How the News Influences Politics and Government*. Westport, Conn.: Praeger, 2007.

Winfield, Betty Houchin. *FDR and the News Media*. Urbana: University of Illinois Press, 1990.

Wittebols, James H. *Watching M*A*S*H, Watching America: A Social History of the 1972–1983 Television Series*. Jefferson, N.C.: McFarland, 2003.

Wood, Gordon S. *The American Revolution: A History*. New York: Modern Library, 2002.

Young, William H. and Nancy K. Young. *The 1950s*. Westport, Conn.: Greenwood Press, 2004.

Zakon, Robert H. *Hobbes' Internet Timeline*. http://www.zakon.org/robert/internet/timeline/.

Zelizer, Barbie. *Covering the Body: The Kennedy Assassination, the Media, and the Shaping of Collective Memory*. Chicago: University of Chicago Press, 1992.

Zeller, Bob. *The Blue and Gray in Black and White: A History of Civil War Photography*. Westport, Conn.: Praeger, 2005.

Zinn, Howard. *SNCC: The New Abolitionists*. Boston: Beacon Press, 1964; reprint, Cambridge, Mass.: South End Press, 2002.

Zuckoff, Mitchell. *Ponzi's Scheme: The Story of a Financial Legend*. New York: Random House, 2005.

Index

A

Abbott, Robert S., 149–50, 207
ABC World News Tonight, 6
abolitionist press, 78–82
Abu Ghraib, 6
Account of the Province of Carolina, An, 13
Adams, Eddie, 92, 233–34
Adams, John, 1, 11, 22, 53
Adams, John Quincy, 70
Adams, Samuel Hopkins, 135
Adams, Samuel, 1–2, 11, 21, 258
Addison, Alexander, 7
advertising. nineteenth century, 122–23
    post World War II, 237–38
Agence France-Presse, 284
Akron (Ohio) Beacon Journal, 267–68
Albany Congress, 28
Aldrin, Buzz, 243
Alien and Sedition Acts of 1798, 52–53, 57–58
Alien and Seditions Acts of 1917, 195
Alien Registration Act, 199, 207
All in the Family, 249
All the President's Men, 245
All-American Newsreel, 208
Alsop, Joseph and Stewart, 209
American Broadcasting Company, 222
American Equal Rights Association, 142
American Marconi Company, 172, 178
American Newspaper Annual, 101
American Revolution, 11, 33–34
Americans All, 208
Anarchiad, The, 46
Andros, Edmund, 17
Anthony, Susan B., 142
Antietam, Battle of, 91
antifederalists, 48–49
Anti-Saloon League, 150
Apollo 11, 244
Appeal to Reason, The, 136
Arbuckle, Roscoe "Fatty," 169–70
Areopagitica, 15
Arizona, U.S.S., 205
Armat, Thomas, 164
Armstrong, Louis, 163
Armstrong, Neil, 243, 280
Arnaz, Desi, 249
Arsenio Hall Show, The, 260
Articles of Confederation, 45–46
Associated Press, 149, 197, 284
Atlanta Constitution, 105–106
Aurora, 43, 52
Ayer, Francis, 122

B

Bache, Benjamin Franklin, 51–53
Bailey, Gamaliel, 80
Baird, John Logie, 182
Baker, Chuck, 263
Baker, Ray Stannard, 133, 146, 171
Baldwin, Herbert, 148–49
Ball, Lucille, 249
Baltimore Afro-American, 163
Baltimore riots of 1812, 58–59

Barnard, George, 90
Barnum, P. T., 181
Barron, C. W., 148
Beatles, the, 241–42
Bell, Alexander Graham, 158
Bennett, James Gordon, 69, 85–86, 87, 88, 142
Berliner, Emile, 160
Bernstein, Carl, viii, 7, 245–47
Bernstorff, Johann Heinrich von, 190, 194
Berry, Chuck, 239
Beverly Hillbillies, The, 249
Bewitched, 249
Biddle, Francis, 207
Bill of Rights, 49
Birney, James, 79
Birth of a Nation, The, 168–69, 206
Black Hand, The, 166
black press, 82–84, 114–17, 205–208
Blackboard Jungle, 239
Blackwell's asylum, 113–14
Blaine, James G., 111–12
Blair, Ezell Jr., 226
Blair, Jayson, 290
blockades, 55–56
Blumer, Herbert, 170
Bly, Nellie, 1–2, 7, 112–14, 123, 144, 248
Bok, Edward, 135
Boston Massacre, 32
Boston Post, 148–49
Boston Tea Party, 32–33
Boston Traveler, 147
Bowles, Samuel, 85–86
Boylston, Zabdiel, 19
Bradford, William, 15
Brady, Mathew, 90–91
Braestrup, Peter, 223
Brick, Al, 205
Brinkley, David, 223
Broadcast Measurement Bureau, 4
Brokaw, Tom, 269–70
Brown v. Board of Education, 229
Brown, John, 77
Brown, Murphy, 268
Bryan, William Jennings, 130, 144
Bryce Committee, 193

Buchenwald, 3, 202–203
Buffalo Springfield, 240
Burk, John Daly, 52, 53
Bush, George H. W., 266
Bush, George W., 261, 279–80, 283, 285–86

C

Cable News Network, 257–58, 270, 272
Cagney, James, 198
Cahn, Lu Ann, 1, 2, 6, 7, 278
Calhoun, John C. 82
Calloway, Cab, 163
Campbell, W. Joseph, 3
Captivity of Mary Rowlandson, The, 16
Carroll, Diahann, 249
Carter, Horace, 224–26
Carter, Horace, viii
Carter, Jimmy, 268
Caruso, Enrico, 160
CBS Evening News, 271
Chaplin, Charlie, 166
Checkley, John, 18–19
Cheney, James, 229
Chesapeake, U.S.S., 55, 56
Chicago Defender, 149–50, 206, 207
Chomsky, Noam, 7
Christianity. as a concept for colonization, 14
cinématographe, 164
Cisneros, Evangelina Cosío y, 118–19
civic journalism, *see* community journalism
civil rights movement, 223, 226–29
Civil War, 221
    and illustrated newspapers, 89
    and photography, 90–91
Cleveland, Grover, 111–12
Clinton, Bill, 247, 260, 263
    and the Lewinsky scandal, 271–72
CNN, 281
Cobb, Irvin, 193
Cockerill, John, 1, 108, 113
Cold War, 209–13, 237, 242–43
Cole, Nat King, 249
Coleman, William, 60–62

Index

colleges, beginning in America, 22–23
Collins, LeRoy, 228
colonization literature, 11–13, 295
Colored American, 84
Columbia Broadcasting Situation, 180, 222
comics, 102
"Coxey's Army," 104–105
Committee on Civil Rights, 208
Committee on Public Information, 195–97
Common Sense, 33
Communism, 209–12, 224
community journalism, 2, 258–59, 267–68
community weeklies Gilded Era, 103–105
Compromise of 1850, 76
Conrad, Frank, 178
Constitution, The, 46–47, 50–51
Cooper, Anderson, 283
Cornish, Samuel, 83–84
Cosby, Bill, 249
Country Joe and the Fish, 240
Court TV, 269
Creedence Clearwater Revival, 240
Creel, George, 191, 195
Crime of Carelessness, The, 166
Crisis, The, 34, 206, 208
Cronkite, Walter, 6, 286, 290
  and the space race, 244
  and Vietnam, 235
Cuban Missile Crisis, 209

D

Daguerre, Louis, 90
Davis, Gray, 262
Davis, Richard, 119
Day, Benjamin, 85
De Bow, James, 81
De Bow's Review, 81
De Forest, Lee, 171, 174–76, 194
De Mott, R. W. 179
Debs, Eugene V., 131
Declaration of Sentiments, 145
DeFranco, Philip, 278
DeFranco, Philip, vii

DeKay, James, 92
Democratic National Convention 1968, 221–22, 247
Dewey, John, 104
Diana, princess of Wales, 270–71
Dickinson, John, 32
Dickson, William, 161
Disney, Walt, 191, 203–205
Division of Press Intelligence, 199
Domino, Fats, 239
Dorsey, Tommy, 163
"Double V" campaign, 207–208, 210
Douglas, Stephen A., 76
Douglass, Frederick, 81, 82, 84
Douglass, William, 18–19
Doyle Dane Bernbach, 237
Drudge Report, The, 271
Drudge, Matt, 6, 273
Du Bois, W. E. B., 150, 206
Duane, William, 43
Duck, Donald, 204
Duran, Francisco, 263
Dylan, Bob, 239–40

E

Ed Sullivan Show, The, 241–42
Edes, Benjamin 11
Edison, Thomas, 160, 163
election of 1796, 53
election of 1800, 53–55
Elkins Act, 136
Ellsberg, Daniel, 248
Emporia Gazette, 190
"Era of Good Feelings," 57
Espionage Act, 195, 207
Executive Order 9981, 208

F

Facebook, 288
Faraday, Michael, 172
Farnsworth, Philo, 180
Federal Bureau of Investigation, 210, 226
Federal Communications Commission, 222
Federal Republican (Baltimore), 58–59

Federalists Papers, 47–50
Felt, Mark, 246
Fenno, John, 50, 53
Fessenden, Reginald, 171, 174–75
"Fireside Chats," 214
first printing press, 13
Fleet, Thomas, 15
Fleming, John, 176
Ford, Gerald, 233, 245, 268
FOX News, vii
Frank Leslie's Illustrated Newspaper, 89
Franklin, Benjamin, 21, 28, 34, 258
    and printing networks, 15–16
    and George Whitefield, 25, 26
Franklin, James, 18
Free Speech (Memphis), 114–15
Freed, Alan, 239
Freedom's Journal, 83–84
French and Indian War, 26–30
    and the growth of newspapers, 27–28
Freneau, Philip, 50–51
Fugitive Slave Act, 76–77

G

Gaine, Hugh, 23, 28, 29
Gallup, George, 278
Ganley, Gladys, 5–6
Gardner, Alexander, 90–92, 159
Gardner's Photographic Sketch Book of the War, 91–92, 159
Garrison, William Lloyd, 77–79
Gates, Daryl, 266
Gaye, Marvin, 241
Gazette of the United States, 50
Gerry, Elbridge, 48
Getting Ideas from the Movies, 170
Giap, Vo Nguyen, 235
Gibbons, Floyd, 196–97, 206, 214
Gill, John, 11
Gleason's Pictorial, 89
Goddard, William, 20
Godey's Lady's Book, 89, 93
Godkin, E. L., 142
Goldman, Ronald, 269
Goodman, Andrew, 229
Goodman, Benny, 163

Goodwin, H.M.S., 172–73
Gore, Al, 279–80
Grady, Henry, 105–107
Graham, Katherine, 248
Grammatical Institute of the English Language, 44
gramophone, 160
"Great American Fraud, The," 135
Great Awakening, The, 24–26
Greeley, Horace, 100
Greeley, Horace, 77, 85–86
Greensboro (N.C.) Record, 226–27
Greensboro, North Carolina, 226–27
Griffith, D. W., 166, 168
Guthrie, Woody, 239

H

Habermas, Jürgen, 21
Hakluyt, Richard, 12
Haldeman, H. R., 246
Hamilton, Alexander, 47–48, 50–51, 57
Handy, W. C., 162
Hanson, Alexander, 58–59
Harper's Weekly, 89
Harris, Benjamin, 16
Hart, Gary, 268
Hauser, Philip, 170
Hays Office, 203
Hays, Will, 170
Hearst, William Randolph, 224, 258
    and journalism of action, 2–3, 99–100, 117–22
    and Cuba and Spanish American War, 118–22
Henry, Joseph, 172
Hepburn Act, 136
Herman, Edward, 7
Hersh, Seymour, 6
Hertz, Heinrich, 158
Hewitt, Don, 230–31, 243, 248, 287
Hidden Persuaders, The, 238
Hill, Isaac, 45
Hiss, Alger, 211
Hodges, Willis, 82
"Hoe-Man in the Making, The," 137
Hoe, Richard, 87

Index 321

Holliday, George William, 266
Hollywood Babylon. 169
Holmes, Oliver Wendell, 101
Holt, John 33
Hoover, Herbert, 178, 180, 214
Hoover, J. Edgar, 207, 213
House Committee on un-American Activities, 211
How They Rob Men in Chicago, 166
Humphrey, Hubert, 221
Hunter, William, 21
Hurricane Katrina, 278, 282–84
 and racial profiling, 284
Hutchins Commission Report, 3

I

Illustrated News, 89
illustrated press, 88–89
impressment, 55–56
Independent Reflector, 16
inoculation controversy of 1721, 18–20
Internet, 272–73, 277–79, 283–90
iPod, 290
Isaacson, Walter, 281

J

Jackson, Andrew, 70, 89
Jackson, Michael, 257, 287
Jaworski, Leon, 246
Jay, John, 47–48
Jay's Treaty, 52
Jazz Singer, The, 161, 170
jazz, 161–63
Jefferson Airplane, 241
Jefferson, The, 249
Jefferson, Thomas, 23, 50–51, 53, 55, 74
 and election of 1800, 53–55
Jenkins, Francis, 164
Jennings, Peter, 6
Johnson, Lyndon Baines, 221, 234–35
JOIN, or DIE, 28, 31
Jolson, Al, 161
Journal of Occurrences, 1, 31
journalism of action, 117–22
Jungle, The, 136–37

K

Kansas-Nebraska Act, 77
Kendall, Amos, 80
Kennedy, John F., 209, 259
 and 1960 election, 230–31
 and assassination, 231–32, 280
 and space race, 243
Kennedy, Robert, 221, 261
Khrushchev, Nikita, 209
kinetoscope, 164–65
King, Larry, 270
King, Martin Luther Jr., 221, 227, 228, 261
King, Rodney, 266–67, 272
Koenig, Friedrich, 86–87
Korean conflict, 189, 191–92, 208–11
Krakatau, 158
Ku Klux Klan, viii 224–26
Kurtis, Bill, 247

L

La Follette, Robert, 130
Laconia, 196–97, 206
Landrieu, Mary, 283
Lasswell, Harold, 197
Late Night with David Letterman, 261, 281
Lauterer, Jock, 2
Lear, Norman, 249
LeBlanc, Maurice, 182
Lee, Henry, 58–59
Lee, Ivy, 145–47
"Letters from a Pennsylvania Farmer," 32
Lewinsky, Monica, 271–72
Liberator, 78–79
Life magazine, 243
Limbaugh, Rush, 258, 262–64, 267, 286
Lincoln, Abraham, 81, 101
Lind, Jenny, 181
literacy rates 1880–1900, 101–102
Little Richard, 239
Livingstone, William, 16, 23
Loan, Nguyen Ngoc, 233–34
Log Cabin, 86

Lovejoy, Elijah, 79–80
Ludlow mining disaster, 145–47
Lumière, Auguste and Louis, 164
Lusitania, 189, 193, 194
lynchings, 115–17
Lyon, Matthew, 52, 53

M

M*A*S*H, 249
MacLeish, Archibald, 191
Madison, James, 46, 47–48
magazines.
    growth 1875–1905, 131–32
    growth 1800–1860, 69
    growth 1885–1905, 101
    and muckraking, 133–41
    number in 1960s and 1970s, 222
Maher, Bill, 282
Maine, U.S.S., 119–20
manifest destiny, 73
Mann Act, 136–37
"Making Steel and Killing Men," 136–37
Marconi Wireless Telegraph Company, 172
Marconi, Guglielmo, 171, 173–74, 194
Markham, Edward, 137
Martineau, Harriet, 78
Mather, Cotton, 17
Maude, 249
Maxwell, James Clerk, 158
McCain, Franklin, 226
McCain, John, 287–88
McCarran, Pat, 211
McCarthy, Joseph, viii, 6, 210–12, 224
McCombs, Maxwell, vii, 2, 222
McCullough, David, 278, 282
McKinley, William, 118–20, 130, 142, 165
McLuhan, Marshall, 4, 221
McMasters, William, 148
McNeil, Joseph, 226
McPherson, Harry, 234
McQuire, Barry, 240–41
McVeigh, Timothy, 263
Méliès, Georges, 166
Mexican War, 74–75

Miller, Glenn, 163
Miller, Samuel, 44
Milton, John, 15
Mindich, David, 2
Mirror of Liberty, 82–83
Missouri Compromise, 73–74
Morse, Samuel F. B., 87
Motion Picture Producers and Distributors of America, 170
motion pictures, 159, 163–71, 181, 203–205
    theaters and attendance in early twentieth century, 167
    and juvenile delinquency, 167–68, 169–71
Mount Tabor Tribune, 225–26
Movies, Delinquency, and Crime, 170
Movietone newsreels, 205
muckrakers, 129–30, 133–41, 145, 166, 194, 258
Munsey, Frank, 224
Murrow, Edward R., viii, 3,6, 201–203, 205, 210, 212–13, 214
"Music Box Memo," 178
Music Television, 257–58, 272, 281
Mutual Film Corporation v. Industrial Commission of Ohio, 169
Muybridge, Edweard, 159
My Lai Massacre, 6
MySpace, 285, 288

N

NAACP, 168, 206
Nader, Ralph, 261
Nast, Thomas, 110–11
Nation, Carrie, 166
National Aeronautics and Space Administration (NASA), 243–44
National American Woman Suffrage Association, 142, 144
National Broadcasting Company, 180, 222
National Era, 76–77, 80, 82, 84
Native Americans, 56
New Age, 115
"New South," the, 105–107
New Spirit, The, 204

Index

New York Associated Press, 88
New York Graphic, 159
New York Journal, 140
New York Times, 140
New York World, 113, 140
New-England Courant, 18–19
newspaper war on Britain, 30–32
newspapers
    reasons for beginning in America, 15–23
    stimulating debate in eighteenth century, 20–23
    growth 1783–1820, 44–45
    rise in advertising, 62–63
    growth of 1820–1860, 69–70, 71
    growth of, 1870–1900, 101–102
    growth 1900–1920, 130–31
newsreels, 204, 208
New-York Evening Post, 60–62
New-York Tribune, 142–43
nickelodeon, 166–67
Niepce, Joseph, 90
nineteenth amendment, 145
Nintendo, 265
Nipkow, Paul, 182
Nixon, Richard M., 233
    and "Checker's Speech," 229–30
    and Watergate, 245–47

O

Obama, Barack, 287–88
Ochs, Adolph, 3
Office of Censorship, 199
Office of War Information, 199, 205, 214
Oglethorpe, James, 13
Okinawa, 200–201
Oswald, Lee Harvey, 231–32

P

Packard, Vance, 238
Paine, Thomas, 33, 34
Palin, Sarah, 278
Palmer, Frederick, 193, 196
Palmer, Volney, 122
Paramount News, 208

Parker, George B., 191
Parker, Kathleen, 285
Parks, Rosa, 229
"'Patent Medicine' Curse, The," 135
Paul, Robert, 163
Payne Fund Studies, 170–71, 181, 203
Pearl Harbor, 189, 197–98, 199, 203, 206, 264
penny press, 84–86
Pentagon Papers, 248
Persian Gulf War, 264–66
Peterson, Ruth, 170
Phillips, David Graham, 129, 137
"philosophical" journals, 92–93
phonograph, 159
photography, 90–92
Pinkham's Vegetable Compound, 135
Pittsburgh Courier, 207
political parties, rise of, 49–55
political patronage, 50–52; end of, 100
Ponzi, Charles, 147–49
population, United States, 1820–1860, 70–71
population, United States: 1625–1700, 13;
populism, 130
Post Office Act of 1792, 87
postal system, 21
Postman, Neal, 5–6
President's Committee on Equality of Treatment and Opportunity in the Armed Services, 208
presidential debates. 1948, 229; 1960, 230–31
presidential election of 2000, 279–280
presidential election of 2008, 287–88
Presley, Elvis, 239
Price, Byron, 198
prohibition, 150, 162
Project for Excellence in Journalism, 277
Publick Occurrences, 16–17
Pulitzer, Joseph, 1, 3
Pulitzer, Joseph, 99–100, 103, 114, 123, 224, 258
    and "New Journalism," 107–109
    buys the *World*, 108

and the 1884 presidential election, 111–12
and yellow journalism, 117–22
and Cuba and Spanish American War, 118–22
Pure Food and Drug Act of 1906, 135, 136
Pyle, Ernie, 199–201

Q

Quayle, Dan, 268–69

R

Radio Act of 1912, 177
Radio Corporation of America, 178, 180
radio, 157–58, 171–79, 180, 181, 190, 214, 224
    competition, 174–78
    and department stores, 178
    and Great Depression, 178–79
    number at beginning of World War II, 201
Ramsey, David, 5
Ramsey, JonBenet, 270–71
Rather, Dan, 247, 281, 285–86
Ray, Charles, 84
Reagan, Ronald, 258, 259
recorded music, 160–63, 181, 223, 239–42, 248
Recording Industry Association of America, 290
Red Channels, 210
Red Record, The, 116
Red Scare 1950s, 211–13
Reid, Whitelaw, 102, 103
religious newspapers, 93
Remington, Frederic, 118
Revere, Paul, 32
Rheingold, Howard, 277
Rhett, Robert Barnwell, 77
Rice, Donna, 269
Richmond, David, 226
Ritchie, Thomas, 56
rock 'n' roll, 223, 239–42
    and juvenile delinquency, 239

Rockefeller, John D. Jr., 146
Rockefeller, John D. Sr., 133–34
Rodríguez, Adolfo, 118
Rolling Stone magazine, 241
Roosevelt, Franklin Delano, 180, 197, 214259
Roosevelt, Theodore, 2, 129–30, 165, 258
Rosedale, Philip, 290
Rowell, George, 122
Russell, Benjamin, 51–52
Russwurm, John, 83

S

San Francisco earthquake, 181
Sandburg, Carl, 146
sanitation and New York City, 60–62
Santosuosso, P. A., 148
Sarnoff, David, 178, 179
Saturday Evening Post, 131
Saturday Night Live, 268
Schenck v. United States, 197
Schlesinger, Arthur M., 5
Schwarzenegger, Arnold, 262
Schwirmer, Michael, 229
Scopes "monkey" trial, 150, 179
Scott, John Morin, 16
Scripps, E. W., 224
Second Life, 290
sectionalism, 72–78
seditious libel, 21–22
See It Now, 212–13
Seeger, Pete, 239
Seneca Falls women's rights convention, 142, 145
Sengstacke, John 207
September 11, 2001, 280–82
    and television coverage, 281
Seven Years' War, 26
Seversky, Alexander P. de, 204
Shapiro, David, 282
Shaw, Bernard, 264
Shaw, Donald, vii, 2, 222
Sherman, William T., 106–107
Ship Act of 1910, 177
Simon and Garfunkel, 240
Simpson, Nicole, 269

Simpson, O. J., 269–70
  and U.S. race relations, 270
Sinclair, Upton, 135–36, 146
sit-ins, 226–28
slavery, 63
slavery, 76–82
smallpox, 18–20
Smith Act, *see* Alien Registration Act
Smith, John, 12
Smith, Shepard, 283
Smith, William Jr., 16
Society for Education among Colored Children, 84
Society of Professional Journalists, 290
Southern Horrors: Lynch Law in All Its Phases, 115
Southern Literary Messenger, 81
space race, 243–45
Square Deal, 130
St. Louis Dispatch, 107–108
St. Louis Post, 107–108
Stamp Act of 1765, 29–30
Standard Oil, 133, 146
Stanton, Elizabeth Cady, 176
Starr, Edwin, 241
Starr, Kenneth, 272
steam press, 87
Steffens, Lincoln, 133
stenographers, 160
Stowe, Harriet Beecher, 80–81
Stubblefield, Nathan, 4, 157, 171
stunt journalism, 112–14
suffrage, 141–45
Sullivan, Timothy, 90
Sun (New York), 85

T

Talbot, Henry Fox, 159
Tarbell, Ida, 133–34
telegraph, 87–88, 92, 101, 157–58
telelectroscope, 165
telephone, 101, 158
television, 179–80, 182, 222–24
  saturation in America 1950s, 212
  cost of sets 1950s and 1960s, 222
  and Vietnam, 236
  and 1960 presidential debates, 230–31
  and advertising 1960s, 237–38
  and rock 'n' roll, 239
  and the Beatles, 241–42
  record sales 1960, 241
  and the space race, 243–44
  and election of 2000, 279–80
  nightly news viewers, 286
Temptations, The, 241
Tesla, Nikola, 171
Tet offensive, 233–36
Thomas, Isaiah, 44
Thomas, Jesse, 74
Thompson, Hunter S., 248
Thompson, James, 206–207
Thurstone, L. L., 170
Time magazine, 284–86
Times-Picayune (New Orleans), 283
Tisdale, Geneva, 227–28
Tocqueville, Alexis de, 4, 5, 69
Tonight Show with Jay Leno, The, 262
Townshend Acts, 31
transatlantic cable, 157–58
Transatlantic Times, 174
"Treason of the Senate, The," 129, 137
Treaty of Ghent, 57
Tregaskis, Richard, 189
Triangle Shirtwaist Fire, 137–41, 166
Trip to the Moon, A, 165–66
Truman, Harry, 201, 208, 209
Tweed, William M. "Boss," 109–10
  investigation by *Harper's Weekly* and *New York Times*, 110–11
twenty-four/seven news cycle, 268–71
Twitter, 285–87, 288

U

Uncle Tom's Cabin, 80–82
Understanding Media, 221
union of colonies, 29–30
United Nations, 211
United States Magazine, 73
United States. population, 1870–1900, 107
  population early twentieth century, 132, 134

education early twentieth century, 132
immigration 1900–1915, 132–33
immigrant population 1917, 190
Usurer, The, 166

V

Van Buren, Martin, 70
Vann, David, 223
Victory Through Air Power, 204, 205
Victrola, 160
video cassette recorder, 258
Vietnam war, 223, 233–36, 264
Virginia Resolutions, 57
Virginia Slims, 238
Virtual Community, The, 277
vitascope, 163–64
Volkswagen Beetle, 237
voting patterns, 1775–1832, 70

W

Wagner, Jacob, 58–59
Walkman, 258
wall of separation, 23
Wall Street Journal, 148–49
Wallace, George, 228, 261
Walters, Barbara, 260
War Hawks, 56
War of 1812, 56–57
Washington Post, 245–47
Washington, George, 33, 43, 52
Waskow, Henry T., 200
Watergate, viii, 245–47
Watterson, Henry, 106
Web logs (blogs), 286–87
Webster, Noah, 44
Wells, Ida B., 114–17, 123
Westliche Post, 107
Westmoreland, William, 235
White, William Allen, 104–105
Whitefield, George, 24–26
Whiteman, Paul, 163
Whiteville (N.C.) News Reporter, 226
Wikipedia, 285
Wilmot Proviso, 76
Wilmot, David, 75

Wilson, Flip, 249
Wilson, Samuel, 13
Wilson, Woodrow, 144, 168, 177, 189, 192, 206
Winfrey, Oprah, 264
Wintrop, John, 13
"wireless," *see* radio
women and the vote, 59–60
woodcuts, 88
Woodson, Carter G., 162–63
Woodward, Bob, viii, 7, 245–47
Woolworth's, F. W., 226–28
Worcester, Massachusetts, women's rights convention, 142
WordPress, 286
World War I, 177, 189–91, 192–97, 206, 214
    and atrocity stories, 193
    and radio, 194
World War II, 180, 189–91, 197–208, 214, 282
    and advertising 237
    and censorship, 198–99
    and radio, 4
World Wide Web, 284–87

X

XYZ Affair

Y

Yalta Conference, 211
yellow journalism, 117–22
Yeltsin, Boris, 269
YouTube, 285, 287, 288

Z

Zenger, John Peter, 21–22
Zworykin, Vladimir, 179–80